CAMBRIDGE

GCSE for **AQA**

GEOGRAPHY
Student Book

Rebecca Kitchen, David Payne, Alison Rae, Emma Rawlings Smith, John Rutter, Helen Young and John Pallister
Series editor: David Payne

CAMBRIDGE
UNIVERSITY PRESS

University Printing House, Cambridge CB2 8BS, United Kingdom

Cambridge University Press is part of the University of Cambridge.

It furthers the University's mission by disseminating knowledge in the pursuit of
education, learning and research at the highest international levels of excellence.

www.cambridge.org
Information on this title:
www.cambridge.org/9781316604632 (Paperback)
www.cambridge.org/9781316604649 (Cambridge Elevate enhanced edition, 2 years)
www.cambridge.org/9781316608784 (Cambridge Elevate enhanced edition, 1 year)
www.cambridge.org/9781316608791 (Cambridge Elevate enhanced edition, School Site Licence)
www.cambridge.org/9781316604663 (Paperback + Cambridge Elevate enhanced edition, 2 years)

First published 2016

Printed in the United Kingdom by Latimer Trend

A catalogue record for this publication is available from the British Library

ISBN 978-1-316-60463-2 Paperback
ISBN 978-1-316-60464-9 Cambridge Elevate enhanced edition (2 years)
ISBN 978-1-316-60878-4 Cambridge Elevate enhanced edition (1 year)
ISBN 978-1-316-60879-1 Cambridge Elevate enhanced edition (School Site Licence)
ISBN 978-1-316-60466-3 Paperback + Cambridge Elevate enhanced edition (2 years)

Additional resources for this publication at www.cambridge.org/education

Cambridge University Press has no responsibility for the persistence or accuracy
of URLs for external or third-party internet websites referred to in this publication,
and does not guarantee that any content on such websites is, or will remain,
accurate or appropriate.

Approval message from AQA

This textbook has been approved by AQA for use with our qualification. This means that we have checked that it broadly covers the specification and we are satisfied with the overall quality. Full details of our approval process can be found on our website.

We approve textbooks because we know how important it is for teachers and students to have the right resources to support their teaching and learning. Please note, however, that the publisher is ultimately responsible for the editorial control and quality of this book.

Please note that when teaching the GCSE Geography (8035) course, you must refer to AQA's specification as your definitive source of information. While this book has been written to match the specification, it cannot provide complete coverage of every aspect of the course.

A wide range of other useful resources can be found on the relevant subject pages of our website: www.aqa.org.uk

Contents

Introduction   6

Section 1: The challenge of natural hazards   8

Chapter 1: Tectonic hazards
1.1 What is the Earth made of?   10
1.2 What are the different types of plate margins
     and their resulting landforms?   12
1.3 Why is volcanic activity found near plate margins?   14
1.4 Volcanic hazards   16
1.5 Why do people live near areas of volcanic activity?   18
1.6 Why do earthquakes occur near plate margins?   20
1.7 What is the impact of earthquakes on people
     and places?   22
1.8 How can the earthquake risk be reduced?   24
     EXAMPLE: The Tohoku earthquake, Japan, 2011   26
     EXAMPLE: The Gorkha earthquake, Nepal, 2015   28
1.9 Issue evaluation: Predicting seismic activity in China   30

Chapter 2: Weather hazards – tropical storms
2.1 The general model of global atmospheric circulation   34
2.2 Where do tropical storms develop?   36
2.3 How might climate change affect tropical storms?   38
2.4 What are the effects of tropical storms?   40
2.5 How can the effects of tropical storms be reduced?   42
     EXAMPLE: Typhoon Haiyan 2013   44

Chapter 3: Extreme weather in the UK
3.1 What extreme weather events affect the UK?   46
3.2 What are the effects of extreme weather?   48
3.3 Is the weather in the UK becoming more extreme?   50
     EXAMPLE: 2013 UK winter storms   52

Chapter 4: Climate change
4.1 What is the evidence for climate change?   54
4.2 What are the causes of climate change?   56
4.3 What are the effects of climate change
     and how do we manage them?   58

Section 2: The living world   60
Ecosystems – an overview   61
Chapter 5: Tropical rainforests
5.1 What are the distinctive characteristics of tropical
     rainforests?   66
5.2 What are the rainforest layers?   68
5.3 Why does deforestation take place in areas
     of tropical rainforest?   70
5.4 What are the impacts of deforestation?   72
5.5 Why is sustainable rainforest management important?   74
     CASE STUDY: The Chocó rainforest   76
5.6 Issue evaluation: Hydroelectric power generation in
     Madagascar's tropical rainforest   78

Chapter 6: Hot deserts
6.1 What are the distinctive characteristics of hot desert
     ecosystems?   82
6.2 What is the hot desert ecosystem like?   84
6.3 What development opportunities are found in hot
     deserts?   86
6.4 What are the challenges of developing hot desert
     environments?   88
6.5 What are the causes and impacts of desertification?   90
6.6 What strategies can be used to reduce the risk
     of desertification?   92
     CASE STUDY: The hot desert environment of Qatar   94

Chapter 7: Cold environments
7.1 What are the distinctive characteristics
     of cold environments?   96
7.2 How are plants and animals adapted to cold
     environments?   98
7.3 What are the challenges of developing cold
     environments?   100
     CASE STUDY: Development opportunities and
     challenges in Alaska   102
     CASE STUDY: The North Slope region, Alaska   104
7.4 How can cold environments be protected?   106

Section 3: Physical landscapes in the UK   108
Physical landscapes in the UK – an overview   109
Chapter 8: Coastal landscapes in the UK
8.1 What happens when waves reach the coastline?   114
8.2 How do physical processes affect the coast?   118
8.3 What landforms are associated with coastal
     erosion?   120
8.4 What landforms are created by coastal
     deposition?   122
     EXAMPLE: The Holderness coast, Yorkshire   123
8.5 Protecting coastlines from the effects of physical
     processes   126
8.6 Protecting coastlines from the effects of physical
     processes – soft engineering   128
     EXAMPLE: Coastal management scheme, Ventnor to
     Bonchurch, Isle of Wight   130

Chapter 9: River landscapes in the UK
9.1 How do river valleys change as rivers flow
     downstream?   132
9.2 What are the processes that affect river valleys?   134
9.3 Distinctive landforms resulting from different
     physical processes   136
9.4 Assessing and managing flood risk   140
     EXAMPLE: Responding to a flood
     event – the Boscastle Flood 2004   142

Chapter 10: Glacial landscapes in the UK
10.1 Why do glacial landforms exist in the UK? — 146
10.2 What are glaciers? — 148
10.3 How do glacial processes shape the landscape? — 149
10.4 What landforms result from glacial erosion in upland areas? — 152
10.5 What landforms result from glacial erosion in lowland areas? — 154
10.6 What landforms result from glacial deposition? — 155
10.7 What economic opportunities do glacial landscapes provide? — 157
10.8 Why do we need strategies for managing glacial landscapes? — 157
EXAMPLE: The Lake District — 158
EXAMPLE: The Lake District National Park — 160

Section 4: Urban issues and challenges — 164
Chapter 11: Changing urban areas
11.1 Living in an increasingly urban world — 166
11.2 What are the causes of urbanisation in LICs and NEEs? — 168
11.3 The growth of megacities — 170
CASE STUDY: Mumbai – a city of national and international importance — 172
CASE STUDY: Opportunities and challenges of urban growth in Mumbai — 174
EXAMPLE: Urban planning in Mumbai — 176
11.4 The UK's urban landscape — 178
CASE STUDY: Birmingham – national and international links — 180
CASE STUDY: How urban change has created opportunities in Birmingham — 182
CASE STUDY: How urban change has created challenges for Birmingham — 184
EXAMPLE: Urban regeneration in Birmingham — 186
11.5 Sustainable urban living — 188
11.6 Managing traffic congestion in cities — 190
11.7 Issue evaluation: City centre regeneration: Birmingham, UK — 192

Section 5: The changing economic world — 196
Chapter 12: The development gap
12.1 How do we measure levels of development? — 198
12.2 Classifying countries according to their level of development — 200
12.3 What are development indicators? — 200
12.4 What are the limitations of development indicators? — 202
12.5 What is the Demographic Transition Model? — 204
12.6 Why is development uneven? — 206
12.7 What are the economic causes of uneven development? — 208

12.8 What are the consequences of uneven development? — 210
12.9 What strategies exist to reduce the global development gap? — 212
EXAMPLE: How tourism in Tanzania is reducing its development gap — 216
CASE STUDY: Rapid economic development in Malaysia — 218
CASE STUDY: How the manufacturing industry can stimulate economic development in Malaysia — 220
CASE STUDY: International aid and Malaysia — 222

Chapter 13: Economic futures in the UK
13.1 What are the causes of economic change in the UK? — 224
13.2 How has industry in the UK changed? — 228
13.3 How has UK infrastructure changed? — 230
EXAMPLE: Teesside: impacts of industry on the physical environment — 233
13.4 Social and economic changes in rural areas — 234
13.5 What strategies have been used to resolve regional differences in the UK? — 236
13.6 The place of the UK in the wider world — 238
13.7 Issue evaluation: Expansion of an industrial estate in Bournemouth — 240

Section 6: The challenge of resource management — 244
Resource management – an overview — 245
Chapter 14: Demand on resources in the UK
14.1 How is the demand and provision of food resources in the UK changing? — 250
14.2 How has the demand for water changed? — 253
14.3 Matching supply and demand – areas of deficit and surplus — 254
14.4 The changing energy mix — 256
14.5 What is the future of energy in the UK? — 258
14.6 Issue evaluation: Opencast coal mining in Northumberland — 260

Chapter 15: Food resources
15.1 Where does our food supply come from? — 264
15.2 Food security — 265
15.3 Where are the world's food resources consumed? — 266
15.4 Is there enough food? — 268
15.5 What factors affect food supply? — 270
15.6 What are the impacts of food insecurity? — 272
15.7 What strategies can be used to increase food supply? — 274
EXAMPLE: Large-scale agricultural development in Kilombero Valley, Tanzania — 276
15.8 How can food supply be increased sustainably? — 278
EXAMPLE: Sustainable food production in Kinshasa — 280

Chapter 16: Water resources
16.1   Where do our water resources come from?   282
16.2   Global patterns of water surplus and deficit   283
16.3   What is water used for?   284
16.4   What are the reasons for increasing water consumption?   285
16.5   Water stress and deficit   288
16.6   Is there enough water for everyone?   288
16.7   What are the impacts of water insecurity?   290
16.8   What strategies can be used to increase water supply?   292
       EXAMPLE: The South-to-North Water Transfer Project, China   294
16.9   How can we move towards a sustainable water resource future?   296
       EXAMPLE: The Kyeni Kya Thwake water conservation scheme in Kenya   298

Chapter 17: Energy resources
17.1   Where does our energy come from?   300
17.2   How are energy consumption and economic development linked?   304
17.3   Why are there areas of energy surplus and deficit?   306
17.4   What factors affect energy supply?   308
17.5   What are the impacts of energy insecurity?   310
17.6   What strategies can be used to increase energy supply?   312
       EXAMPLE: The Gannet oilfield in the North Sea   314
17.7   How can we use energy more sustainably?   316
       EXAMPLE: The Solar Mini Grid Scheme in Melela, Tanzania   318

Section 7: Fieldwork, skills and assessment preparation   320
Chapter 18: Fieldwork
18.1   What is a fieldwork enquiry?   322
18.2   Methods of collecting data   324

18.3   Methods of presenting data   326
18.4   Describing, analysing and explaining data   327

Chapter 19: Skills practice
19.1   Graphical skills – graphs and charts   330
19.2   Graphical skills – population pyramids, choropleth maps   334
19.3   Numerical skills – number, area, scale   336
19.4   Numerical skills – ratio, proportion, sampling   338
19.5   Statistical skills – central tendency, percentage increase and decrease   340
19.6   Statistical skills – how do we find relationships between data?   344
19.7   Cartographic skills – gradients, contours   346
19.8   Cartographic skills – cross-sections, transects   348
19.9   Cartographic skills – coordinates   350
19.10 Cartographic skills – GIS   352
19.11 Atlas skills   354

Chapter 20: Preparing for an assessment
20.1   Structure of the GCSE Geography exams   356
20.2   Types of questions   357
20.3   Answering questions   358
20.4   Approaching extended response questions   360
       Assess to progress question bank   362

Glossary   366

Index   374

Acknowledgements   378

Introduction

This book has been written by experienced teachers to help build your understanding and enjoyment of the geography you will meet at GCSE.

The book is divided into seven sections. Sections 1–6 each cover a section in the GCSE Geography specification. Section 7 helps you with planning and carrying out fieldwork and developing geographical skills.

At the start of each chapter, there is a box of **Learning outcomes**. These explain what you will learn in the chapter. At the end of the chapter, you can return to these and check that you are confident with each statement.

Throughout the book there are **Case studies** and **Examples**. Case studies are in-depth studies of a country, city or recent event that demonstrates the required content of the geographical topic that you are studying. They require you to understand the context of the country, city or event. Examples are similar but smaller in scale, focusing on a specific location, event or situation and containing less detail. Try to learn these case studies and examples as they may be useful when answering exam questions.

There are other features throughout the book to help you build knowledge and improve your skills:

 Fact file

Fact files list the key facts about an event or topic.

 Key terms

Important geographical terms are written in orange. You can find out what they mean in **Key terms** boxes and also in the **Glossary** at the back of the book.

 Skills link

Skills links show where there are opportunities to practice geographical skills. For help on developing these skills, see Chapter 19 Skills practice.

 Fieldwork

Fieldwork boxes show where there are opportunities for fieldwork. For help on carrying out fieldwork, see Chapter 18 Fieldwork.

 Did you know?

Did you know? boxes contain interesting geographical facts.

 Tip

Tip boxes provide useful information or helpful hints.

WORKED EXAMPLE

Worked examples guide you through student sample answers to help you understand methods of answering questions.

 Discussion point

Discussion points contain thought-provoking questions to stimulate discussion.

 Further research

Further research boxes suggest opportunities for investigating topics further.

ACTIVITY

Activities contain questions to test your understanding of topics.

 Assess to progress

Assess to progress boxes contain practice questions so you can test your learning.

The **Cambridge Elevate enhanced edition** of this Student Book allows you to annotate, highlight and add bookmarks and weblinks. Videos and animations help to demonstrate important concepts and processes. You can also download worksheets to help you with activities. Look out for the following icons in the book, which show you where you can find extra resources in the Elevate enhanced edition.

 Visit Cambridge Elevate to view a video or animation.

 Visit Cambridge Elevate to download a worksheet.

 Visit Cambridge Elevate for some useful additional information or weblinks.

IMPORTANT NOTE:

AQA has not approved any Cambridge Elevate content.

SECTION 1

The challenge of natural hazards

1 Tectonic hazards

In this chapter you will cover:

- plate tectonics
- the global distribution of earthquakes and volcanoes
- the different types of volcanoes and how they are formed
- the different hazards and benefits associated with volcanoes
- how the movement of plates leads to earthquakes
- the primary and secondary effects of volcanic activity and earthquakes
- the measurement of earthquakes
- how the effects and responses to earthquakes vary between areas of contrasting levels of wealth
- how the earthquake risk can be reduced

2 Weather hazards – tropical storms

In this chapter you will cover:

- global atmospheric circulation patterns
- the global distribution of tropical storms
- how tropical storms are formed
- the effects of tropical storms and how these effects are managed
- how climate change may affect the formation and effects of tropical storms

3 Weather hazards – extreme weather in the UK

In this chapter you will cover:

- the extreme weather events that affect the UK
- the effects of extreme weather and how they can be reduced
- if the weather in the UK is becoming more extreme
- the causes, impacts and responses for an extreme weather event

4 Climate change

In this chapter you will cover:

- the evidence for climate change
- whether climate change is man-made or a natural process
- the possible consequences of climate change
- ways to help combat the effects of climate change

Natural hazards – an overview

Natural hazards are naturally occurring events that pose major threats to people and property around the world. These include tectonic events, weather events and climate change. The effect of climate change on the frequency and strength of storm hazards and rising sea levels pose an increasing threat to the millions of people who live in vulnerable coastal areas.

As the global population continues to grow, increasing numbers of people are living in areas at risk from the effects of natural hazards. Hazard risk is often higher in lower income countries and newly emerging economies, which might not have the money to build protective infrastructure such as flood defences and earthquake-proof buildings and can also find it more difficult to recover and rebuild after a natural hazard.

Understanding the processes involved in the development of natural hazards is an important step in working towards reducing their threat.

Looking at examples of past hazard events and considering the impact they had on people and property gives valuable information about the risks linked to particular hazards. This knowledge can then be used to put in place effective planning and preparation strategies in order to reduce the risks of future hazard events.

Figure S1.1 Hazards associated with volcanic eruptions are easy to identify: lava flows, ash falls and gas clouds. However, there are also secondary hazards to consider, such as landslides and even climate change. How might a volcanic eruption lead to an outbreak of cholera?

1 Tectonic hazards

In this chapter you will learn about...

- the global distribution of earthquakes and volcanoes
- how the movement of plates leads to earthquakes and volcanic eruptions
- the benefits of living in tectonically active areas
- the primary and secondary effects of earthquakes and volcanic eruptions
- how people respond to tectonic activity
- how the effects of and responses to earthquakes vary between areas of contrasting levels of wealth.

Key terms

crust: the outermost layer of the Earth

mantle: a layer of rock between the core and crust made of molten rock

magma: heat from the Earth's core is hot enough to melt rock in the mantle; this molten (liquid) rock is called magma

core: dense hot rock at the centre of the Earth

1.1 What is the Earth made of?

The Earth is made up of several layers (Figure 1.1). The top layer, called the **crust**, is between 5 km and 90 km in thickness. Underneath this is the **mantle**, a mass of hotter material which includes **magma**. This becomes denser towards the centre of the Earth, which is called the **core** and is so dense it is like solid rock.

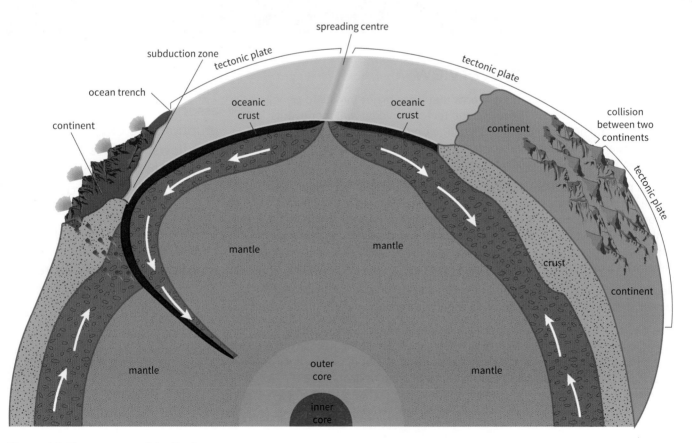

Figure 1.1 The structure of the Earth.

Plate tectonic theory has shown that the Earth's crust is not a solid mass but is split into smaller crustal plates which 'float' on the semi-molten upper mantle. There are two types of crustal plate: continental crust and oceanic crust (Table 1.1). The places where the plates meet are called **plate margins**. In these areas the Earth is particularly unstable and **tectonic hazards** (earthquakes and volcanic activity) are common.

	Continental crust	Oceanic crust
Thickness (km)	30–90	5–10
Oldest rock (years)	4 billion	200 million
Main rock type	granite	basalt

Table 1.1 Characteristics of continental and oceanic crust.

Scientists once believed that **convection currents** in the mantle, generated by high temperatures, were the main force driving the movement of tectonic plates. It is now thought that **tectonic plate** movement is driven by the weight of denser, heavier tectonic plates sinking into the mantle at ocean trenches. This drags the rest of the plate with it and is called **slab pull theory**. In some places the plates are moving towards each other and the Earth's crust is being destroyed. In other places the plates are moving apart and new crust is being created (Figure 1.2).

Key terms

plate margins: the place where tectonic plates meet and the Earth is particularly unstable

tectonic hazards: natural hazards caused by movement of tectonic plates (including volcanoes and earthquakes)

convection currents: circular movements of heat in the mantle; generated by radioactive decay in the core

tectonic plate: a rigid segment of the Earth's crust which floats on the heavier, semi-molten rock below

slab pull theory: a theory that outlines how large and dense tectonic plates sinking into the mantle at ocean trenches drives tectonic plate movement.

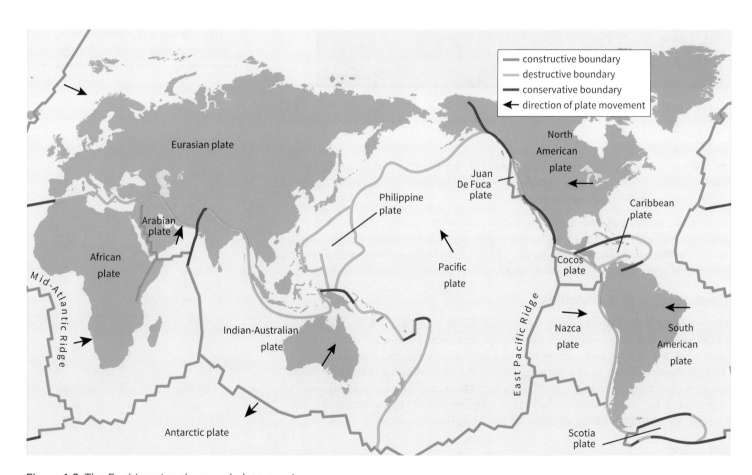

Figure 1.2 The Earth's major plates and plate margins.

1.2 What are the different types of plate margins and their resulting landforms?

Destructive plate margins

Destructive plate margins occur where two plates move towards each other (converge). This happens at **subduction** and collision zones.

Subduction zones

Where an oceanic plate converges with a continental plate, the denser oceanic plate is forced under the lighter continental plate and sinks into the mantle; a process known as subduction. This creates a deep ocean trench on the edge of the continental plate. Where plates subduct, heat in the mantle melts the crust to form magma. Magma rises to the Earth's surface because it is hotter and less dense than the surrounding rock. Close to the surface, the pressure decreases and gases dissolved in the magma build up to eventually form volcanoes that erupt violently (Figure 1.3).

Key term

subduction: the process by which one tectonic plate moves under another tectonic plate and sinks into the mantle

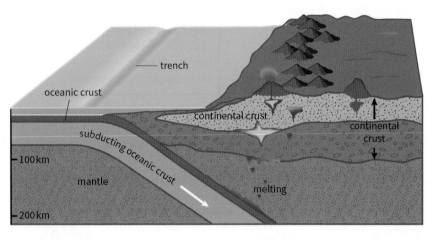

Figure 1.3 A destructive plate margin.

Discussion point

The theory of plate tectonics introduced by Alfred Wegener suggests that North and South America once fitted together with Europe and Africa – what evidence supports this theory?

Collision zones

Where two continental plates collide they push the Earth's crust upwards forming fold mountains. The Himalayas are being created by continental plates carrying India colliding with the Eurasian plate.

Where two oceanic plates converge, subduction of the cooler, denser plate results in volcanic eruptions that form a chain of volcanic islands, known as island arcs. The Mariana Islands in the western Pacific Ocean were formed as a result of the Pacific plate being forced under the Mariana plate.

Constructive plate margins

Constructive plate margins occur where two plates move away from each other (diverge). Where two oceanic plates diverge, the crust fractures. Initially, magma rises from the mantle and creates new sea floor along the mid-oceanic ridge. This process is known as sea-floor spreading. As more magma rises, submarine (underwater) volcanoes form, and if they grow enough, they can rise above sea-level to form volcanic islands (Figure 1.4). The islands of Iceland and Saint Helena are both formed by volcanic on the Mid-Atlantic Ridge.

Rift valleys

Where two continental plates are moving apart a rift valley can form. Rift valleys are relatively narrow compared to their length, with steep sides and a flat floor.

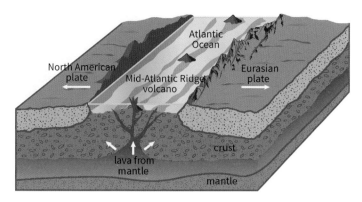

Figure 1.4 A constructive plate margin.

As the sides of a rift valley move farther apart, the floor of the valley sinks lower. Millions of years ago, Africa was attached to the Arabian Peninsula. The African plate moved away from the Arabian plates and the rift valley was flooded by the Indian Ocean, creating the Red Sea (Figure 1.5).

Conservative plate margins

Conservative plate margins occur where two plates move past each other in opposite directions or they move in the same direction at different speeds. The movement of plates is not always smooth and sometimes plates stick together and pressure builds up over time. The eventual release of this pressure results in earthquakes. In California, USA, the cities of San Francisco and Los Angeles, are located on or near the San Andreas fault (part of a conservative plate margin), and consequently prone to earthquakes. There is no volcanic activity at conservative plate margins as the crust is not being created or destroyed.

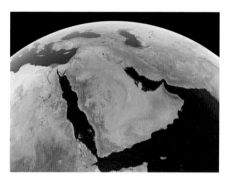

Figure 1.5 Satellite image of the Red Sea.

ACTIVITY 1.1

1 Define the terms:
   a subduction zone
   b sea-floor spreading
2 Study Figure 1.1.
   a Draw a cross-section through the Earth and label each layer.
   b On your cross-section, add a definition for each layer.
3 Study Table 1.1.
   a How does continental crust differ from oceanic crust?
   b Why is continental crust much older than oceanic crust?
4 a Explain how convection currents cause plate movement.
   b Draw a simple annotated diagram to explain how volcanoes are formed at destructive plate margins.
5 Study Figure 1.2.
   a Name five areas in the world where tectonic activity is common.
   c Name five areas where tectonic activity is uncommon.

Use the resources on the **National Geographic website** to find out more about plate tectonics (www.cambridge.org/links/gase40001)

Download Worksheet 1.1 from Cambridge Elevate for additional questions.

1.3 Why is volcanic activity found near plate margins?

There are more than 500 active volcanoes in the world and more than half of these encircle the Pacific Ocean. Volcanic eruptions occur where molten rock and ash erupts from inside the Earth. Most volcanic activity occurs near plate margins where the Earth's crust is unstable. However, in places where the Earth's crust is thin magma can reach the surface. These places are known as volcanic hotspots, for example the lava flows in Hawaii (Figure 1.6).

An inactive volcano is said to be **dormant**. If it has not erupted in the last 10 000 years, it is thought to be **extinct**. Evidence of past volcanic activity can be seen in the form of hot springs and geysers where underground water is superheated by the hot rocks close to the Earth's surface.

Volcanoes are closely associated with disturbed crust found at constructive and destructive plate margins (Figure 1.7). Around 80 per cent of the world's active volcanoes are composite volcanoes (Figure 1.8) which occur along destructive plate margins. The longest of these is known as the Pacific 'Ring of Fire' and marks the location where the Pacific plate is subducted under several continental plates including the Eurasian and North American plates. Shield volcanoes occur along constructive plate margins (Figure 1.9).

Figure 1.6 Lava flow in Hawaii.

Key terms

dormant: has not erupted in living memory, but it could become active in the future

extinct: has not erupted in historic times, in the last 10 000 years

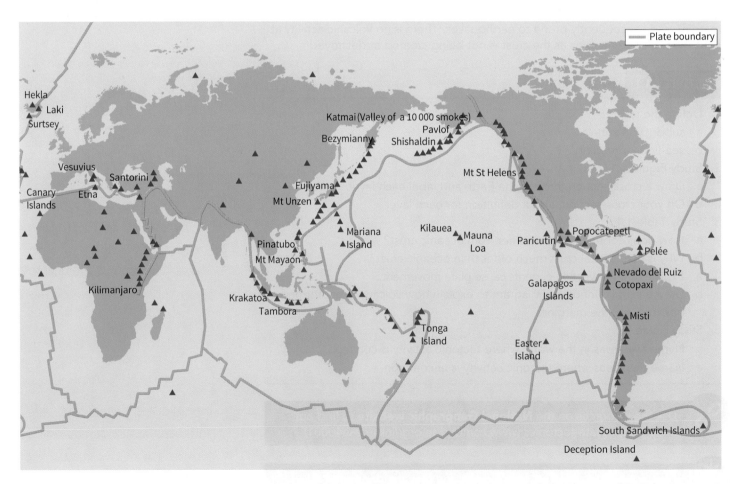

Figure 1.7 The global distribution of active volcanoes.

Volcano shape	composite volcano	shield or basic volcano
Form of volcano	steep slopes tall cone with narrow base made of alternative layers of lava and ash	gentle slopes low cone with wide flanks made of numerous lava flows
Cross-section of volcano	**Composite volcano** – layers of ash and lava, viscous lava travels short distances Figure 1.8 The characteristic features of a composite volcano.	**Shield volcano** – layers of lava, runny lava travels long distances Figure 1.9 The characteristic features of a shield volcano.
Lava	volcanic gases do not escape the lava easily viscous, less hot and slow flowing flows for short distances violent, infrequent eruptions	volcanic gases remain dissolved in the lava low viscosity – hot and runny flows over long distances mild, frequent eruptions
Formation	Where the plates collide at a destructive margin, the cooler, denser plate is subducted into the mantle. Heat melts the plate and generates magma. Being less dense than the surrounding rock, the viscous magma forces its way to the Earth's surface and erupts violently.	Where the plates diverge at a constructive margin, the crust fractures and magma rises to the surface. This adds new crust to the sea floor and submarine volcanoes form. Some volcanoes grow tall enough to rise above sea-level and form volcanic islands.
Examples	Krakatoa in Indonesia Mount St Helens, USA (Figure 1.10)	Mauna Loa in Hawaii Heimaey in Iceland

Table 1.2 Comparing composite and shield volcanoes.

Figure 1.10 Mount St Helens, after it erupted on 18 May 1980.

Skills link

Locating plate margins, volcanoes and earthquakes on a map is an important geographical skill. You can develop this skill by mapping all the named locations in this chapter on a world map.

Did you know?

Scientists now believe that the mantle contains more water than all the oceans combined.

1.4 Volcanic hazards

Volcanic eruptions are hazardous. Volcanic bombs, pyroclastic flows, ash and gas clouds can all kill people when volcanoes erupt suddenly. As the time between eruptions can be long, many people can live safely near volcanoes, never experiencing any of the hazards shown in Table 1.3.

Volcanic eruptions have primary and secondary hazards. **Primary hazards** are directly caused by the volcano while **secondary hazards** are caused indirectly, as a result of the primary hazards.

Primary hazards	Secondary hazards
Volcanic bombs: solid lumps of lava which fly through the air; some can be the size of a house (Figure 1.11)	**Lahars:** volcanic mud flows which travel along river valleys (Figure 1.13)
Pyroclastic flows: very hot (around 800 °C) flows of gas and ash reaching speeds of 700 km/h (Figure 1.12)	**Glacier bursts:** large floods caused by the melting of ice beneath a glacier
Lava flows: streams of molten rock which slow as they cool	**Landslides:** rock and earth which tumble down a slope, triggered by tectonic activity
Ash falls (<2 mm in size): small erupted material which can travel long distances	**Tsunamis:** giant sea waves generated by undersea volcanoes (Figure 1.14)
Gas clouds (including carbon dioxide, hydrogen sulphide and sulphur dioxide) can suffocate people	**Climate change:** volcanoes send ash into the atmosphere. This reflects radiation from the Sun back to space, causing cooling

Table 1.3 Primary and secondary volcanic hazards.

Key terms

primary hazards: those caused directly by the hazard, such as lava flows, ash falls as a result of a volcano erupting

secondary hazards: hazards caused as an indirect result of the primary hazard; for volcanoes these include landslides and tsunamis

Discussion point

Are primary hazards or secondary hazards more dangerous to people and property?

Skills link

You might be asked to draw a sketch or a simple diagram of a geographical feature. Drawing simple labelled diagrams quickly and accurately can be a useful revision technique.

Figure 1.11 Boiling lava and volcanic bombs on the slopes of the crater of the volcano cone Tolbachik, Kamchatka, Russia.

Figure 1.12 Pyroclastic flow descending down Tar River Valley, Soufrière Hills volcano, Montserrat, Caribbean.

Figure 1.13 Lahar damage to road from Mount Merapi volcano eruption, Yogyakarta, Indonesia.

Figure 1.14 Devastation on the West Coast of Aceh, Sumatra, Indonesia following a tsunami.

Primary or immediate effects	Secondary or longer-term effects
• people killed and injured • farmland and buildings destroyed • communications damaged or disrupted (transport, water, electricity and gas supply)	• spread of disease with no clean water or broken sewers • hospitals overwhelmed by people needing attention • shortage of food, water, shelter and medicine/healthcare • loss of farmland and local businesses reduces income and food production • economic impact from air travel disruption and the cost of rebuilding

Table 1.4 Primary and secondary effects of volcanic eruptions.

Immediate and long-term responses to a volcanic hazard

After successive eruptions of the Soufrière Hills volcano, large areas of Montserrat have been destroyed by pyroclastic flows, ash and lava (Figure 1.12). Islanders responded immediately by setting up an exclusion zone in the volcanic region, abandoning the capital city of Plymouth and evacuating to safety in the north of the island or overseas.

In the longer-term the Montserrat Volcano Observatory (MVO) was built to monitor volcanic activity. New roads, houses and a port have been constructed at Little Bay in the north for returning residents. The tourist industry has expanded, with the volcano as the prime attraction, replacing jobs lost in the farming industry. Residents are now better prepared to cope when the Soufrière Hills volcano violently erupts again.

 Further research

Use the **Global Volcanism Program website** (www.cambridge.org/links/gase40002) to find out more about the volcanoes mentioned in Table 1.2. Start by downloading the Google Earth data layer to explore where active volcanoes are located.

ACTIVITY 1.2

1 Are these statements true? Give reasons for your answer.
   a Shield volcanoes are more dangerous than composite volcanoes.
   b Volcanoes are only hazardous for people living close to them.
2 Study Figures 1.8 and 1.9.
   a Draw simple labelled diagrams showing both composite and shield volcanoes.
   b Outline how and why composite and shield volcanoes are different and how and why they are similar.
3 Study the photograph of Mount St Helens in Figure 1.10.
   a Draw a sketch of the photograph of Mount St Helens.
   b Add labels for each of the volcano features you can see.
   c What features of the volcano can you not see in the photograph?

1.5 Why do people live near areas of volcanic activity?

Over 500 million people live near active volcanoes. There are a number of reasons for this, including the economic opportunities that such areas can provide and the feeling of relative safety because the volcanic activity is being monitored. Since July 1995, the active Soufrière Hills volcano on the Caribbean island of Montserrat has been monitored and an early warning system is in place.

Economic opportunities

Tourism

Volcanic areas offer a variety of tourist attractions. Yellowstone, a volcanic **caldera** in the USA, draws 3 million tourists a year who come to view the geysers and other geothermal activity. Yellowstone National Park provides over 2000 rooms for guests, and jobs for 5000 permanent and seasonal workers.

Farming

Farming in southern Italy can be difficult as the limestone bedrock is nutrient poor. The region around Naples is famous for vines, olives, tomatoes and fruit trees grown on the volcanic soils from the many eruptions of Mount Vesuvius (Figure 1.15). Coffee production in Nicaragua is also only possible in the volcanic soil around Estelí and Jinotepe. Over time, weathering of mineral-rich volcanic rock produces rich fertile soil which is good for growing crops and provides a valuable income for farmers.

Geothermal energy

Iceland, a volcanic island located on the Mid-Atlantic Ridge, uses geothermal heat to supply a quarter of domestic electricity needs, as well as heat and hot water to 90 per cent of homes. Cucumbers, tomatoes and peppers are grown in geothermally heated greenhouses (Figure 1.16) and steel is produced using excess renewable energy.

How can monitoring reduce the risks from volcanic eruptions?

With 750000 people living in the shadow of Mount Fuji and the **megacity** Tokyo located just 100km away, it is no wonder that the Japanese carefully monitor tectonic activity. By monitoring and measuring volcanic activity, scientists can warn when the risk of an eruption is high. This gives people time to evacuate. The National Research Institute for Earth Science and Disaster Prevention and the Japan Meteorological Agency collect data (Figure 1.17) about Mount Fuji and 46 other active volcanoes through a network of Volcano Observation and Information Centres located across Japan.

Prediction, planning and preparation

The three Ps (prediction, planning and preparation) can reduce the impact of volcanic eruptions.

Prediction involves trying to forecast when a volcano will erupt. Indicators of an imminent eruption include the time between volcanic activities, changing ground conditions and rising steam and gases. A major volcanic eruption is often preceded by a series of earthquakes. In 1976, a false evacuation of 70000 people from Guadeloupe Island in the Caribbean proved costly when La Soufrière volcano produced only minor explosions. Mount Fuji last erupted in 1707 and there are signs that it is still active, but it is difficult to predict when it might erupt again.

Planning helps communities to respond and recover from natural disasters. It includes drawing up emergency evacuation plans and using hazard maps to prevent building in high risk areas as well as setting up warning systems and volcano shelters for use in the event of a sudden eruption.

Figure 1.15 Farming on the fertile slopes of Vesuvius.

Figure 1.16 Greenhouses in Iceland heated using geothermal energy.

Watch the video on Cambridge Elevate which explains some of the reasons why people live near volcanoes in Iceland.

Discussion point

What are the social, economic and environmental reasons that people continue to live in areas of volcanic and earthquake activity?

Key terms

caldera: a large, basin-like depression formed as a result of the explosion or collapse of the centre of a volcano

megacity: a city that has 10 million or more people

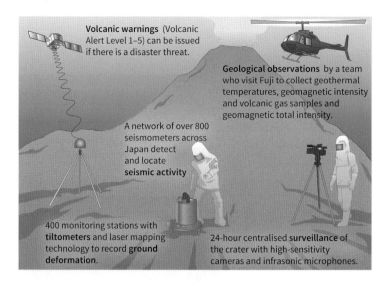

Figure 1.17 Techniques used to monitor Mount Fuji.

Preparation involves educating people about how to evacuate to safety and what to pack in a survival kit if they cannot get out of danger. Planning and preparation are common in higher income countries (**HICs**). Japan holds nationwide disaster drills on 1 September every year to mark the anniversary of the 1923 Tokyo earthquake.

Disasters occur when natural hazards threaten people's lives and property. The risk grows as the frequency or magnitude of hazards increases and people's vulnerability increases or their capacity to cope decreases. This can be presented as the **hazard risk** equation:

$$\text{RISK}(R) = \frac{\text{MAGNITUDE OF HAZARD}(H) \times \text{VULNERABILITY}(V)}{\text{CAPACITY TO COPE}(C)}$$

Increasing numbers of people are affected by hazards because more people are living in hazardous areas. Rapid urbanisation and poverty in lower income countries (**LICs**) means that increasing numbers of people are living in poorly constructed homes. People in richer countries are less vulnerable to natural disasters as they have the money available to monitor volcanoes and plan their response to disasters.

ACTIVITY 1.3

1 Study Figure 1.17.
   a Why is volcano monitoring so important in Japan?
   b Why is it that not all volcanoes in the world are monitored?
   c Outline how each technique used to monitor Mount Fuji can help to predict an imminent eruption.
2 a Outline the advantages and disadvantages of living in an area of tectonic activity such as Iceland.
   b Why is it not possible to prevent all loss of life or damage to property when a volcano erupts?
   c A local person who lives in an active volcanic region was asked why he lived there. His reply was 'because it is worth the risk'. Explain his response.
   d Study the hazard risk equation. Why has the risk of natural disasters increased over the last 30 years?
3 What items might you include in an Emergency Kit in the event of a nearby volcano erupting? Justify your choices.

Discussion point

Should a permanent 15 km exclusion zone be placed around all volcanoes to reduce the risk of death in the event of an eruption?

Further research

Use the **US Geological Survey website** (www.cambridge.org/ links/gase40003).

- Outline the different tectonic hazards associated with particular volcanoes.
- Explain why more people die in some volcanic eruptions than others.
- Investigate how active volcanoes are being monitored.

Key terms

HIC: a higher income country is defined by the World Bank as a country with a gross national income per capita above US$12 735 in 2014

hazard risk: the probability or chance that people will be seriously affected by a natural hazard

LIC: a lower income country is defined by the World Bank as a country with a gross national income per capita below US$1 045 in 2014

1.6 Why do earthquakes occur near plate margins?

Earthquakes are a sudden or violent movement (shockwave) within the Earth's crust caused by the release of built-up pressure. The initial movement may only last for a few seconds but it is usually followed by a series of aftershocks. These may be felt for months after the initial shockwave. The point within the Earth where the shockwave starts is known as the focus. The point on the Earth's surface above the focus is known as the epicentre and is where the earthquake is most strongly felt.

Where do earthquakes occur?

Every year there are over 20 000 earthquakes. Most of these are small in magnitude and go undetected, but some are very powerful and destructive. There are around 20 major earthquakes each year worldwide. These are strong enough to cause considerable damage to buildings and danger to life.

If you map the distribution of recorded earthquakes (Figure 1.18) you can see that they are concentrated in long, narrow belts that follow plate margins. One belt arcs all the way around the Pacific plate and is where 75 per cent of the world's seismic energy is released. Indonesia, Japan, China, Fiji and the Philippines are located along this 40 000 kilometre active seismic zone and experience some of the largest and most frequent earthquakes.

Discussion point

Is the death toll and damage to property always greatest at the epicentre of an earthquake?

Watch the animation on Cambridge Elevate to see how earthquake magnitude is measured on a seismometer.

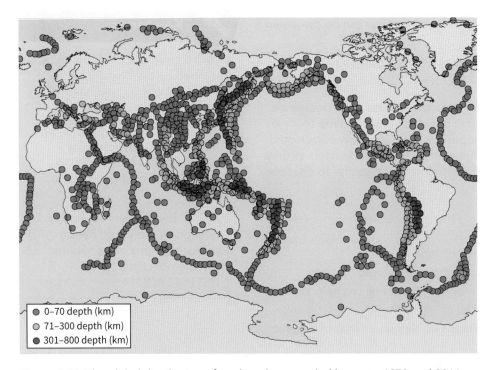

- ● 0–70 depth (km)
- ○ 71–300 depth (km)
- ● 301–800 depth (km)

Figure 1.18 The global distribution of earthquakes recorded between 1973 and 2011.

Why do earthquakes happen?

Over 90 per cent of earthquakes occur where plates are moving together at destructive plate margins or slide past each other at conservative plate margins. Where plates are moving apart, limited pressure builds up and earthquakes are characteristically low magnitude. The movement of one plate past another is not always smooth and sometimes plates 'stick' as they move past each other. Intense pressure builds up and rocks deform (bend and twist) under stress. As the pressure is released there is a sudden movement, creating seismic waves which radiate outwards from the focus through the Earth's crust. Shallow-focus earthquakes are more dangerous than those with a deeper focus, as more power is unleashed closer to the Earth's surface.

How are earthquakes measured?

Earthquakes can be measured using two different methods. The **Richter scale** measures the magnitude or energy released, on a scale of 1 to 10. The Richter scale is logarithmic, so an earthquake measuring 7.0 is ten times more powerful than one measuring 6.0.

A seismometer is the equipment used to detect the ground motions produced by an earthquake, at a specific location. A suspended pen draws a line on a rotating drum. If the ground moves so does the pen and the relative motion between the pen and earth is recorded.

The **Mercalli scale** describes the intensity and likely effects of an earthquake. The resulting damage is compared with descriptors on a scale of 1 (weak) to 12 (complete destruction). It is a subjective judgement and has limited use in areas where there are few buildings. See Figure 1.19 in Section 1.7.

Key terms

Richter scale: a scale used to measure the magnitude of earthquakes

Mercalli scale: a scale used to describe the damage caused by an earthquake

Did you know?

Earthquakes that measure more than 9.0 on the Richter scale are very rare. The most powerful earthquake ever recorded hit Chile in 1960; it measured a massive 9.5 on the Richter scale. The most powerful earthquake this century hit Sumatra, Indonesia in 2004; it measured 9.1 on the Richter scale and killed over 230 000 people.

ACTIVITY 1.4

1 Using data from the US Geological Survey showing recent earthquakes (Table 1.5):
   a Plot the location of the earthquakes listed in the table on a world map.
   b Explain the relationship between earthquake location and plate margins.
   c Suggest possible reasons for the lower than expected death toll in Pakistan and Peru.
2 a Explain the advantages and disadvantages of the Richter scale and the Mercalli scale.
   b How many times more powerful is an earthquake measuring 8.0 on the Richter scale compared to one measuring 6.0?

Earthquake	Magnitude (Richter scale)	Death toll
Wenping, China, 2014	6.2	729
Awaran, Pakistan, 2013	7.7	825
Negros, Philippines, 2012	6.7	113
Tohoku, Japan, 2011	9.0	15 853
Haiti, 2010	7.0	220 000
Sumatra, Indonesia, 2009	7.5	1 117
Sichuan, China, 2008	7.9	87 150
Near the coast, Peru, 2007	8.0	514
Java, Indonesia, 2006	6.3	5 749
Pakistan, 2005	7.6	80 361

Table 1.5 Data on recent earthquakes from the US Geological Survey.

Download Worksheet 1.2 from Cambridge Elevate for help with Activity 1.4, question 1.

1.7 What is the impact of earthquakes on people and places?

What are the primary effects of earthquakes?

The immediate damage (**primary effects**) caused by an earthquake can include the destruction of roads, bridges and buildings. People can be killed or trapped under fallen rubble. The number of deaths and scale of damage is affected by physical factors and human factors (see Table 1.6).

Key term

primary effects: the immediate damage caused by a tectonic hazard. It can include death and destruction of property

Physical factors	Human factors
Location of epicentre: if near urban areas, more people and more buildings are affected	**Population density:** more deaths and damage occur in earthquakes which strike densely populated areas
Depth of the focus: friction causes less energy from deep-focused earthquakes to reach the Earth's surface	**Level of development:** wealthy countries can afford to reduce the impact of earthquakes through prediction, protection and preparation
Time of day/week: more deaths can occur if people are inside poorly designed buildings; fewer deaths occur if people are outside and far from these buildings	**Prediction:** trying to forecast when and where an earthquake might strike can provide time for people to evacuate to safety
Geology: sedimentary rocks such as clay and sand will amplify shockwaves and cause buildings to collapse	**Building standards:** using earthquake-resistant design reduces the number of deaths

Table 1.6 Why more people die in some earthquakes than others.

I. Instrumental	Generally not felt by people, picked up on seismometers.
II. Weak	Felt only by people that are sensitive, especially on the upper floors of buildings.
III. Slight	Felt quite noticeably by people indoors, especially on the upper floors of buildings. Many do not recognise it as an earthquake.
IV. Moderate	Felt indoors by many to all people, and outdoors by few people. Some awakened. Dishes, windows and doors disturbed.
V. Rather Strong	Felt by nearly everyone. Dishes and windows may break and liquids may spill out of glasses.
VI. Strong	Felt by everyone, many frightened and run outdoors, some heavy furniture moved or overturned. Damage slight to moderate.
VII. Very Strong	Difficult to stand. Damage light in building of good design and construction, considerable damage in poorly built or badly designed structures.
VIII. Destructive	Damage slight in structures of good design. Brick buildings receive moderate to extremely heavy damage. Possible fall of monuments and walls.
IX. Violent	General panic. Damage slight to moderate in well-designed structures. Some buildings may be shifted off foundations. Walls can fall down or collapse.
X. Intense	Many well-built structures destroyed, collapsed, or moderately to severely damaged. Most other structures destroyed. Large landslides.
XI. Extreme	Few, if any structures remain standing. Numerous landslides, cracks and deformation of the ground.
XII. Catastrophic	Total destruction – everything is destroyed. Objects thrown into the air. The ground moves in waves or ripples. Landscape altered, or levelled by several metres. Even the routes of rivers can be changed.

Figure 1.19 The Mercalli scale, which describes the intensity and likely effects of an earthquake.

Destruction of buildings

It is often said, 'it is not earthquakes that kill people – buildings do'. When a 6.6 magnitude earthquake hit the historic city of Bam, Iran, in 2003, the death toll was high because the mud brick houses collapsed as people lay sleeping. This happened again when a 7.0 magnitude earthquake struck Haiti (the poorest country in the western hemisphere) in 2010. Poorly constructed slum housing in the capital Port-au-Prince collapsed as walls gave way (Figure 1.20). Over 295 000 buildings were destroyed killing 220 000, injuring 300 000 and leaving 1.5 million people homeless.

Population density

People are vulnerable to disasters if they live in earthquake zones in densely populated areas. An earthquake with magnitude over 8.0 on the Richter scale can completely destroy communities near the epicentre. When a 7.9 magnitude earthquake struck Sichuan, China in 2008, a high death toll was inevitable as 15 million people lived close to the epicentre. The earthquake killed 87 150 people.

What are the secondary effects of earthquakes?

Unforeseen after-effects (**secondary effects**) of earthquakes include:

- **Landslides:** common when earthquakes occur in mountainous areas. After the Nepal earthquake, 2015, landslides blocked roads and prevented aid reaching remote Himalayan villages.
- **Fires:** gas pipes can break and fires take hold in urban areas. Over 300 fires burnt for days in Kobe, Japan, after an earthquake struck in 1995.
- **Disease:** the 2010 earthquake in Haiti led to shortages of fresh water and damaged sanitation systems resulting in the spread of cholera to 740 000 people, leaving 8600 dead.
- **Food shortages:** after an earthquake, ports and airports are often shut, affecting food supplies. Damaged roads also prevent the movement of food.
- **Tsunamis:** giant sea waves caused by earthquakes on the sea floor. After the 9.0 magnitude Tohoku earthquake struck Japan in 2011, 30-metre waves wiped out coastal communities and caused a nuclear meltdown in the coastal power station.

Figure 1.20 Port-au-Prince in Haiti, 2010.

 Key term

secondary effects: the unforeseen consequences of tectonic hazards such as fires, spread of disease and food shortages

---

ACTIVITY 1.5

1 Study Figure 1.20.
   a Describe the primary effects of the Haiti earthquake shown in the photograph.
   b If you were an aid organisation seeing this image, what emergency aid would you send to Haiti?
2 a Describe the secondary effects of earthquakes.
   b A news reporter said, 'The outbreak of cholera in Haiti was inevitable'. Explain whether you agree with him.
3 Study Table 1.6.
   a Explain why earthquakes in LICs usually have a greater human impact than similar magnitude earthquakes in HICs.
   b Using examples, explain why more people die in some earthquakes than others.

1.8 How can the earthquake risk be reduced?

Why do people live in areas at risk of an earthquake?

There are a number of reasons why people remain living in areas where an earthquake could strike. Major earthquakes happen relatively infrequently so many generations of a family can take the risk that a major earthquake will not occur in their lifetime. For example the highly unstable San Andreas Fault had a violent earthquake in San Francisco in 1906 (Figure 1.21) but did not have another one until 1989. Other people, living in earthquake zones like Japan, think that with good planning and preparation the risk of death is greatly reduced. Many of the Earth's natural resources of oil, minerals and geothermal energy are concentrated near plate margins and so the economic benefits can outweigh the risks. In Nepal, a million tourists a year go trekking and mountain climbing and contribute US$1.8 billion to the country's economy. Employment opportunities for the local Sherpa population as guides and porters are the biggest income stream for the remote region.

Figure 1.21 The 1906 San Francisco earthquake.

Reducing the risks of an earthquake

Earthquake prediction is not advanced. Over the last 40 years scientists have tried to observe possible earthquake precursors such as unusual animal behaviour, changes in radon gas emissions and disturbances in the magnetic field in order to warn people of an impending earthquake. For each scientific report that supports a particular precursor, there is another study that disputes it. Instead of monitoring earthquake precursors, scientists have turned to earthquake prediction. By mapping past seismic activity, scientist can better predict where and when an earthquake might strike. In the seismic gap model, a location with recent seismic activity has a lower probability of an impending earthquake strike than one without it. The 2015 Nepal earthquake (see the Gorkha earthquake example) had been widely predicted by scientists, as the region west of Kathmandu had not experienced a major earthquake since 6 June 1505.

As we cannot accurately predict when earthquakes will strike, people living in known earthquake zones may respond to the threat of disaster by being prepared. Methods of reducing risk can involve protection (educating people or improving building design) and planning (emergency evacuation plans, information management and warning systems).

Earthquakes are a regular occurrence along the San Andreas Fault, California. In 1906, a 7.8 earthquake struck San Francisco killing 3000 people and triggering fires that ravaged the city. In response to this disaster, authorities prepared people and emergency services (Figure 1.22). San Francisco now has strict building laws, with earthquake-resistant buildings such as the Transamerica Pyramid (Figure 1.23 and 1.24). Earthquake-resistant buildings protect people and reduce damage to property in the event of an earthquake. Old masonry buildings have been modified and strengthened (retrofitted) to make them more resistant to seismic activity. When a shallow but powerful 6.5 magnitude earthquake struck California in 2003 there were only two deaths.

Figure 1.22 Preparation: methods of reducing the impact of earthquakes.

Figure 1.23 The Transamerica Pyramid, San Francisco.

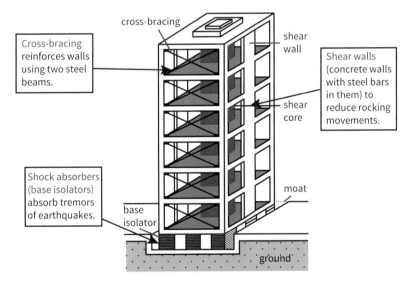

Figure 1.24 Protection: earthquake-resistant building designs.

Further research

Find out how the Torre Mayor building in Mexico City, Mexico and the Transamerica Pyramid in San Francisco, USA are designed to resist earthquakes. Suggest why different earthquake-resistant designs are more suitable for higher income or lower income countries.

ACTIVITY 1.6

1 Study the text and Figure 1.17 in Section 1.5.
   a Draw a table to show the different means of monitoring and predicting volcanoes and earthquakes.
   b For each technique used to monitor volcanoes, explain the reasons for each technique and how it can predict a volcanic eruption.
   c To what extent do you think it is easier to predict a volcanic eruption than an earthquake?

The Tohoku earthquake, Japan, 2011

What caused the earthquake?

- The tectonic plates in the area around Japan are complex. The Pacific plate, moving westwards at a rate of 83 mm a year, is being forced under the minor tectonic plate on which the north of Japan sits (the Okhotsk plate) at a subduction zone known as the Japan Trench.
- The earthquake was shallow-focused at a depth of approximately 30 km.
- Faulting occurred along 300 km of the Japan Trench and uplifted the fault by over 30 m. It was the most powerful earthquake ever recorded in Japan and lasted over 3 minutes.

The impact of the earthquake was increased by the following factors:

- Japan's coastal area is flat, low-lying and the soft soil amplifies shockwaves.
- There are deep coastal bays that amplify tsunami waves.

What were the primary effects?

The March 2011 earthquake was preceded by a series of large foreshocks over the previous two days. The Japan Meteorological Agency sent out an earthquake warning on national television and radio, giving precious time for people to evacuate.

The Tohoku earthquake

- The earthquake shifted the Earth on its axis by 10 cm.
- The coastal areas of Sendai, located 130 km from the epicentre, were all but destroyed.
- Skyscrapers in Tokyo (370 km from the epicentre) swayed, but damage was limited due to earthquake-resistant design.
- Most casualties and damage occurred in Iwate, Miyagi and Fukushima where 127 000 buildings were completely destroyed and a million were damaged across north-east Japan.
- Oil refineries in Ichihara and Sendai were ablaze.
- One school in Ishinomaki, Miyagi lost 84 of 121 students and teachers.
- Around 4.4 million households in north-east Japan were left without electricity and 1.5 million without running water.
- 1100 sections of train track were damaged. Only four trains were derailed because all high-speed trains had been automatically stopped by an early warning system.

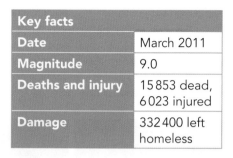

Key facts	
Date	March 2011
Magnitude	9.0
Deaths and injury	15 853 dead, 6 023 injured
Damage	332 400 left homeless

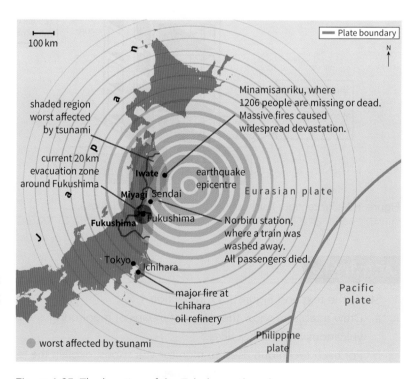

Figure 1.25 The location of the Tohoku earthquake.

What were the secondary effects?

At least 1800 houses were destroyed in Fukushima when the Fujinuma dam failed. The earthquake triggered a tsunami that took 10 minutes to reach the coast with waves up to 38 m high (Figure 1.26). The waves travelled 10 km inland. In total, 56 bridges and 26 railways were washed away along the east coast of Honshu from Chiba to Aomori. The earthquake knocked out the power supply to the Fukushima Daiichi Nuclear Power Plant. The tsunami then flooded back-up generators and triggered nuclear meltdown in three reactors. Further afield, the tsunami caused several massive slabs of ice to calve (break away) from the Sulzberger Ice Shelf, Antarctica and damage to property in California, Hawaii and Ecuador. Nissan's UK Sunderland plant shut down due to a parts shortage from its Japanese factories. The damage caused was in excess of $300 billion, making this the costliest natural disaster in world history.

Figure 1.26 Tsunami waves engulfing Sendai International airport.

Immediate responses

People got outside or ran for the hills when they heard earthquake and tsunami warnings. The government was overwhelmed by the scale of the disaster and quickly asked for international help. Over 116 countries and 28 international organisations responded. The military were forced to bury the dead in mass graves to prevent the spread of disease. AT&T (an American telecommunications corporation) maintained wireless and telephone networks to help people communicate free of charge. Japan declared a state of emergency following problems at six nuclear reactors. Around 140 000 residents were evacuated from a 20 km radius around the Fukushima plant (Figure 1.27).

Figure 1.27 Damage caused by the Tohoku earthquake.

Long-term responses

All 55 nuclear reactors across Japan were taken offline for safety reasons. This resulted in power blackouts across Japan. Toyota and Sony stopped production and other companies decided to permanently relocate abroad, even though all of Japan's ports reopened to limited ship traffic by the end of March. The railway network was fully functional by September. Japan's food exports were limited due to radiation fears. A year later, homes needed to be constructed for the 330 000 victims who were still homeless.

In the longer-term, reconstruction of the coastal areas will need to take place. This will include the re-building of road, rail and port infrastructure, as well as houses and community facilities. Questions over the future location of nuclear power plants and the height of new tsunami defences will need to be answered.

Visit the **NOAA website** for more information about the Tohoku earthquake. (www.cambridge.org/links/gase40004)

The Gorkha earthquake, Nepal, 2015

What caused the earthquake?

- Nepal is being uplifted by the collision of the Eurasian and Indian plates.
- Kathmandu Valley is located on the Himalayan fault line, which slipped by around 3 metres. Shaking lasted for 30 seconds.
- The earthquake was shallow-focused at a depth of approximately 15 km.
- Kathmandu sits on 600 m of soft lake infill in the Kathmandu Basin. The soft soil amplified the shockwaves (Figure 1.28).

Key facts	
Date	25 April 2015
Magnitude	7.8
Deaths and injury (estimates)	9 100 dead, 18 000 injured
Damage (estimate)	100 000 left homeless

Did you know?

The timing of an earthquake can have a huge impact on the death toll. It is thought that fewer people died in the Gorkha earthquake because it struck during Saturday lunchtime when people were out in the fields and not indoors in poorly constructed buildings.

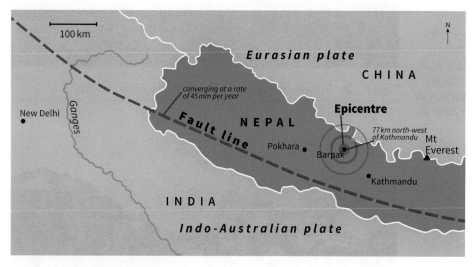

Figure 1.28 The location of the Gorkha earthquake.

What were the primary effects of the earthquake?

Close to the epicentre, fewer than 50 houses out of 1400 remained standing in the village of Barpak and almost no houses were left in the trekking village of Langtang. UNESCO World Heritage sites in the Kathmandu Valley, including the royal palaces of Kathmandu (Figure 1.29), Patan and Bhaktapur were reduced to rubble. Some 24 000 classrooms collapsed, affecting the education of over 1 million Nepalese children. Many deaths occurred in the tightly packed and narrow lanes of Kathmandu where buildings had been constructed out of rubble from the 1934 earthquake. Severe damage occurred to roads, houses and infrastructure (gas, electric, water, sanitation systems and communication cables).

What were the secondary effects?

An avalanche on Mount Everest (Figure 1.30) destroyed the Khumbu Icefall and engulfed base camp killing at least 19 climbers and making it the deadliest day on the mountain in history. Landslides in Langtang Valley blocked roads and hampered the aid effort. The 2015 monsoon triggered more landslides on the unstable slopes of the Kathmandu Valley. Those that did survive the quake were left with no power, communications or clean water, living in temporary shelters and with a greater risk of disease. Food shortages were widespread, affecting 1.5 million people. It is estimated that reconstruction costs could exceed $10 billion.

Immediate responses

- For the first 24 hours survivors were left to fend for themselves, moving rubble with bare hands to try to reach relatives and friends.

Figure 1.29 The collapse of the temple at Durbar Square, Kathmandu.

Figure 1.30 Climbers being evacuated after the avalanche on Mount Everest.

- With only nine functioning helicopters, the Nepal Army could only reach a few of the badly injured victims stranded in remote, mountainous villages.
- Relief was sent rapidly, as the government quickly asked for help. This included a 60-strong emergency team from China and a 30-bed mobile hospital from Pakistan. India sent its military to assist with rescue efforts.

Long-term responses

- Many tourists cancelled their trips to Nepal in 2015 following the earthquake, reducing a vital income stream.
- As one of the poorest countries in the world, Nepal will struggle to get back on its feet, relying on $450 million of international aid to rebuild (Figure 1.31).

ACTIVITY 1.7

1 Describe the causes of the Tohoku and the Gorkha earthquakes. Include information about convection currents, plate margins and the focus of the earthquakes.
2 Describe how and explain why the effects of an earthquake differ in countries at different stages of development.
3 Copy Table 1.7 onto a large piece of paper to summarise the key facts about the two earthquakes.

	Tohoku earthquake	Gorkha earthquake
Cause		
Primary effects		
Secondary effects		
Immediate responses		
Long-term responses		

Table 1.7 Earthquake comparison summary.

Tip

Know the differences between the primary and secondary effects of the Tohoku and the Gorkha earthquakes. Understand how the responses to the earthquakes were different in the immediate and longer term.

Discussion point

Should tourists and climbers on Mount Everest pay hazard insurance, just in case there is an earthquake during their visit and they need to be evacuated?

Figure 1.31 Relief aid arriving by aeroplane in Kathmandu.

Further research

Search for the BBC report 'Why are some tremors so deadly?' which compares the 2015 Nepal earthquake with earthquakes in Chile.

Tip

In the exam **read the question** carefully. It might be helpful to underline the command word and highlight key words, before you start to write your answer.

Visit the **Earthquake Report website** for more information about the Gorkha earthquake.

Assess to progress

1 Explain how risks from a tectonic hazard can be reduced. `6 MARKS`

2 What is:
  a a constructive plate margin `2 MARKS`
  b a destructive plate margin `2 MARKS`
  c a conservative plate margin? `2 MARKS`

3 Using examples, explain why people choose to live in areas at risk from tectonic hazards. `6 MARKS`

4 a Why do volcanoes erupt at subduction zones? `4 MARKS`
  b Why do earthquakes happen at conservative plate margins? `4 MARKS`

5 a How does the Richter scale measure earthquakes? `2 MARKS`
  b How does the Mercalli scale measure earthquakes? `2 MARKS`

1.9 Predicting seismic activity in China

In this issue evaluation you will consider the effectiveness and relative advantages and disadvantages of different methods of earthquake monitoring, prediction and mitigation.

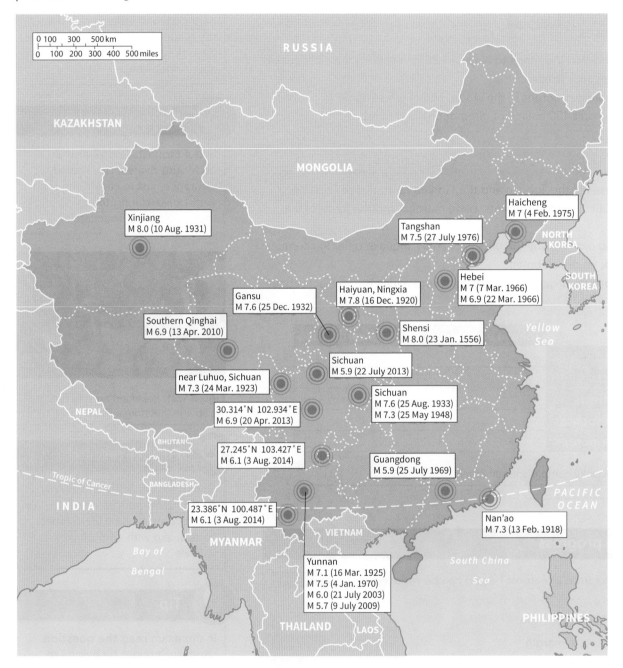

Figure 1.32 The location of major earthquakes in China.

China's attempts to predict and monitor earthquakes

- Seismologists in Nanjing, the capital of eastern Jiangsu province, have set up seven observation centres at zoos and animal parks in the region to watch for changes in behaviour in the animals, which might be a sign of an imminent earthquake tremor.

- The 1975 Haicheng earthquake was successfully predicted and was based on reports of changes in groundwater and soil elevations.
- China's Seismological Society suggested, after the 2008 Sichuan earthquake, that the same fault line was likely to rupture again. They used seismograph data to make this prediction.

Predicting seismic activity is notoriously difficult, if not impossible. All of the methods described above are highly experimental, and, in the case of seismograph data, are only available after the earthquake event. However, China is developing a network of Seismo-Electromagnetic Satellites (CSES) as part of its earthquake monitoring network. Strong seismic activity can cause anomalies in the Earth's atmosphere and magnetic field which has the potential to be detected by satellites. Indeed, two days before Japan's March 2011 earthquake, Chinese researchers detected abnormal electromagnetic signals in the area using ground-based systems. When a disturbance is recorded it is important to compare data at a number of locations to ensure that it is not caused by another factor. This is the reason that China's satellite network is more ambitious than existing systems, developed by countries such as France, because it uses several satellites. This means that locations are monitored more frequently meaning that data is likely to be more reliable. China is working with other countries including Italy, Russia and Ukraine to develop the system as these countries already have similar, although less ambitious, systems.

Is a satellite network the best option?

However, the technique is still experimental and poses a risk of false alarms and therefore predictions are not being made public in case they are wrong. Critics have also suggested that the $1 billion predicted spend on the system would have been better spent on building earthquake-proof buildings in China's seismic zones.

Many of the fatalities in the Sichuan earthquake occurred due to collapsing buildings and falling masonry (Figure 1.33); in total 1.5 million houses collapsed and 6 million were damaged. Many of Sichuan's public buildings, including schools and hospitals were not earthquake-proof. Parents of pupils who were killed blamed poor construction as many government buildings surrounding schools which collapsed managed to withstand the shock. In the year following the earthquake, 2009, the Chinese government spent $1218 million strengthening schools in earthquake prone areas and this seems to have had an impact. When an earthquake struck Sichuan in April 2013 none of the school buildings constructed since 2008 collapsed and those that experienced similar levels of shaking typically had less damage.

Figure 1.33 A collapsed building following the Sichuan earthquake in May 2008.

Factfile

Country: China

Location: Between the Earth's two largest seismic zones – the Pacific and the Indian – it is compressed by the Pacific, Indian and Philippine plates (Figure 1.32).

Number of earthquakes since 1900 of magnitude 6 and above: over 600

Number of fatalities since 1900: 627 000 (23 per cent of all earthquake fatalities worldwide)

Deadliest Chinese earthquake in the 21st century: 12 May 2008, Sichuan. This earthquake was of a 7.9 magnitude and 69 180 were killed and over 5 million people were made homeless.

Factfile

Impacts of the Sichuan earthquake

- Many rivers became blocked by landslides as a result of the quake.
- Chengdu International airport was shut down for several hours.
- A cargo train carrying 13 tanks of petrol derailed and caught on fire.
- In Beichuan County, 80 per cent of the buildings collapsed.
- In Shifang, two chemical plants collapsed leading to the leakage of 80 tonnes of ammonia.
- Mobile and internet communications were cut.
- The Zipingpu Hydropower plant was damaged and the Tulong reservoir upstream was in danger of collapse.
- 12.5 million livestock, mainly birds, were killed.

Are new earthquake-proof buildings effective?

'If the buildings were older and built prior to the 1976 Tangshan earthquake the chances are that they weren't built adequately for earthquake forces. It is only following this earthquake event that China put in place a design code to try to make buildings as earthquake-proof as possible.'

Reginald DesRoches, professor of civil and environmental engineering at Georgia Tech.

Before the 2008 earthquake, building codes in China's rural west were poor. Buildings were made from adobe (Figure 1.34), which is a type of clay used as a building material, or watered down concrete and not well equipped to withstand even moderate tremors.

Figure 1.34 A house in rural China made from adobe.

However, China has moved quickly and most cities contain modern, steel high-rise buildings and a larger percentage of the population lives in earthquake-proof buildings (Figure 1.35). Building codes in China are now well defined and comparable with Japan and the US. However, trying to enforce these codes is more problematic. Engineers and builders need to follow the codes which has not always happened. Contractors feel under pressure to complete projects ahead of schedule and may cut corners while builders have been known to substitute cheap materials in order to cut costs. As a consequence, earthquake-proof buildings are expected to last 25 to 30 years in China compared with 70 to 75 years in the US. Following the 2008 earthquake, 40 000 people were resettled into a new city called Yongchang which is 10 km from the epicentre. However, cracks have already appeared in the brand new homes raising questions about whether they will be able to withstand future tremors.

There is a saying that 'earthquakes don't kill people, buildings do'. Therefore should the money being spent on China's earthquake monitoring satellites be spent instead on retrofitting existing buildings or constructing new, earthquake-proof ones?

Figure 1.35 An example of how China has made its buildings earthquake-proof.

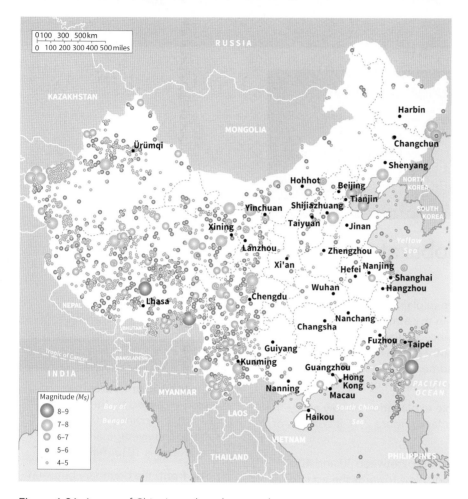

Figure 1.36 A map of China's earthquake records.

For more information see the **BBC website** (www.cambridge.org/links/gase40005)

Assess to progress

1 Study Figure 1.36 which shows earthquakes that have occurred in China between 1949 and 2000. Which two of the following statements are true? 2 MARKS
   - Most of China's earthquakes occur in the east of the country.
   - Some earthquakes in China reach a magnitude of 8 to 9.
   - In the west of the country the earthquakes are clustered together.
   - There are no earthquakes in the east of the country.
   - Earthquakes in China range from magnitude 5 to magnitude 9.
2 Suggest why the buildings in China are so vulnerable to earthquake tremors. 2 MARKS
3 Should China invest money in satellite technology for earthquake monitoring? Justify your answer. 9 MARKS + 3 SPaG MARKS

Tip

- For question 2, it could be helpful to think about what the buildings are constructed from and how seismic waves work.
- For question 3, there are a number of things you could consider here. Think about how effective satellite technology is compared with other forms of prediction, monitoring and mitigation. You could also consider whether all of China is affected by earthquakes and therefore whether it's fair or not to invest such vast sums of money.

2 Weather hazards – tropical storms

In this chapter you will learn about...

- global atmospheric circulation patterns
- the global distribution of tropical storms
- how tropical storms are formed
- the effects of tropical storms and how these effects are managed
- how climate change may affect the formation and effects of tropical storms.

2.1 The general model of global atmospheric circulation

The general atmospheric circulation model tries to explain how the energy which controls weather and climate is transferred in the atmosphere. It also shows how surface and high-level winds are influenced by areas of low and high **atmospheric pressure**. All air is under pressure but if it is rising then the air is said to be unstable and the pressure is lower. In some circumstances, air pressure can be extremely low and this might influence the formation of tropical storms.

The three-cell model outlined below gives a simple explanation of how air moves at both high altitudes and at the Earth's surface (Figure 2.1). It consists of three cell types.

Hadley cell – The largest cell extends from the equator to between 30 and 40° north and south. Within the Hadley cell, surface winds blow towards the equator and then rise as they are heated. At high altitudes this air then moves towards the poles, decending between 30 and 40° north and south to form areas of high pressure. It is under these conditions of high pressure that many of the world's great deserts, including the Sahara, are formed.

Polar cells – Polar cells extend from the poles to between 60 and 70° north and south. The cold, dense air descends at the higher latitudes and moves towards lower latitudes. As it does so, the air warms and begins to rise at an area known as the **polar front**.

Ferrel cells – Ferrel cells lie between the Hadley cell and the Polar cell. In these latitudes the cooler air from the poles meets the warmer air from the tropics, causing the air to rise. This often occurs around the latitude of the UK and partly explains the unsettled weather experienced in the UK.

The rising air in the Hadley cell at the equator forms an area of low pressure at the surface. The intensity of the Sun's rays means that air rises rapidly and huge cumulonimbus rain clouds are formed, associated with thundery downpours.

Where the warm air rises at the polar front, another area of low pressure is formed.

Key terms

atmospheric pressure: the pressure caused by the weight of air at any point on the Earth's surface; the average air pressure at sea level is 1013 millibars

polar front: the boundary between a Polar cell and Ferrel cell

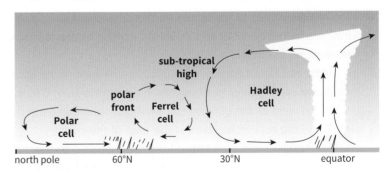

Figure 2.1 The three-cell model of global circulation.

How does the global circulation influence surface wind patterns?

The effects of the three-cell model are seen at the surface in the way they affect wind patterns. In theory, air should move from high pressure to low pressure (just like it does when a tyre is deflated!), such as from the sub-tropical high pressure areas towards the equator. However, the rotation of the Earth produces the **Coriolis effect**, which alters the direction of the winds and deflects them to the right in the northern hemisphere and to the left in the southern hemisphere. This leads to the prevailing westerly/south-westerly winds experienced in the UK. Together, pressure differences and the Coriolis effect produce the common wind patterns shown in Figure 2.2.

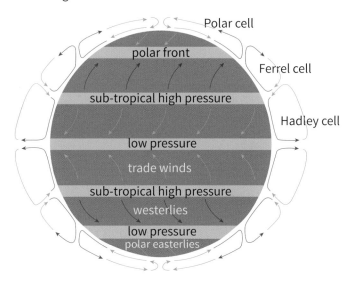

Figure 2.2 Global atmospheric circulation and surface wind patterns.

- Trade winds blow at the surface from the sub-tropical high pressure areas towards the equator. The north-east trade winds are formed in the northern hemisphere and the south-east trade winds in the southern hemisphere.
- Mid-latitude westerlies blow towards the poles from the sub-tropical high pressure areas. These winds bring series of **anticyclones** and **depressions** that influence much of the weather in the UK.
- Polar easterlies flow from the polar high pressure areas towards the mid-latitudes.

ACTIVITY 2.1

1 Draw a labelled diagram to show the three-cell model of global atmospheric circulation.
2 Describe the main surface wind patterns in both the northern and southern hemispheres.

 Download Worksheet 2.1 from Cambridge Elevate for help with Activity 2.1, question 2.

3 Why is the sinking air associated with high pressure conditions more stable than the rising air found in low pressure areas? Think about how this will influence the formation of tropical storms.

Did you know?

High in the atmosphere are bands of very fast winds called jet streams (Figure 2.3). Windspeeds can reach up to 320 km/h. This fast-flowing air has a strong influence over the weather conditions in the UK and other mid-latitude countries.

Key terms

Coriolis effect: the effect, caused by the rotation of the Earth, which deflects winds to the right in the northern hemisphere and the left in the southern hemisphere

anticyclone: a large-scale circulation of winds around a central region of high atmospheric pressure; the circulation is clockwise in the northern hemisphere and anti-clockwise in the southern hemisphere

depression: where warm and cold air meet, usually at mid-latitudes over the UK – the warm air is less dense than the cold air and so rises above it, creating low pressure on the ground; weather associated with a depression includes rain and strong winds

Figure 2.3 A jet stream.

Further research

Find out more about how the global circulation and jet streams influence the weather and climate of the UK.

2.2 Where do tropical storms develop?

Tropical storms are intense areas of very low atmospheric pressure, which generally develop between the tropics where ocean temperatures are higher. On average, there are between 80 and 100 tropical storms each year.

Tropical storms are known by different names in different parts of the world:

- hurricanes in the East Pacific and Atlantic oceans
- cyclones in the Arabian Sea and Bay of Bengal off the coasts of India and Bangladesh, and in the South West Pacific and Western Australia
- typhoons in Japan, China and other countries in Eastern Asia.

However, they all share similar characteristics: they are all areas of very low air pressure and generally develop within the tropics between 5° and 20° north and south of the equator (Figure 2.4). Closer to the equator the rotation of the Earth does not produce sufficient 'spin' for a tropical storm to develop.

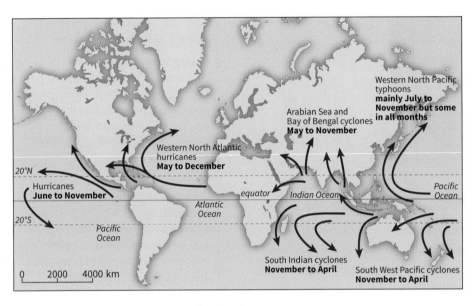

Figure 2.4 Worldwide distribution of tropical storms.

What causes tropical storms to develop?

Meteorologists are not sure exactly why tropical storms develop. They do know that tropical storms contain enormous amounts of energy, generated from the heat of the atmosphere and moisture from the oceans. From the general atmospheric circulation it is known that the areas of tropical storm formation are characterised by low pressure and rising air. This results in a strong, upward movement of very moist, warm air (known as severe air instability). Other conditions have been identified as common characteristics in the development of tropical storms. These include:

- large areas of tropical ocean where the water temperature is greater than 26.5 °C
- a considerable depth of warm water – usually at least 70 m
- a strong upward movement of very moist, warm air (known as severe air instability)
- predominant winds generally blowing in the same direction.

As a tropical storm begins to develop, warm air rises in a spiralling motion, drawing up huge amounts of evaporated water which cools and condenses quickly to form towering banks of cloud. As this happens, a vast amount of

Did you know?

Tropical storms have individual storm names, for example Typhoon Haiyan, Hurricane Katrina, Cyclone Aila.

Skills link

Developing your cartographic skills is an important part of the AQA Geography course. You need to be able to recognise and describe distributions and patterns such as those shown in Figure 2.4.

Did you know?

The naming of individual tropical storms follows a strict procedure controlled by the World Meteorological Organisation. Lists of alternating male and female names are used on a six-yearly rotation. If any one storm proves especially damaging or deadly, the name is retired from the list and never used again.

Key term

meteorologist: a scientist who studies the causes of particular weather conditions

Visit Cambridge Elevate for an animation showing the formation of a tropical storm.

energy is produced. This energy is what powers a storm. As a storm builds more and more water vapour is drawn upwards. The clouds grow larger, wind speed increases and the likelihood of very heavy rainfall increases.

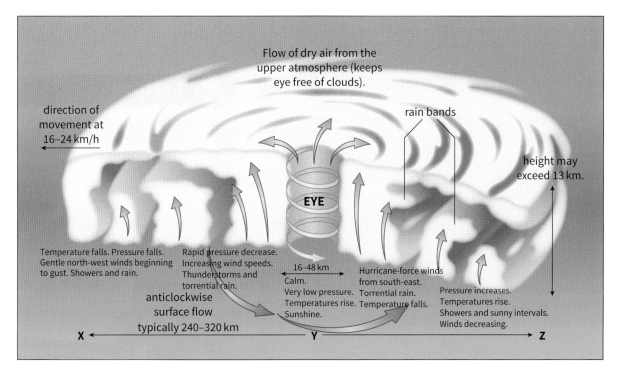

Flow of dry air from the upper atmosphere (keeps eye free of clouds).

rain bands

direction of movement at 16–24 km/h

height may exceed 13 km.

EYE

Temperature falls. Pressure falls. Gentle north-west winds beginning to gust. Showers and rain.

Rapid pressure decrease. Increasing wind speeds. Thunderstorms and torrential rain.

16–48 km
Calm.
Very low pressure.
Temperatures rise.
Sunshine.

Hurricane-force winds from south-east. Torrential rain. Temperature falls.

Pressure increases. Temperatures rise. Showers and sunny intervals. Winds decreasing.

anticlockwise surface flow typically 240–320 km

X ←   Y   → Z

Figure 2.5 The structure of a tropical storm.

Satellite images show very clearly the spiralling winds of a tropical storm surrounding an easily identifiable centre known as the **eye**. Here, denser air sinks towards the ground and conditions are calm and clear (Figure 2.5). This calmness can be very deceptive for people affected by tropical storms who, after a period of intense rainfall, may believe that the storm has passed. However, huge banks of cloud known as the **eye wall** surround the eye (Figure 2.6) so, once the eye has passed over an area, wind speeds and rainfall increase again.

Once formed, a tropical storm increases in power as it moves over the ocean towards land. Upon reaching land the storm starts to weaken. In some areas, such as the Caribbean, the storm may decrease in intensity as it passes over an island, only to increase again as it reaches the ocean once more. The average lifespan of a tropical storm is between 7 and 14 days.

Figure 2.6 Hurricane with clearly visible eye just off the coast of Florida, USA.

ACTIVITY 2.2

1 Annotate a world map to show the global distribution of tropical storms. Use the different terms for the storms in the appropriate locations.

 Download Worksheet 2.2 from Cambridge Elevate for help with Activity 2.2, question 1.

2 a Describe the conditions required in order for tropical storms to develop.
  b Suggest why these conditions are required for their development.
3 Draw a labelled diagram showing the structure of a tropical storm.

 Key terms

eye: the centre of a tropical storm where sinking air causes relatively calm, clear conditions

eye wall: the towering banks of cloud bearing heavy rainfall which surround the eye

2.3 How might climate change affect tropical storms?

It is clear that one of the conditions required in order for tropical storms to develop is a relatively high ocean temperature. Recent evidence suggests that there has been a marginal increase in ocean temperatures over the past 20 years (see Chapter 4). Consequently, according to the USA's National Oceanic and Atmospheric Administration (NOAA), **climate change** might have the following effects on tropical storms:

Key term

climate change: the global increase (or decrease) in temperature and its effect on the world's climate

- a wider distribution area as the latitude of oceans with a temperature of 26.5 °C or more spreads further north and south of the equator
- the intensity of tropical storms increasing by 2–11 per cent as higher ocean temperatures generate more energy
- the actual number of tropical storms may remain largely the same but there might be a greater number of extremely intense storms
- an increase of up to 20 per cent in the amount of rainfall within 100 km of the eye of the storm.

Recent research into the links between climate change and tropical storms has found evidence for some of these changes. Scientists studying Typhoon Haiyan, which struck the Philippines in 2013, found that the increased sea temperatures and ocean heating along its path increased the strength of the typhoon, making its destructive power greater (Figure 2.7). Similarly, unusually warm waters increased the effects of Hurricane Sandy, which affected the Caribbean and north-eastern USA in 2012. In one report, scientists said:

"It is possible that subways and tunnels may not have been flooded without the warming-induced increases in sea level and storm intensity and size, putting the potential price tag of human climate change on this storm in the tens of billions of dollars."

Source: 'Attribution of climate extreme events', Kevin E. Trenberth, John T. Fasullo, Theodore G. Shepherd. *Nature Climate Change*, 22 June 2015

Figure 2.7 Damage caused by Typhoon Haiyan (Philippines 2013).

What are the reasons for the changes in tropical storm activity?

Many climate change models, based on current scientific evidence, show that sea surface temperatures (SST) in the areas where tropical storms develop are set to increase markedly over the next 100 years. This warming will occur to a greater depth of water than at present, leading to the wider distribution and intensity of tropical storms.

The increase in SST could also mean an increase in the volume of rising water vapour which powers a tropical storm. Higher temperatures will lead to more evaporation and higher wind speeds may also increase the destructive potential of each storm.

One further factor that needs to be taken into account is the potential for global sea-level rise (Figure 2.8) and the effect this may have when combined with the storm surge associated with tropical storms. Many storms hit low-lying coastal areas such as the eastern seaboard of the USA, Pacific islands and countries such as the Philippines and Bangladesh. Already these areas are regularly affected by flooding due to tropical storms. With the increasing intensity of storms, higher sea levels and lower atmospheric pressure, storm surges could be considerably higher, leading to an increasing flood risk.

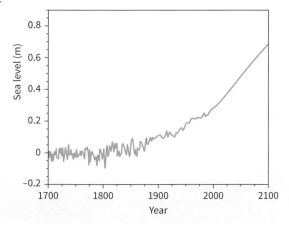

Figure 2.8 Sea-level change (1700–2100) as identified by the Intergovernmental Panel on Climate Change (IPCC) (using lowest estimates).

An increasing threat

Although there are variations in the number, frequency and energy (Figure 2.9) of tropical storms, it does appear that tropical storms are becoming more destructive. In the USA, six of the ten most powerful recorded hurricanes in history have occurred since 1990. One reason for this recorded increase may be the advances in storm detection. Data collection has become increasingly reliable and has improved markedly following technological developments in satellite monitoring. The increased destruction may also be a result of more people now living in more expensive homes near the coast (Figure 2.10) so the economic cost of storm damage has increased.

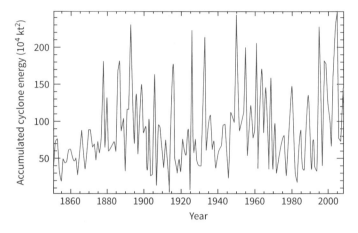

Figure 2.9 Hurricane activity and energy in the Atlantic Basin (1850–2010).

Figure 2.10 Expensive housing on low-lying land in Florida.

ACTIVITY 2.3

1 What effect might climate change have on the distribution and intensity of tropical storms?
2 Why will low-lying coastal areas be particularly vulnerable to the changes in tropical storm intensity brought by climate change?

 Further research

The phenomenon of El Niño may also have an effect on the distribution and intensity of tropical storms and how they may change in the future. Find out more using textbooks and the internet.

2.4 What are the effects of tropical storms?

Tropical storms are capable of causing considerable destruction and can have a significant impact on the physical and human environment. The immediate risks of tropical storms are caused by:

- strong winds – with wind speeds upwards of 200 km/h (125 mph) there can be widespread damage to buildings, infrastructure (including power lines, roads, ports and airports), trees and crops
- **storm surges** – the low air pressure means that sea levels are very high. Strong on-shore winds create huge waves which move towards coastal areas, causing extensive coastal flooding
- heavy rainfall – torrential rain can cause widespread flooding
- the possibility of localised tornadoes.

NOAA defines these as primary effects – effects that result directly from the event itself. There are also secondary effects (effects that result from the primary effects) as the hurricane leads to the formation of other hazards. There may be mudslides and **landslides** as unstable slopes become completely saturated with rain and flood water. When the storm has passed and the floods are subsiding, polluted water supplies can cause further deaths through diseases such as cholera. People may be forced to seek make-shift shelters because of the damage to their homes, and basic services will be affected as infrastructure awaits repair.

How are tropical storms measured?

The **Saffir-Simpson scale** is used to measure the strength of tropical storms. Table 2.1 shows the different types of damage that may occur according to the category of the storm.

Key terms

storm surge: when low-pressure storm conditions cause the sea to rise

landslide: the movement of earth or rock from a slope as a result of it becoming unstable (usually from heavy rainfall)

Saffir-Simpson scale: the five-point scale used to classify tropical storms according to their strength

Look on Cambridge Elevate for a video on the impact of Cyclone Aila, which struck south-western Bangladesh on 25 May 2009.

Category	Wind speed	Type of damage
1	119–153 km/h 74–95 mph	Very dangerous winds – some damage to well-constructed houses, roofs and gutters. Large branches will snap. Power outages with damage to power lines and poles.
2	154–177 km/h 96–110 mph	Extremely dangerous winds – extensive damage to roofs and houses. Shallow trees will be uprooted and block roads. Near-total power loss for a minimum of several days.
3 (major)	178–208 km/h 111–129 mph	Devastating damage to well-built houses. Many trees uprooted and blocking numerous roads. Electricity and water unavailable for days, if not weeks.
4 (major)	209–251 km/h 130–156 mph	Catastrophic damage – severe damage to houses, especially roofs and walls. Fallen trees and power lines will leave residential areas isolated with no power for weeks or months.
5 (major)	252 km/h + 157 mph +	Catastrophic damage – widespread destruction with a high proportion of homes and infrastructure destroyed. Most of the area uninhabitable for weeks or months.

Table 2.1 The Saffir-Simpson scale (*Source*: National Hurricane Center).

Responses to tropical storms

Immediate responses

When a tropical storm hits an area, the population and government need to respond immediately. Immediate responses include putting the emergency services on alert, evacuating the area, and rescuing and treating any victims of the storm.

In richer countries there will be detailed plans already in place and forecasting and efficient communication systems mean that damage and loss of life will be kept to a minimum. The population will be alerted through media services including the internet. Before severe storms the police, army and other organisations will often conduct street-by-street reconnaissance trips, ensuring that people comply with evacuation orders or have sufficient supplies if they are staying put.

In lower income countries the lack of infrastructure such as decent roads and communication systems make it more difficult to alert people to the dangers of the coming storm and people are more reliant on the help that comes after the storm has hit. In many countries help will arrive from friendly governments and charities such as Médecins Sans Frontières who provide medical support and MapAction who will help with targeting support where it is most needed. With government agencies often struggling to cope, these organisations are invaluable in helping to find and treat those in affected areas.

Long-term responses

Once the immediate danger is over, longer-term responses can begin. These include rebuilding damaged housing and infrastructure, and often setting up or improving protective systems such as levées. Again, in richer countries, the need for intensive help is often less than in LICs. Houses and other buildings are often built more solidly and therefore damage is reduced to a minimum. Governments can respond quickly to make sure infrastructure such as bridges and roads are repaired quickly.

In LICs it may be months or years (or longer) before things are returned to the way they were before the storm. There may be long-term health concerns and, if people have lost their homes and are gathered in temporary housing, this can include an increasing incidence of diseases such as typhoid, cholera and malaria. With fewer resources available it can take a long time for roads to be rebuilt or for people to move back to their home areas and start their lives again.

41

2.5 How can the effects of tropical storms be reduced?

Efforts to reduce the damaging effects of tropical storms include monitoring storms, predicting where storms will hit, protecting at-risk assets and planning how to deal with the storm impacts.

Monitoring

Satellite monitoring at agencies such as the National Hurricane Center in Florida and the Joint Typhoon Warning Center in Hawaii identifies the formation of tropical storms and tracks their paths and strength. They classify storms according to the Saffir-Simpson scale by measuring their strength and the types of damage that they produce.

Prediction

Meterologists can use the monitoring information to provide warnings, which give an opportunity for local areas to prepare. However, despite advances in scientific methods, accurate prediction is not always easy because the strength and path of a tropical storm can change quickly.

International scientific weather agencies regularly update tropical storm forecasts. They supply forecasts to national weather services and media channels. The forecasts tend to cover a large area known as a **cone of uncertainty** (Figure 2.11) – the area most likely to be affected by the tropical storm.

Protection and planning

As methods of forecasting have improved, so have the ways in which people prepare for a storm. In many countries schoolchildren are taught about the dangers of tropical storms and given lessons about what to do if a storm hits. Governments produce posters, leaflets and information for the media, and people are encouraged to prepare disaster kits.

Key term

cone of uncertainty: the area defined by forecasters where a hurricane may cause damage

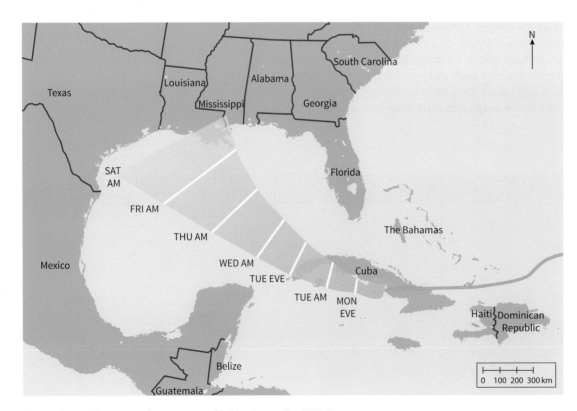

Figure 2.11 The cone of uncertainty for Hurricane Ike (2008).

Figure 2.12 Tropical storm shelter.

In some coastal areas, building regulations mean houses have windproof roofing tiles and stormproof windows. In lower income countries, storm shelters (Figure 2.12) are common and can ensure the survival of whole villages.

Governments have plans in place to act swiftly when a storm hazard is likely. Evacuation plans may be carried out and emergency services, voluntary organisations and the armed forces put on alert. These organisations need to plan for the immediate effects of the storms including injuries, death, loss of homes, and damage to infrastructure, food and water supplies. Long-term response plans also need to be put in place. These will include rebuilding roads and rehousing people who have lost their homes. There may be long-term health effects including psychological problems and depression which need both medical attention and counselling.

Lower income countries are often less well equipped to plan for the effects of tropical storms, resulting in greater loss of life and a much longer time to recover. They will also be much more reliant on foreign aid to cope in the event of a major disaster.

Assess to progress

Test your progress by answering the following questions.

1 Identify **three** conditions that are needed for a tropical storm to develop. `3 MARKS`

2 Using the information in Table 2.1, describe **three** kinds of damage caused by a category 4 tropical storm. `3 MARKS`

3 Describe **two** features of a storm surge. `4 MARKS`

4 Assess the extent to which the primary effects are more significant than the secondary effects of a tropical storm in a low or high income country.
Use a named example in your answer. `9 MARKS + 3 SPaG MARKS`

ACTIVITY 2.4

1 Identify the main socio-economic and environmental impacts of tropical storms.

2 Suggest five pieces of advice that might be given to children in preparation for a tropical storm.

3 Explain how prediction and preparation can reduce the risks from tropical storms.

Look on Cambridge Elevate for some useful weblinks:
- the BBC (www.cambridge.org/links/gase40006), National Geographic (www.cambridge.org/links/gase40007), and the *Guardian* (www.cambridge.org/links/gase40008) have lots of information about hurricanes
- the Joint Typhoon Warning Center (www.cambridge.org/links/gase40009) and National Hurricane Center (www.cambridge.org/links/gase40010), in the USA, provide warnings of tropical storms already occurring or on the way

Tip

Many longer questions ask you to give an example – make sure you locate any example used and learn some detailed facts about it.

Typhoon Haiyan, 2013

When the 270 km/h gusts of Typhoon Haiyan swept through the central Philippines (Figure 2.13) on Friday, 8 November 2013 it was one of the strongest storms ever to make landfall. By the time it had passed, more than 6000 people were dead and the storm had affected an estimated 11 million people.

Did you know?

In tropical countries there are many unexpected dangers in the aftermath of a storm – snakes may shelter in houses or piles of debris, while malaria-carrying mosquitoes breed in car tyres and other rubbish filled with stagnant water.

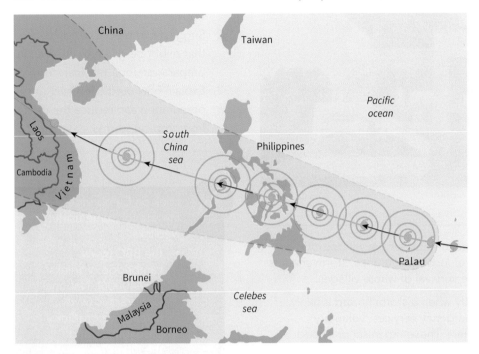

Figure 2.13 Track of Typhoon Haiyan from the Philippines towards Southeast Asia.

Figure 2.14 Damage to property caused by Typhoon Haiyan in the Philippines.

The government of the Philippines was well aware a big storm was on the way and had issued warnings and evacuated thousands of people from the areas that were expected to be worst hit. However, many did not heed the warnings or, in the rural areas of the country, were not aware of them. There were also concerns that the government did not emphasise the danger of the likely storm surge which, when combined with the high tides, measured over 5 metres high. Many low-lying areas were completely flooded, with houses and buildings flattened (Figure 2.14). The infrastructure in these areas, including roads and power lines, was completely destroyed.

In a relatively poor country such as the Philippines, the effects of the 'super typhoon' were magnified. Without money to feed the many millions of now homeless people, or to provide adequate medical care or fresh water, the country relied on foreign aid. Although aid arrived, it was often impossible to move the supplies (Figure 2.15) to where they were needed outside of the major cities and airports because of the destruction of the roads and other infrastructure. For many in the most remote areas, the delays in the distribution of essential supplies caused prolonged suffering and more fatalities.

While in many cases there is a huge initial response to disasters such as Super Typhoon Haiyan, the long-term effects can be more difficult to deal with. Infrastructure may remain in disrepair, economic activity is severely affected and health suffers for months after the storm has passed.

	Typhoon Haiyan
Location	Philippines, Southern China, Vietnam
Maximum windspeed	275 km/h (171 mph)
Deaths	6 201 in the Philippines (1 785 missing)
Homeless	5 million
Number of people affected	16 million (13 million in Philippines alone)
Effects on infrastructure	550 000 homes destroyed; 580 000 homes severely damaged; lack of safe drinking water, sanitation, food and fuel
Effects on the economy	Destruction of infrastructure including roads led to problems in crop production and distribution
	Almost 6 million workers in the Philippines affected by loss of income
	Total cost unknown
Recovery	Six months to several years in places

Table 2.2 The effects of Typhoon Haiyan.

In the case of Haiyan at least 6 million people lost their jobs and 30 000 fishing boats were destroyed. Schools and hospitals were damaged beyond repair leading to increased long-term economic effects. UK-based charity Oxfam estimated it would be at least three to five years before life returned to normal for most people.

Figure 2.15 Aid supplies waiting for distribution in the Philippines.

ACTIVITY 2.5

1 With reference to specific locations, describe the track of Typhoon Haiyan shown in Figure 2.13.
2 Draw a table to show the short-term and long-term effects of Typhoon Haiyan.
3 Many people in the Philippines ignored the warnings given about the approach of Typhoon Haiyan. What do you think could be done to make more people take notice?

In this chapter you will learn about...

- what is meant by the term 'extreme weather'
- the types of extreme weather that the UK is likely to experience
- the effects of extreme weather and how they can be reduced
- the causes, impacts and responses for an extreme weather event.

Key terms

climate: an area's average weather conditions measured over a number of years; this is measured in a variety of terms, e.g. average precipitation, maximum and minimum temperatures, sunshine hours, humidity

weather: the day-to-day condition of the atmosphere, for example, the weather may be sunny, windy, rainy and so on

UK (United Kingdom): the UK comprises the countries of England, Wales, Scotland and Northern Ireland

depression: (in meteorology) where warm and cold air meet, usually at mid-latitudes over the UK – the warm air is less dense than the cold air and so rises above it, creating low pressure on the ground; weather associated with a depression includes rain and strong winds

3.1 What extreme weather events affect the UK?

The UK's weather and climate

The **climate**, or normal pattern of **weather** for the **UK** can be described as temperate maritime (Figures 3.1 and 3.2). Temperate climates are those which have moderate temperatures and lie between the temperature extremes of cold, polar areas and warm, tropical areas. Temperate climates can be divided into those that are continental (influenced by land) or maritime (influenced by sea). The UK's position as an island means that it has a maritime climate. This means that the UK has cool summers and mild winters. It also has a smaller annual temperature range than continental climates.

High and low pressure systems bring day-to-day changes to the UK's weather. Low pressure occurs where air rises. The air cools and condenses to form clouds and so the weather that low pressure brings is cloudy, rainy and windy. Low pressure systems are called **depressions**. High pressure occurs where air is sinking. The skies are clear and so the weather is sunny and calm in summer and very cold in winter. High pressure systems are known as **anticyclones**.

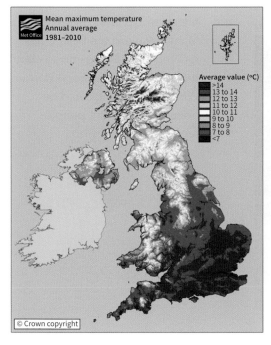

Figure 3.1 UK average temperature.

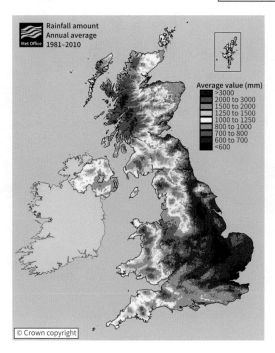

Figure 3.2 UK average rainfall.

Examples of extreme weather in the UK

Extreme weather can be defined as 'weather that is different from the normal pattern that would be expected and includes severe gales, thunderstorms and hailstorms'.

Extreme weather events in the UK tend to be caused by extremes of low or high pressure. Extremes of low pressure bring storms, strong winds and, particularly in winter, heavy snowfall and blizzards. In contrast, extremes of high pressure bring long periods of drought in summer months (Figure 3.3) but can bring bitterly cold, icy conditions in winter (Figure 3.4).

An example of extreme weather caused by an anticyclone occurred in August 2003. The UK and parts of Europe experienced a heatwave where temperatures reached 38.5 °C, the warmest for 500 years. Rainfall levels were also much lower than average throughout June, July and August.

Figure 3.3 Low water levels in a reservoir on the Isle of Wight.

Figure 3.4 Blizzard in Surrey, UK.

In March 2013, parts of the UK experienced bitterly cold temperatures, strong winds, heavy rain and flooding caused by a depression. Snow depths of over 40 cm were reported at sites in the Midlands that would normally have an average of 6.4 cm in March and the strong winds and cold temperatures meant that much of the snow drifted and did not thaw for several days (Figure 3.4).

Key term

anticyclone: a large-scale circulation of winds around a central region of high atmospheric pressure; the circulation is clockwise in the northern hemisphere and anti-clockwise in the southern hemisphere

Discussion point

What extreme weather have you experienced in the UK? Why was the weather event extreme? What happened?

Did you know?

The lowest temperature ever recorded in the UK was in Scotland. On three separate occasions, in 1895 (Braemar, Aberdeenshire), 1982 (Braemar, Aberdeenshire) and 1995 (Altnaharra, Highlands) the temperature reached −27.2 °C.

ACTIVITY 3.1

1 Use the data in the table below to draw a climate graph for the UK.

Month	Jan	Feb	Mar	Apr	May	Jun	Jul	Aug	Sep	Oct	Nov	Dec
Average temp. (°C)	6.4	6.6	8.9	11.4	14.7	17.3	19.4	19.1	16.5	12.8	9.1	6.7
Average rainfall (mm)	121.7	88.6	95.1	72.7	70.0	73.4	78.1	89.5	96.4	127.1	121.2	120.2

Table 3.1 Climate data for the UK.

2 a Describe the pattern of temperature and rainfall. Refer to the highest and lowest values, the total volume of rainfall and the temperature range.

  b What is the link between the temperature and rainfall data? Suggest reasons for this link.

3 Look at either Figure 3.3 or Figure 3.4.

  a Write down annotations for the photograph which describe in detail the effects of extreme weather in the UK.

Skills link

- For further guidance on graphs, see 19.1, Graphical skills – graphs and charts.
- For further guidance on photograph annotation, see 18.3 Methods of presenting data.

3.2 What are the effects of extreme weather?

The precise effects of extreme weather depend very much upon the weather event itself. High temperatures will have very different effects to high rates of **precipitation**. However, because extreme weather is out of the ordinary, people are often unprepared and so the effects can be severe. Effects can be social, economic and environmental.

The effects of heatwaves

The heatwave of August 2003 was linked to the deaths of over 20 000 people across Europe, of whom 2000 were in the UK. Many others became ill through heat stroke and dehydration. Water supplies were affected and a hosepipe ban was introduced in an attempt to conserve water. In addition, railway tracks buckled and some road surfaces melted in the heat making travel difficult. In London, underground trains overheated making conditions unpleasant. Many crops and farm animals died which led to an increase in food costs (Figure 3.5). It is estimated that throughout Europe the cost of the heatwave to farmers was over 13 billion euros.

While many of the effects of extreme summer weather are negative there are also some positive effects. For example, during the 2003 heatwave tourism experienced a boost as people took advantage of the warm conditions and stayed in the UK rather than holidaying abroad. Sales of food associated with warm weather, such as ice cream and cold drinks, also increased.

Key term

precipitation: any form of water, both liquid and solid, which falls from the sky; rain, snow, sleet and hail are all examples of precipitation

Tip

The effects of a heatwave can be sorted into categories as follows.

Social:
- illnesses and deaths
- hosepipe ban
- transport problems

Economic:
- deaths of crops and livestock leading to increased food costs
- transport problems causing problems for businesses
- increased tourism and sales of warm weather items

Environmental:
- deaths of crops and livestock

Figure 3.5 Drought conditions affecting crops.

The effects of storms and snowstorms

Storms are characterised by strong winds and torrential rainfall. Winds in excess of 70 mph can cause electricity pylons and trees to blow over causing damage to cars and buildings (Figure 3.6). The torrential rainfall can lead to widespread flooding, particularly if the ground is already saturated. During winter the precipitation often falls as snow to cause a snowstorm. The widespread snow of March 2013 saw extreme effects. The weight of the snow and ice caused both branches of trees and power cables to break leading to traffic chaos and power cuts throughout the UK; 137 000 people were without power in Northern Ireland. The drifting snow meant that it was very difficult for engineers to reach remote locations to repair the power cables. Many were not repaired for several days.

How can the effects of extreme weather be reduced?

The Met Office is responsible for monitoring and forecasting the weather in the UK and it issues warnings if weather is likely to be extreme. The purpose of this is to provide information so people can plan and prepare. Other agencies, such as the Environment Agency and local councils use this information to put control measures such as flood defences in place. Effective monitoring, planning and preparation reduces the negative effects of extreme weather as much as possible.

Figure 3.6 An electricity pylon blown down in high winds.

ACTIVITY 3.2

1   Investigate an extreme weather event in the UK from the last ten years. A good starting point is the **Met Office website** (www.cambridge.org/links/gase40011): search for 'Weather case studies'.
  a   Describe the event and suggest why it occurred.
  b   Examine the effects of the event. Start by sorting the effects into those that are positive and those that are negative. Then sort these into those which are social, those which are economic and those which are environmental.
  c   Explain what was done to try to reduce the effects of the event.
2   Carry out some research into how the Met Office prepares people and businesses in the UK for extreme weather events.
  a   Search for '**Be prepared**' and '**Weather warning guide**' as a starting point.
  b   Create a fact sheet and/or a video to explain how the Met Office specifically prepare for an extreme event.
3   Watch the weather forecast for a week and keep a diary which notes what the forecast reports and what the actual weather conditions are for each day.
  a   How accurate are the forecasts for the week?
  b   Why do you think this is?

 Further research

Use the **Environment Agency website** (www.cambridge.org/links/gase40012) to investigate how they prepare for extreme weather events.

 Download Worksheet 3.1 from Cambridge Elevate for help with Activity 3.2, question 3.

3.3 Is the weather in the UK becoming more extreme?

If the weather in the UK is becoming more extreme, we would expect extreme weather events to become more frequent. We might also expect these events to last for a longer period of time or for the impacts to be more severe than in the past.

Has the weather in the UK always been variable?

Throughout history there have been examples of extreme weather in the UK. The Great Storm of 1703 (Figure 3.7) is thought to be one of the greatest natural disasters to have affected the south of England: 2000 chimney stacks collapsed in London and the lead roofing blew off Westminster Abbey. It is thought that between 8000 and 15000 lives were lost. Most of these were seamen who were returning from the War of Spanish Succession.

Although the average summer temperature in the UK is between 15°C and 16°C, heatwaves have also been a feature of the weather in the past. From 31 August to 3 September 1906, temperatures were over 32°C across most of the UK. Bawtry in South Yorkshire recorded the hottest ever September day in the UK with a temperature of 35.6°C.

Did you know?

During the Little Ice Age, which occurred between 1350 and 1850 in northern Europe, temperatures were significantly cooler than at present. In London temperatures were so cold that the River Thames froze over in winter and frost fairs were held on its icy surface.

Figure 3.7 A drawing of storm conditions off the coast of Scarborough, UK in 1703.

Is extreme weather in the UK becoming more frequent?

While this evidence suggests that the UK has always experienced extreme weather there is also evidence to suggest that events are becoming more frequent. Weather patterns over the North Atlantic affect conditions in the UK; five out of the ten most extreme North Atlantic winter weather patterns occurred within the last decade. There is also evidence from climate models run by the Met Office to suggest that, despite summers in future years becoming drier, there is also likely to be an increase in intense summer rainfall events which could lead to flash flooding.

A particularly good example of weather conditions fluctuating from hot to cold and wet to dry occurred during 2012. Hosepipe bans were in force in March due to drought conditions. Yet there was torrential rain, cooler temperatures and widespread flooding, particularly in the latter half of the year, which made 2012 the second wettest year since records began (Figure 3.8).

Discussion point

How has the monitoring of the weather improved? Why is it more accurate and reliable?

Figure 3.8 Flooding in the UK.

Figure 3.9 The position of the jet stream over Canada.

These volatile conditions were blamed on the unusual position of the **jet stream** (Figure 3.9). In summer it usually moves north of the UK, but in the summer of 2012 it stayed in the south and therefore brought extreme weather. It is thought that the main reason for this is the global change in climate. Scientists are concerned that the UK is likely to see more extreme weather events in the future due to global warming. While temperatures may increase, flooding events are also likely to become more frequent as the position of the jet stream becomes more erratic.

 Look on Cambridge Elevate for some useful weblinks:

Visit the **Met Office website** (www.cambridge.org/links/gase40013) for a more in-depth look at the jet stream.

The **BBC website** (www.cambridge.org/links/gase40014) and the *Guardian* website (www.cambridge.org/links/gase40015) have good summaries of the extreme weather events which occurred in 2012.

The **Met Office website** (www.cambridge.org/links/gase40016) and the environmental section of the *Guardian* website (www.cambridge.org/links/gase40017) have summaries of the extreme weather events which occurred in 2014 putting the UK in a global context.

ACTIVITY 3.3

1 Search on the **Met Office website** (www.cambridge.org/links/gase40018) for 'UK climate - Extremes - Met Office' which shows UK climate records.
   a Do these figures suggest that the UK's weather is becoming more extreme?
2 Using newspaper clips or headlines:
   a Create a timeline of extreme weather events since 2000. The **Met Office website** (www.cambridge.org/links/gase40019) is a good place to start for examples of extreme events which could be included; search for 'Past weather events'.
   b What evidence does your timeline add to the argument that UK weather is becoming more extreme?
3 What is likely to happen to the frequency and impact of extreme weather events in the UK in the future? Why do you think this?

Key term

jet stream: strong winds (around 200mph) that circle the Earth between 5 and 10 miles up in the atmosphere

 Watch a video on Cambridge Elevate about some of the impacts of the 2007 floods on a family in Tewkesbury.

Discussion point

Do you think that the UK weather is becoming more extreme?

Further research

Look for an article on the **Met Office website** (www.cambridge.org/links/gase40020) called 'Global circulation patterns' which explains global atmospheric circulation and the UK's place within this.

2013 UK winter storms

From early December 2013 to early January 2014 the UK experienced a succession of severe winter storms. The storms initially brought strong winds, particularly to Scotland and northern England. However, as rainfall totals increased, large rivers, including the River Severn and River Thames, began to flood. Finally, towards the end of the year and into January, the strong winds, river flooding and high spring tides led to storm surges along the coast.

The causes of the 2013 winter storms

It is thought that the main cause of the 2013 storms was a particularly deep depression which formed over the Atlantic Ocean (Figure 3.10). On four occasions the pressure reached 950 **mb** which is very low and, on 24 December a pressure of 936 mb was recorded, the lowest pressure recorded in the UK since 1886. The reason for this low pressure was that temperatures in the south of the UK were around 1.5 °C warmer than average during December 2013. This strongly contrasted with the bitterly cold temperatures coming from the north.

The impacts of the 2013 winter storms

The 2013 winter storms had significant social impacts. The strong winds at the beginning of December led to travel chaos, particularly in Scotland. The rail network shut down, planes from Glasgow, Edinburgh and Aberdeen were cancelled and there were many road accidents resulting in two fatalities. Over 100 000 homes also lost electricity, mainly due to trees falling onto power lines (Figure 3.11).

The storm then moved towards the south of the UK. Again there was travel disruption but this was made worse by widespread flooding which affected large areas of Dorset, Hampshire, Surrey and Kent; 50 000 people had no electricity over Christmas and several rivers burst their banks (Figure 3.12). One man was swept away by the floodwaters in Devon.

The storms also had severe economic impacts. In March 2014, the Association of British Insurers (ABI) estimated that insurance claims from the storms would cost around £1.1 billion, including £446 million for homes and businesses that were flooded. Travel disruption would have also affected business deliveries and staffing.

The environmental impacts of the storms included increased erosion of beaches and damage to farmland.

Responses to the 2013 winter storms

The management strategies put in place to respond to the 2013 winter storms took effect at both a national and a regional scale. For example, on a national scale the UK's **COBRA** committee met twice to discuss plans for responding to the storms and the Environment Agency issued 40 severe flood warnings indicating that there was danger to life.

At a regional scale, local emergency services were in operation throughout the period in order to minimise the impacts of the storm event. On the east coast of England 15 000 people were advised to evacuate their homes.

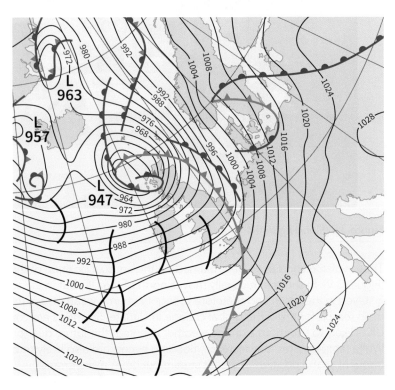

Figure 3.10 A synoptic chart to show the pressure patterns over the UK in December 2013.

Figure 3.11 An uprooted tree as a result of strong winds.

Figure 3.12 A river that has burst its banks in Essex, UK.

Figure 3.13 Cliff collapse as a result of storm surges on the Norfolk coast.

Figure 3.14 The Thames Barrier flood defence.

Emergency rest shelters were set up for those that were evacuated, although some people stayed with relatives. Some people refused to leave their houses, saying that they were going to protect their homes and possessions. In late January 2014, Somerset County Council and Sedgemoor District Council both declared a 'major incident', requesting extra support including help from the armed forces.

Modern flood defences, such as the Thames barrier, were also deployed during the period of the storms.

The management strategies that were used meant that the impacts of the winter storms were less severe than a similar weather event on the east coast in 1953 in which over 300 people were killed.

ACTIVITY 3.4

1   Create an annotated map of the UK which clearly shows the different impacts of the 2013/2014 winter storms. You could colour code your annotations to show whether they are cause, impact or response and include facts and figures in your annotations.

Download Worksheet 3.2 from Cambridge Elevate for help with Activity 3.4, question 1.

2   Search the **National Trust website** (www.cambridge.org/links/gase40021) for an article called 'How the 2013/14 winter storms affected the coast' which highlights how the 2013/2014 winter storms affected the coast. Find a suitable photograph and describe examples of the impacts.

3   Search the **BBC website** (www.cambridge.org/links/gase40022) for an article called '10 key moments of the UK winter storms' published on 17 February 2014 which looks at ten key moments of the UK winter storms. Read the article and then use the text, maps or images to pick out five interesting things which you were not aware of before.

Assess to progress

1   'The weather of the UK is becoming more extreme.'
    Use evidence to support this statement.

 6 MARKS

Key terms

mb (millibars): a unit of atmospheric pressure

COBRA: stands for Cabinet Office Briefing Room A; a cabinet committee which meets to discuss the response to crises

Discussion point

Most of the impacts of extreme weather events such as the 2013 winter storms are negative. Can you think of any positive impacts for weather events such as this?

Look on Cambridge Elevate for some useful weblinks:

- The Environmental management section for the government website has more information about the Thames Barrier flood defence (www.cambridge.org/links/gase40023).

- The Met Office has produced a Recent Storms Briefing which contains a more in-depth look at the 2013 winter storms (www.cambridge.org/links/gase40024).

Tip

The Assess to progress question asks you to use evidence to support the statement so, in this instance, do not worry about including evidence that counters the argument.

- You could consider extreme temperatures and rainfall in recent years.
- Give specific examples as evidence – facts and figures are important.

4 Climate change

In this chapter you will learn about...

- what the evidence for climate change is
- whether climate change is man-made or a natural process
- the possible consequences of climate change
- ways to help combat the effects of climate change.

4.1 What is the evidence for climate change?

There is clear evidence that the Earth's climate has gone through considerable periods of change during the **Quaternary period**. Scientific research and visual evidence in the landscape shows the planet has experienced both cold, glacial periods and warmer, drier periods. Some scientists believe that the mass extinction events during the Quaternary period, which saw large mammals such as woolly mammoths wiped out, were the result of a 6 °C increase in temperature between 15 000 and 10 000 years ago.

Since the last ice age, which ended around 12 000 years ago, there has been an upward trend in global temperatures despite the fluctuations in the climate. Data now suggests there has been a more consistent increase in temperatures since 1950, and especially in the last 20 years (Figure 4.1). It is this recent, rapid warming of the Earth's climate that has become known as **global warming**.

Scientists have gathered a large amount of evidence for climate change. Some shows very long-term trends stretching back thousands of years. Other, more recent data have been gathered as scientific instruments and monitoring have become more sophisticated.

Long-term evidence

Since 1988, scientists working in Antarctica and Greenland have carried out drilling experiments that have provided evidence of past climates dating back almost one million years. **Ice cores** of up to three kilometres deep have been extracted from the glacial ice sheets. These cores show a series of rings, which each equate to one year's worth of snowfall. By analysing the concentration of gases in the rings, including water vapour, carbon dioxide, oxygen and methane, scientists have been able to build up a picture of past climates. Analysis of the pollen trapped in the cores (from plants and trees that previously grew in these areas) has also helped improve the accuracy of estimates of historic temperatures.

Another method of dating climate events is the analysis of fossil plankton from the Baltic Sea and Black Sea. Higher temperatures lead to increased precipitation and decreasing salinity, which has a direct correlation with the amount of plankton found in cores from the sea bed.

The science of dating events by looking at evidence from tree ring growth has also been used to study climate change because the rings are wider in warm and wet years. Using rings from ancient trees such as the Bristlecone Pines (Figure 4.2) which live for 6000 years or more in California, gives a lot of information that can be used to show how the climate has changed.

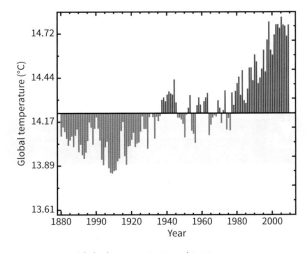

Figure 4.1 Global temperature change.

Figure 4.2 Bristlecone Pine.

Historical evidence from the past 200 years

Direct instrument readings

Measurements from thermometers show that, since the 1970s, there has been a rapid global temperature increase of around 0.55 °C. The warmest years on record have occurred in the last two decades with 2005 and 2010 only being beaten by the temperatures in 2014 when the average land and ocean temperatures were 0.69 °C above the long-term average.

Glacial retreat

Photographs of glaciers taken in the Alps and other high mountain ranges since the mid 1800s highlight changes in the Earth's climate. They show that glaciers have been melting as temperatures increase and the snout melts (Figure 4.3). Recent computer models also suggest up to 25 per cent of the world's current mountain ice could disappear by 2050.

Cover of Arctic sea-ice

There is currently around 50 per cent less sea-ice over the Arctic Ocean than there was 30 years ago. Sea temperatures appear to be getting warmer (Figure 4.4) with the permanent ice cover decreasing at around 9 per cent per decade. There are fears the Arctic may become ice-free by 2100.

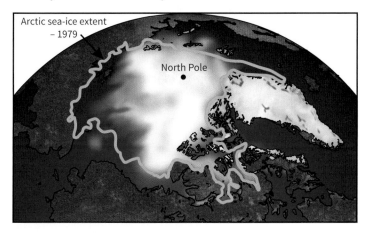

Figure 4.4 Sea-ice thinning and retreat over the Arctic Ocean.

Spring, autumn and winter changes

In the northern hemisphere, certain species of plants and birds seem to be appearing earlier in the year than they did 30 years ago. People talk about much harsher winters in the past and are often convinced that snowfall is less than it used to be. Useful as it is, this kind of anecdotal evidence is not very scientific but can be used as a starting point for more research.

ACTIVITY 4.1

1   What visual evidence do we have that global warming has taken place?
    a   Draw a timeline from 10 000 years ago to the present day.
    b   Add four different types of evidence to your timeline showing how far back they give us information about the Earth's changing climate.
2   Study the photographs showing glacial retreat in Figure 4.3.
    a   What evidence do they provide for climate change?
    b   Suggest what further information could be used to help decide if climate change has taken place.

Watch the video on Cambridge Elevate which shows how glaciers have helped our understanding of climate change.

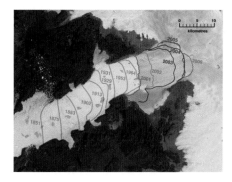

Figure 4.3 Glacial retreat of the Jakobshavn glacier, Greenland.

Discussion point

How convincing is the evidence that exists to support climate change? Could the changes outlined just be natural variations?

Further research

The vast majority of scientists and politicians now agree that recent climate change is a problem caused by the actions of humans that needs to be tackled. There are some people who believe that the evidence for climate change is inconclusive or that it is simply part of the natural cycle.

Make sure that you understand both of these viewpoints.

Download Worksheet 4.1 from Cambridge Elevate for help with Activity 4.1, question 1.

4.2 What are the causes of climate change?

There is clear evidence that climate change has been taking place for many thousands of years and so there must be natural processes that influence it happening. Increasingly, however, the data shows that more recent, short-term global warming has been the result of human development.

Long-term influences on climate change

- Over periods of thousands of years there are small changes in the way the Earth moves around the Sun. (These are called Milankovitch cycles; for more information, go to Cambridge Elevate.) We get warmer temperatures as we get closer to the Sun and cooler periods as we head further away.
- Increased **sunspot** activity leads to more solar radiation and increased temperatures. This was responsible for the warm years of the 1940s although, since 1970, evidence suggests the Sun has cooled slightly.
- Volcanic eruptions send vast amounts of dust into the air. The dust from huge eruptions like those of Krakatoa (1883) and Mount Pinatubo (1991) forms a blanket around the Earth and temperatures decrease as less sunlight gets through.
- More recently, a possible link has been identified between changes in global ocean currents such as those associated with El Niño (Chapter 2) and climate change. There is ongoing research to try and determine what the link is.

Short-term influences on climate change

Perhaps the main natural process associated with climate change is the **greenhouse effect**. Greenhouse gases in the atmosphere – carbon dioxide (CO_2), methane, nitrous oxides and others – keep global temperatures around 30°C warmer than they would otherwise be (Figure 4.5). They are all produced naturally but are being added to by human activity. This **enhanced greenhouse effect** is what many scientists believe is causing global climate change.

How does human activity increase the amounts of greenhouse gases?

Fossil fuels

The use of **fossil fuels** releases large amounts of carbon dioxide and other gases into the atmosphere. Coal, oil and natural gas were all formed over many millions of years but since the mid 1700s, they have been burnt at an increasing rate by industry and power stations, cars, planes and other types of motor vehicles (Figure 4.6). It is estimated that, since around 1750, the amount of CO_2 in the atmosphere has increased by 30 per cent.

Further research

Find out more about the impact of Milankovitch cycles on climate change. As an extension task you could also find out more about the life of the man himself – a very interesting scientist indeed.

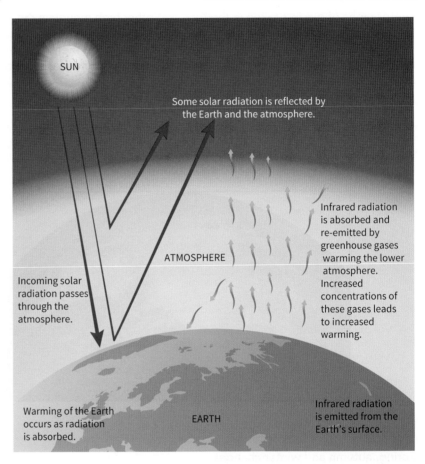

Figure 4.5 The greenhouse effect.

Figure 4.6 Greenhouse gas emissions from a power station.

Changes in agriculture

Over the last few decades there has been a worldwide increase in agriculture to feed the world's growing population. The crops that are grown need chemical fertilisers and pesticides which add greenhouse gases, including methane and nitrous oxides, to the atmosphere. Methane is also produced by cattle and by decomposing vegetation in padi fields.

Figure 4.7 Evidence of burning in tropical rainforests.

Halocarbons

Production of halocarbons in solvents and fridges was banned in 1989 but still has an historic effect on greenhouse gas concentrations.

Deforestation

Deforestation is another source of the increase in CO_2. The burning of tropical rainforests (Figure 4.7), such as the Amazon, has added to the amount of the greenhouse gas. Perhaps more importantly, it has decreased the amount of the gas used up by plants during **photosynthesis**.

ACTIVITY 4.2

1   Look again at Figure 4.1.
    a   See if you can relate any of the causes of climate change in the text above to specific dates on the graph. Some internet research into, for instance, volcanic eruptions such as Mount St Helens or Mount Pinatubo, may help you here.
2   Study Figure 4.5.
    a   With the help of a diagram, explain how the natural greenhouse effect works.
    b   Add information on:
        i   the gases involved
        ii   the human activities that produce them
        iii how they cause an enhanced greenhouse effect.

Download Worksheet 4.2 from Cambridge Elevate for help with Activity 4.2, question 1.

Watch an animation on Cambridge Elevate showing how the greenhouse effect works.

Did you know?

Methane is 21 times more effective at trapping heat in the lower atmosphere than carbon dioxide so the bodily emissions from the world's estimated 1.5 billion cows are a major contributor to the greenhouse gases in our atmosphere.

Key terms

sunspot: dark spots on the surface of the Sun that sometimes discharge big bursts of heat given out by the Sun

greenhouse effect: the gases in the atmosphere which trap outgoing radiation and keep the Earth at a temperature at which humans can live

enhanced greenhouse effect: the increase in the effects of global warming due to human activities

fossil fuels: energy from plant and animal remains, such as coal, oil and natural gas, including shale gas

photosynthesis: the chemical process where plants convert carbon dioxide to oxygen

4.3 What are the effects of climate change and how do we manage them?

Computer models have been used to predict temperature changes and then to look at the potential problems these changes may bring. Nobody is certain what the changes will be but many of them are related to possible sea level rise due to water expanding as it gets hotter (known as thermal expansion). In some parts of the world the consequences are already very real and have had a devastating impact on people.

In the worst possible case, some scientists think global temperatures could increase by between 4 and 6°C by 2100. The changes this increase could bring include:

- more frequent **droughts** and floods and extreme weather events (Figure 4.8)
- widespread coastal flooding
- the extinction of some native plants and animals
- increased demand for water
- warmer winters (meaning cold-related deaths are less likely), but there may be more heat-related death and illness in the hotter summers
- more 'tropical' diseases spreading north (and south) from the tropics.

The impact of climate change will not be the same across the world. Poorer parts of the world will be more severely affected and places that are already affected by storms (such as coastal areas) may find conditions more challenging.

How to manage climate change

The ways to tackle climate change fall into two categories:

- **mitigation:** where we try to stop the change from happening
- **adaptation**: where we make adjustments to our environment as changes take place.

Mitigation

Most of the methods of mitigation are very difficult to introduce as they need the agreement of lots of different governments. This is done through the **Intergovernmental Panel on Climate Change (IPCC)** but progress has been very slow. Some of the ideas put forward include:

- increases in renewable energy including wind, solar and tidal power
- **carbon capture** (Figure 4.9) to store carbon dioxide produced by industry and power stations
- changes in agriculture away from the most polluting forms such as cattle farming and those which release greenhouse gases from fertilisers and pesticides
- large-scale tree planting programmes to act as 'carbon sinks' which soak up large amounts of carbon dioxide during photosynthesis.

These solutions are expensive for individual countries so international agreements such as the **Kyoto Protocol** are needed.

Local-level mitigation is also taking place as individuals and small communities:

Figure 4.8 Flooding due to increasingly wild weather in southern England.

Key terms

droughts: long periods of time without rainfall

Intergovernmental Panel on Climate Change (IPCC): an international body of scientists which assesses all of the evidence of human-induced climate change

carbon capture: where carbon dioxide produced by power stations is stored underground

Kyoto Protocol: an international treaty to reduce greenhouse gas emissions; the original treaty ran from 1997 to 2012 but the Doha Amendment extends the protocol until 31 December 2020

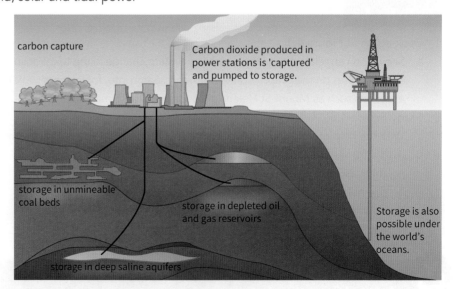

Figure 4.9 Carbon capture.

- cut down on food waste and increase recycling
- introduce energy-saving measures at home
- walk or cycle instead of taking the car
- buy local food to cut down on **food miles**
- build small-scale renewable energy plants.

Some methods may appear to have little impact on a global scale but many believe the more people that adopt them, the more effective they will be.

Adaptation

Some countries have been forced to adapt to climate change. Malé, the capital of the low-lying Maldives, is constantly under threat from storms. The government has built sea walls all around the island (Figure 4.10).

Changing crops grown by farmers, in some countries, to those better adapted to warmer and wetter conditions is one option. Governments also need to look at better management of their water resources to ensure evaporation and loss are kept to a minimum. Some countries are also installing desalinisation plants (see Chapter 16) to cope with water supply problems.

ACTIVITY 4.3

1 Read through the list of possible consequences of climate change and write down the ones that would affect the area where you live.
2 Some of the impacts of climate change would particularly affect certain areas. Research the impacts on parts of Africa (economically vulnerable) and the Arctic (environmentally vulnerable).
3 Many countries will be affected by sea level rise brought by climate change. Find out which areas are most at risk around the globe, including densely inhabited coastal areas and oceans with low-lying islands.

Download Worksheet 4.3 from Cambridge Elevate for help with Activity 4.3, question 2.

Assess to progress

1 'The world's climate is changing.' Use evidence to support this statement. **6 MARKS**
2 Outline one reason why the concentration of methane in the atmosphere has increased over the last 100 years. **2 MARKS**
3 Complete the paragraph below. **3 MARKS**
   International ideas to help combat climate change include using more .... energy, changes in agriculture and carbon .... Agreements such as the .... Protocol are needed to make sure these ideas work.
4 Give two natural causes of climate change. **2 MARKS**

Tip

It is important that you know about carbon dioxide as a primary greenhouse gas but don't limit yourself to this – you need to know about all the main gases involved.

Key term

food miles: the distance food is transported between producer and consumer

Figure 4.10 Malé, capital of the Maldives.

Find out more, visit
- the USA's **National Oceanic and Atmospheric Administration climate change website** (www.cambridge.org/links/gase40025)
- the **Intergovernmental Panel on Climate Change website** (www.cambridge.org/links/gase40026)
- the climate guide on the **UK Met Office website** (www.cambridge.org/links/gase40027)
- **United Nations Climate Change** (www.cambridge.org/links/gase40028) newsroom
- the **about education website** (www.cambridge.org/links/gase40029) for information on Milankovitch cycles
- the **WWF website** (www.cambridge.org/links/gase40030) for information about how WWF is tackling climate change
- the **New Scientist website** (www.cambridge.org/links/gase40031) for information on the science behind climate change

SECTION 2

The living world

You will learn about:
- what an ecosystem is
- why the distribution and characteristics of ecosystems vary

5 Tropical rainforests

In this chapter you will cover:
- how plants and animals have adapted to the conditions in tropical rainforests
- the causes and effects of deforestation in tropical rainforests
- sustainable ways to manage rainforests
- how the Chocóan rainforest is managed

You will study one of the following topics and will learn about:

6 Hot deserts

In this chapter you will cover:
- how plants and animals have adapted to the harsh conditions found in hot deserts
- the economic opportunities that the hot desert environment provides
- why the exploitation of the hot deserts is so challenging
- the causes and impacts of desertification
- the strategies used to reduce the risk of desertification
- Qatar as a case study

7 Cold environments

In this chapter you will cover:
- where cold environments are located
- what the distinctive characteristics of cold environments are
- the economic opportunities that can be found in cold environments
- the challenges for developing cold environments
- some strategies for managing cold environments
- the North Slope region of Alaska as a case study

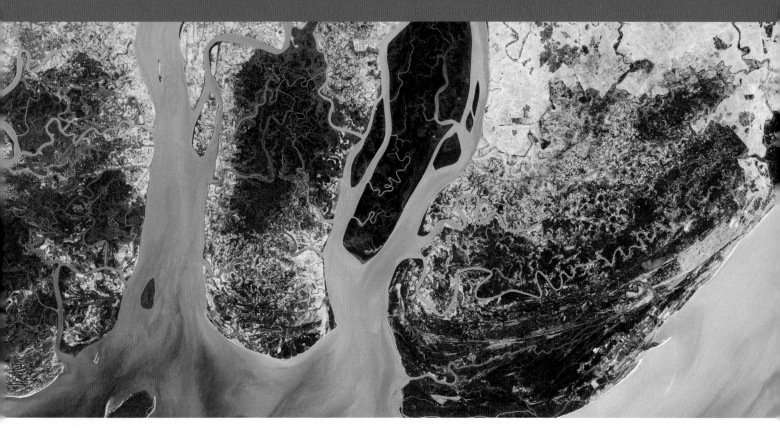

Ecosystems – an overview

Ecosystems are natural systems in which the living parts, such as plants and animals, and the non-living parts, such as climate, rock and soil have to work in harmony in order for the ecosystem to survive. Over the last 200 years human development has put increasing pressure on natural environments as the demand for food and resources has increased. Millions of hectares of forest have been cleared to make way for farming, resource exploitation and human habitation.

Some of the most fragile global environments, including tropical rainforests, hot deserts and cold environments, are under threat because of the economic opportunities that they offer. In order to ensure that these environments are protected and conserved for future generations it is vital to understand how they work and fully appreciate the impact that development is having on them. This knowledge can then be used to ensure that some of the world's most fragile and valuable environments are managed increasingly sustainably.

What is an ecosystem?

An ecosystem is a community of plants and animals within a physical environment. There are both **biotic** (living) and **abiotic** (non-living) elements in an ecosystem. Over time an ecosystem becomes stable, known as being in **equilibrium**. Equilibrium can be disturbed by physical events like climate change, by a volcanic eruption and by human events such as people clearing forest.

Figure S2.1 Mangrove forests (the deep green areas in the photo) protect inland areas from erosion and flooding. In the Irrawaddy River Delta, areas of mangrove forest have been cleared to provide space for rice cultivation. What effect could this have on the area?

 Key terms

biotic: the living part of an ecosystem, e.g. plants and animals

abiotic: the non-living part of an ecosystem, e.g. climate and soils

equilibrium: the balance between all parts in a system

How do components in an ecosystem interact?

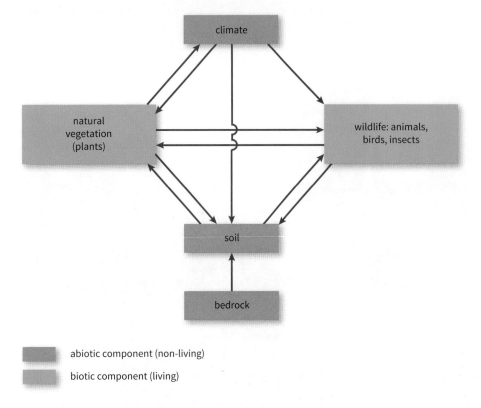

abiotic component (non-living)

biotic component (living)

Figure S2.2 Interaction within ecosystems.

Figure S2.2 shows the general components of an ecosystem and how they interact. Bedrock weathers and dead plants rot to make soil, creating plant food. Rainfall and temperature control which plants and animal wildlife can thrive. Plants (**producers**) are the first food source for animals.

Balance within ecosystems

Change to any part of an ecosystem effects some or all of the other parts. This can be reversible over time – the Krakatoa eruption (Indonesia, 1883) destroyed the rainforest, yet within 100 years the ecosystem had recovered. Mount St Helens in the USA erupted in 1980 and is in the process of recovery today. Reduced or changed vegetation affects the food supply that is available for wildlife; loss of vegetation leads to poorer soil.

Human action can be more devastating to ecosystems than natural events. Logging in the Amazon rainforest removes trees, which reduces available **habitats** and therefore reduces plant and wildlife populations. Soils are made more vulnerable to erosion by heavy rainfall, and as topsoil is lost so are the nutrients required to support new plant growth. Too many elements of the ecosystem have been removed to allow regeneration.

 Key terms

producer: plants that provide food for herbivores at the beginning of the food chain

habitats: the environments that plants and/or animals live in

biomes: global-scale ecosystems

ACTIVITY

1 Write a description of the flow diagram in Figure S2.2 to teach someone who knows nothing about ecosystems exactly how they operate.

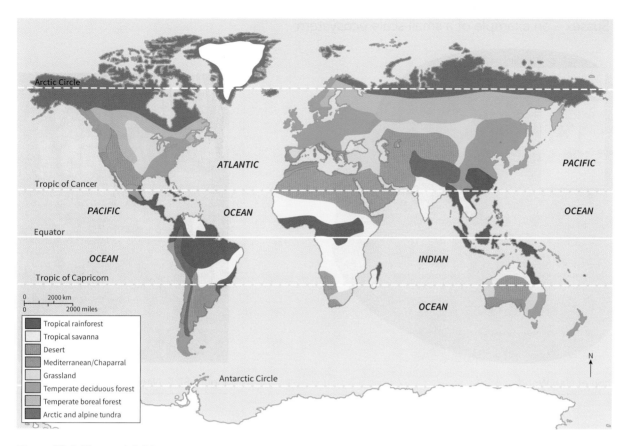

Figure S2.3 The world's biomes.

The distribution and characteristics of global ecosystems

Figure S2.3 shows the distribution of global **biomes**. Latitude is the main factor in the distribution of global ecosystems:

- Tropical rainforests are found close to the equator and are located in wetter areas, near the coast.
- Hot deserts are located closer to the Tropics of Cancer and Capricorn. The greatest extent is in the Sahara desert which extends across the widest part of Africa.
- Cold environments are found at high latitudes approximately 60°N and S of the equator. However, they are also found at high altitudes where alpine ecosystems can be found.

The characteristics of each biome are influenced by a number of factors. Climate is the main factor that influences the type and extent of the vegetation. Other factors influencing vegetation include: latitude, and therefore temperature; how close to the coast they are; and ocean currents.

Temperature and precipitation work together. In wetter regions, the higher the temperature is, the taller and denser is the forest (e.g. Indonesia). Dry or very seasonal areas are dominated by grassland (e.g. East African savanna). Coastal areas are wetter so woodland thrives, as in the UK.

Ocean currents can limit or enhance rainfall. Cold currents, like the Peruvian current, cut off moist onshore winds, reducing rainfall and creating the Atacama desert. The Gulf Stream (North Atlantic Drift) brings warm water to the British Isles, maintaining ecosystems which are not typical of that latitude.

Nap Wood, East Sussex – an example of a small-scale ecosystem

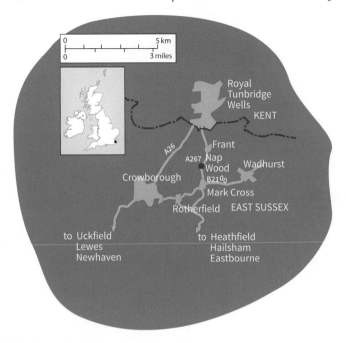

Figure S2.4 The location of Nap Wood.

Figure S2.5 Nap Wood.

Nap Wood's classification

The UK's climatic classification is 'cool temperate western margin'. It lies in the temperate zone and the northerly part, making it 'cool'. The 'western margin' refers to being located on the western edge of continental Europe.

Hundreds of years ago, before extensive human activity had an impact on the landscape, most of the British Isles was covered with mixed **deciduous woodland**. Nap Wood, located in East Sussex, is typical of this ecosystem (Figure S2.4). Today it is protected by the National Trust. Nap Wood is protected for its complex ecosystem as well as its history.

Nap Wood's ecosystem

Temperate deciduous woodland has clear layers. There are four layers compared with a tropical rainforest's five. The tallest trees, oak and beech, can reach 30m, with smaller silver birch and younger specimens growing below the **canopy**. Growth of a **shrub layer** of brambles is limited, but **ground cover** is rich in bracken, grass and low-growing green plants. Nap Wood is important for spring plants like its famous bluebells (Figure S2.5) and wood anemones in May.

Nap Wood has thick layers of litter and humus. Leaves falling in autumn take more than a year to decompose, but the result is rich soil to feed the next generation of vegetation. National Trust policy allows dead wood to rot naturally (Figure S2.5) so that the ecosystem's **decomposers** turn it into future plant food. Figure S2.6 shows the nutrient cycle that occurs in Nap Wood.

The food web in Figure S2.7 shows more detail about the relationships between producers and consumers in a deciduous woodland ecosystem. The food web is made up of multiple, interconnecting food chains. For example, in the food chain consisting of:

common oak → acorns → squirrel

Energy flows from the producer (oak tree) to the consumer (squirrel).

Key terms

deciduous woodland: trees that lose their leaves in winter

canopy: cover of treetops, limiting sunlight

shrub layer: bushes under woodland

ground cover: low-growing plants

decomposers: mosses, lichens and bacteria which decompose and recycle dead vegetation

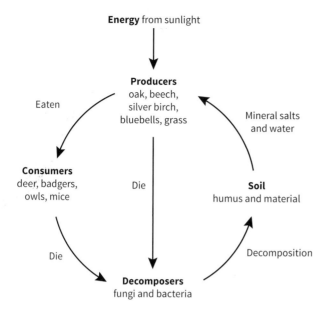

Figure S2.6 Nutrient cycling in the Nap Wood ecosystem.

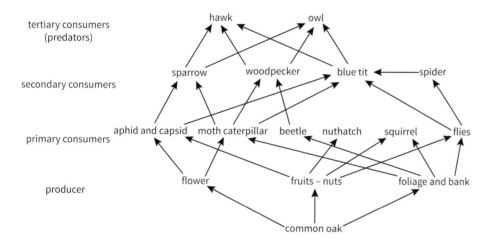

Figure S2.7 A foodweb for a deciduous woodland ecosystem.

Nap Wood's human use

The National Trust website shows a deep medieval pathway through Nap Wood used for driving pigs from farm to market. This path was hollowed out over centuries by the feet of livestock and people. Many trees have been cut back to their base to allow thin branches to grow for fencing. This process is called pollarding.

ACTIVITY

1 On a copy of Figure S2.6, Nap Wood, add annotations to identify and describe the characteristics of this ecosystem.
2 After you have worked through Chapter 5 and learned about rainforest layers, draw a similar diagram to represent the layers in temperate deciduous woodland. Figure S2.6 will help you.

- National Trust information on the history and environment of Nap Wood (see www.cambridge.org/links/gase40032)
- The High Weald Organisation – list of what you would see on a walk in Nap Wood (see www.cambridge.org/links/gase40033)
- The Woodland Trust – photos of Nap Wood (see www.cambridge.org/links/gase40034)

 Discussion point

What do you think would happen to the size of the squirrel population if the nuthatch population suddenly declined? What if the nuthatch population suddenly increased?

 Tip

When annotating, remember to write beside the photograph but locate your annotations on it using arrows.

 Discussion point

Using an atlas, identify the ecosystems where you live. Discuss in small groups and list the components of one of your systems.

5 Tropical rainforests

In this chapter you will learn about...

- how plants and animals have adapted to the conditions in tropical rainforests
- the causes and effects of deforestation in tropical rainforests
- sustainable ways to manage rainforests
- how the Chocóan rainforest is managed.

 Further research

Visit the **Eden Project** (www. cambridge.org/links/gase40035) website to see rainforest plants growing and to learn more about the conditions in which they thrive.

5.1 What are the distinctive characteristics of tropical rainforests?

Tropical rainforests are mainly located between the Tropics of Cancer and Capricorn. They are found in South America, Africa, southern Asia and northern Australasia (Figure 5.1).

Tropical rainforest **biomes** have over 2000 mm of rainfall annually. Temperatures are generally between 20 °C and 35 °C (Figure 5.2). High temperatures cause rapid evaporation, which increases humidity levels during the day. This leads to **convectional storms**. Heat and rainfall provide ideal conditions for plant growth.

 Key terms

biome: a large ecosystem

convectional storms: heavy rain falling as a result of high temperatures

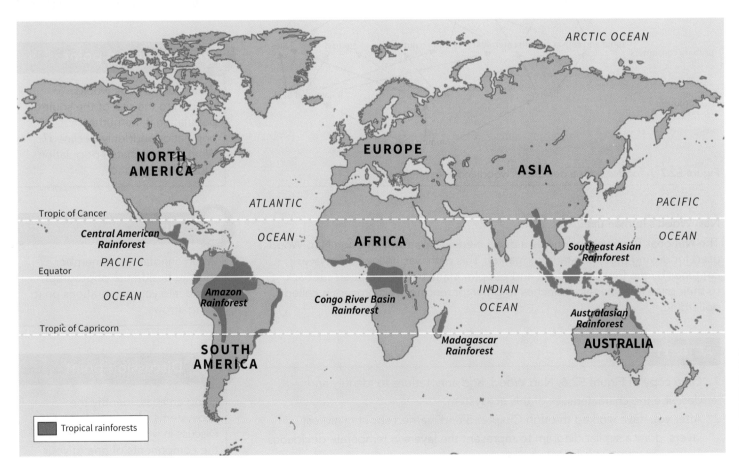

Figure 5.1 The location of tropical rainforests.

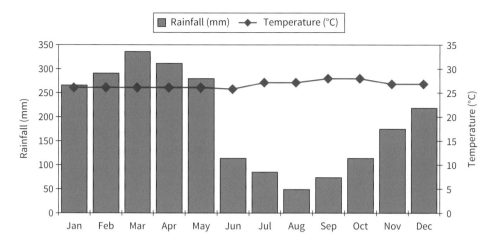

Figure 5.2 A climate graph for Manaus, Brazil. Manaus is a city located in the Amazon region.

In areas of tropical rainforest, there is little variation in the hours of daylight throughout the year and no significant seasonal differences. As a result, trees lose leaves throughout the year.

The hot, wet conditions create ideal conditions for plant growth and form a unique environment (Figure 5.3). The tropical rainforests have a rich **biodiversity**; they support the largest range of plant and animal life of any biome. Estimates suggest that 50 per cent of all life on the planet's land masses is found within the rainforests. Insects, spiders, birds, mammals and reptiles have all adapted to the physical conditions found in the rainforest. There is a greater density of species within rainforests than is found in any other ecosystem.

The living (biotic) features of the rainforest such as plants and animals and the non-living (abiotic) features such as temperature and rainfall are interdependent. Any change to one affects the other. For example, plants and animals need the light, heat and rainfall to thrive. If plants are removed, it affects conditions in the rainforest because, since soils are often poor, much of the energy is stored in the vegetation. This means that removing vegetation from the biome also removes nutrients. This, in turn, affects the water and nutrient cycles which maintain conditions in the rainforest.

Key term

biodiversity: the range of plants and animals found in an area

Figure 5.3 Lush, dense vegetation in the rainforest.

5.2 What are the rainforest layers?

Figure 5.4 shows how rainforests can be categorised into several layers.

- The ground layer receives very little light. Hot, wet conditions cause rapid decomposition of fallen leaves. Saprophytes are found in this layer. They are plants, fungi or micro-organisms that tolerate low light levels and get nutrients from breaking down dead matter. Anteaters are mammals that live in the ground layer of some rainforests in Central and South America. They have long tongues that can gather up to 35 000 ants and termites each day and sharp claws to tear open anthills. Anteaters have a good sense of smell to find their food on the dark forest floor.

- The shrub layer consists of tree saplings and shade-tolerant ferns. Orchids thrive in the damp conditions here. The largest beetle in the world, the Titan Beetle, lives in the shrub layer. The beetle is well adapted to this layer as it feeds on decaying material and is capable of flying.

- The under canopy receives about 5 per cent of the sunlight falling onto the forest. It is hot, damp and shaded from the wind. Vines, called lianas, climb

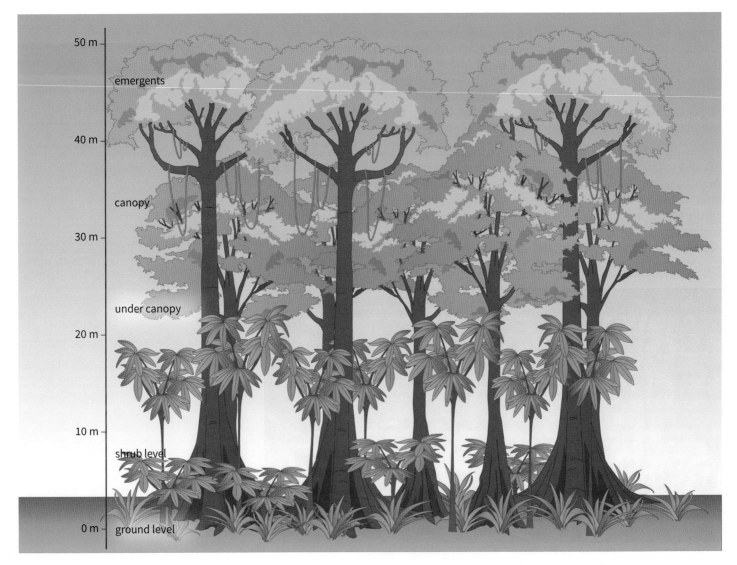

Figure 5.4 The layers of the rainforest.

through trees to reach the light. Amphibians such as frogs and toads are found here because it is close to water supplies on the ground. The flying frog has web-like feet, which allow it to glide through the air to escape predators.

- The canopy trees grow to around 20 to 40 m in height. They compete for the sunlight overhead. Leaves have **drip tips**, which prevent them from staying wet and rotting. Strong light levels mean that there is an abundance of life here, e.g. monkeys, sloths and pythons. The spider monkey has long limbs to move through the trees and sharp nails for peeling off the bark to eat the sap underneath. Epiphytes are plants that grow on branches and obtain their nutrients from water and the air.
- **Emergent trees** grow over 50 m high as they out-compete neighbouring trees. They can withstand winds and high temperatures. Birds such as parrots and eagles thrive here. Parrots have sharp, curved beaks to crack open nuts and access fruit growing on the trees. They have also developed four toes on each foot, which helps to clamber through dense rainforest foliage.

Rainforest soils are often poor because nutrients are washed out by **leaching**. Falling leaves and branches create a thick layer of **litter** on the top of the soil. Trees rely on the rapid breakdown of this litter to provide them with nutrients. Larger trees often have buttress roots, which stabilise them in high winds and spread over great distances to absorb sufficient nutrients.

Key terms

drip tips: leaves that are shaped in a way that allows excess water to run off them

emergent trees: the very tallest trees that grow higher than the rainforest canopy

leaching: when nutrients wash out of soil

litter: fallen plants and leaves on the forest floor

ACTIVITY 5.1

1. Draw a diagram to describe and explain the characteristics of rainforest layers.
2. Explain how plants and animals have adapted to the conditions found in tropical rainforests.
3. How might the characteristics of the rainforest change if the following happened:
   a. large-scale deforestation
   b. global warming
   c. disease affecting the saprophytic organisms?

Did you know?

More than 2000 tropical forest plants have anti-cancer properties, but less than 1 per cent of rainforest species have been analysed for their medicinal value.

Further research

Carry out research into how the Strangler Fig has adapted to rainforest conditions.

5.3 Why does deforestation take place in areas of tropical rainforest?

What is deforestation?

The United Nations Food and Agriculture Organization (UN FAO) defines deforestation as 'the conversion of forest to another land use or the long-term reduction of tree canopy cover below the 10% threshold'. If the amount of forest left is more than 10 per cent of the original area, it is called forest degradation.

Global rates of deforestation remain high, although the situation has improved slightly since the 1990s (Figure 5.5). Tropical rainforests suffer greater losses than other forest ecosystems. Globally, the equivalent of 50 football fields of tropical rainforests have been lost every minute since the year 2000. There were originally 6 million square miles of tropical rainforests globally, but only 2.4 million square miles remain.

Causes of rainforest destruction

There are many causes of rainforest destruction, including:

- **logging for timber and pulp** (Figure 5.6). The construction boom in China has been linked to increased logging in Myanmar and Indonesia.
- **commercial farming**. Plantations of crops such as palm oil and soybeans, cattle ranching and large-scale **shifting cultivation** have destroyed vast areas of rainforest.

Key term

shifting cultivation: when the land is farmed for a period of time and then farmers move to another location and the land is left to recover

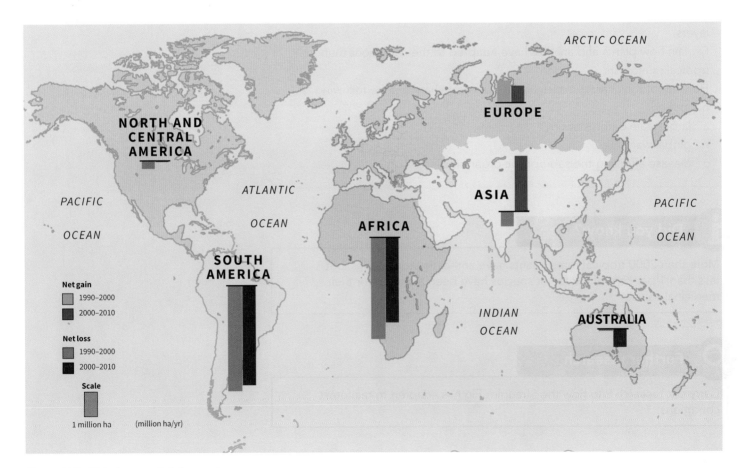

Figure 5.5 Global rates of deforestation.

Figure 5.6 Logging timber.

Figure 5.7 Slash and burn – destroyed tropical rainforest in Amazonia Brazil.

Figure 5.8 Mining for coal in Borneo.

- **subsistence farming**. This is shifting cultivation on a small scale so it is usually sustainable. Small areas of forest are cleared and so recovery is possible. **Slash and burn** is a technique whereby forest is cut down and ignited. The ash provides nutrients for the soil (Figure 5.7).
- **mining and mineral extraction** (Figure 5.8). Coal, gold, copper and even diamonds have been mined in rainforests. More than 70 per cent of the Peruvian Amazon is being made available for potential oil and gas extraction.
- **hydro-electric power** (Figure 5.9). Large dams have flooded areas of rainforests to provide electricity supplies. Many more are planned for Brazil and Malaysia.
- **settlement**. As the global population increases, rainforests are used for construction materials or building plots. Indonesia's transmigration policy encouraged citizens to move from crowded islands to areas of rainforest.
- **road building** (Figure 5.10). New roads and infrastructure projects support the developments in the rainforests. The Trans-Amazonian highway cuts through the Amazon rainforest. It was built in the 1970s and since then new developments have grown alongside it.
- **natural causes**. Forest fires, volcanic activity and tropical storms can cause rainforest loss. Lava flows destroyed rainforests in Cameroon in 1999.

Note that population growth can worsen the effect of many of these; a larger population means an increased demand for resources.

Key term

slash and burn: where forest is cut down and cleared by burning; the ash from the fire provides nutrients to the soil

Figure 5.9 A hydro-electric power station in Mexico.

Figure 5.10 Building roads through the Amazon rainforest.

5.4 What are the impacts of deforestation?

Deforestation has economic benefits and can help lower income countries (LICs) to bring in much-needed income. However, the exploitation of rainforests can have significant impacts on the local and wider environment. The removal of rainforest vegetation leads to a loss of biodiversity, as vegetation is removed and habitats are lost (Figure 5.11). Most rainforest plants and animals cannot survive without the dense rainforest canopy. Soil erosion occurs as land is left unprotected from the heavy rain (Figure 5.12).

Deforestation can lead to local changes in climate as the forests collect rainwater and allow it to evaporate back into the sky. In this way, they create their own local weather conditions. On a global scale, tropical rainforests are 'the lungs of the Earth' as they absorb carbon dioxide and emit oxygen; the removal of rainforests therefore leads to an increase in carbon dioxide, which is a greenhouse gas (see Chapter 4).

ACTIVITY 5.2

1   Use Table 5.1 to answer the questions that follow:

Year	Annual forest loss (sq. km)
1988	21 050
1989	17 770
1990	13 730
1991	11 030
1992	13 786
1993	14 896
1994	14 896
1995	29 059
1996	18 161
1997	13 227
1998	17 383
1999	17 259
2000	18 226
2001	18 165
2002	21 651
2003	25 396
2004	27 772
2005	19 014
2006	14 285
2007	11 651
2008	12 911
2009	7 464
2010	7 000
2011	6 418
2012	4 571
2013	5 891

Table 5.1 Deforestation figures for the Amazon.

a   Draw a bar graph to show rates of forest loss in the Amazon rainforest between 1988 and 2013.

b   Describe the pattern of forest loss shown in the graph.

c   Which years have seen high rates of deforestation? Why might this be?

 Download Worksheet 5.1 from Cambridge Elevate for a set of graph axes for Activity 5.2, question 1.

2   The text describes eight causes of rainforest destruction.

a   Select one or two that you consider to be the most destructive.

b   Explain the reasons for the choices that you made.

c   Are any of the eight causes linked, in that one leads to another?

3   Create a list of the impacts of rainforest destruction. Categorise the list into social, economic and environmental impacts.

 See Cambridge Elevate for an **interactive map of deforestation** between the year 2000 and 2013.

 Watch the animation on Cambridge Elevate to find out more about how deforestation affects convectional rainfall.

 Skills link

For further guidance on drawing and interpreting bar charts, see 19.1 Graphical skills – graphs and charts

 Discussion point

In LICs rainforests are exploited in order to generate income to aid development. Do we have the right to tell other countries what to do with their own resources?

Why does deforestation need to be managed?

Figure 5.11 Aerial view of deforestation.

Figure 5.12 Scarred earth where tropical rainforest has been destroyed by human development in Borneo, Malaysia.

5.5 Why is sustainable rainforest management important?

Why are rainforests important?

Rainforests are global concerns. They regulate global climate, producing oxygen and storing water. Around 25 per cent of our medicines originate from the rainforests, treating illnesses and conditions such as leukaemia and high blood pressure. Rainforests also supply products, such as rubber, bananas and coffee.

On a local scale, rainforests create their own climate. Trees protect soil and supply resources and food. Rainforests provide habitats for 50 per cent of all life on the land and are home to **indigenous** groups of people (Figure 5.13).

The world's rainforests are monitored by a number of international organisations, including:

- The UN FAO, which produces a yearly report on the state of the world's forests that monitors deforestation and records recent policies and decisions made about these biomes.
- Global Forest Watch works with the UN FAO to produce an interactive online mapping system. This allows anyone to view changes to forests.
- Other organisations raise money for conservation projects. The World Wildlife Fund (WWF) and The Rainforest Foundation have both carried out projects in the Democratic Republic of Congo, where rainforests are threatened by roads, industry and agriculture.

Strategies to manage rainforests

Selective logging and replanting

Loggers remove the most valuable trees in the forest, such as mahogany, without damaging surrounding trees. Where trees are removed, the area is replanted as **secondary forest**. These areas initially have lower biodiversity than **primary forests**.

Conservation and education

The Rainforest Alliance's 'Adopt a Rainforest' campaign has enabled the purchase and **sustainable management** of forests in Colombia, Guatemala and Honduras. Global programmes raise funds, and local programmes encourage indigenous people to protect their environment.

Ecotourism

Ecotourism earns money, while providing an incentive for forest preservation. Ecotourism developments are small scale to minimise environmental damage (Figure 5.14). They employ local people and management decisions are based on community agreement. The Santa Lucía Cloudforest Reserve in Ecuador is owned and managed by local communities, supported by organisations such as Rainforest Concern UK. The reserve covers 730 hectares, 80 per cent of which is primary forest. It employs local people and has replanted 6000 trees over 20 hectares.

International agreements

The International Tropical Timber agreement came into force in 2011. It ensures that wood from tropical areas is legally sourced and sustainable. The New York Declaration on Forests (2014), signed by companies like Kellogg's and Barclays, aims to halve forest loss by 2020 and halt it by 2030. The United Nations Sustainable Development Goals (SDGs) were established at the end of 2015. These are targets relating to future international development and the protection of all forest is a key priority.

Key terms

indigenous: native people who originate from a particular place

secondary forest: replanted trees

primary forest: native trees that are undisturbed

sustainable management: management that meets the needs of the present without compromising the ability of future generations to meet their own needs; it takes into account the environment, the needs of present and future generations and the economy

Figure 5.13 Indigenous people in the Brazilian rainforest.

Figure 5.14 An ecolodge in Costa Rica.

Debt reduction

Countries can be encouraged to conserve areas of rainforest in exchange for a reduction of their national debt. 'Debt-for-nature swap' schemes can alleviate poverty and protect rainforests. The US government has reached an agreement to reduce Peru's debt in exchange for conservation measures in areas of rainforest.

ACTIVITY 5.3

1 a Draw a mind map to show reasons to conserve rainforests.
  b Describe the advantages and disadvantages of the five rainforest management strategies.
  c Which do you think is the most effective strategy? Justify your choice.

2 This review was left on a travel website:

> This was our first visit to the Santa Lucía reserve. We loved the peaceful surroundings of the forest. We chose ecotourism as it's important to visit places without damaging them. The food was excellent; grown onsite and cooked by locals. The lodges were comfortable and the showers were heated using solar panels. We went hiking, swam in local rivers and saw interesting archaeological sites. It was expensive, but we're saving up to visit again!

  a Give three pieces of evidence to suggest that the Santa Lucía Cloudforest reserve is an example of sustainable development.
  b 'Ecotourism can put pressure on fragile environments.' Explain this statement.
  c One person responded to this post saying that the idea of conservation had been misunderstood because flights to Ecuador will have created pollution and so the holiday was not an example of 'ecotourism'. Do you agree?

 Discussion point

Can rainforests be exploited and protected at the same time?

 Download Worksheet 5.2 from Cambridge Elevate for help with Activity 5.3, questions 1b and 1c.

The Chocó rainforest

The Chocó rainforests stretch through Panama, Colombia and Ecuador (Figure 5.15). They contain vast stretches of coastal rainforests and are one of the world's wettest and most diverse habitats. The Chocó rainforests contain thousands of undiscovered plant and animal species. At least 66 per cent of the Chocó rainforest has already been destroyed.

Threats to the Chocó rainforests

One of the main threats to the Chocó rainforests comes from palm oil plantations. Colombia and Ecuador produce 1.5 million tonnes of palm oil every year. In order to create a plantation, the forests are cleared, and oil palm trees are planted (Figure 5.16). The oil is used in supermarket products including foods, cosmetics and medicines. It is also used for animal feed and **biofuels**.

Other threats to the Chocó rainforests come from logging for valuable tropical hardwood trees, and the use of heavy machinery and the release of toxins like mercury when mining for gold. The forest is also cleared to make way for new settlements to accommodate a growing population and for farming on a commercial and subsistence level. The Pan American Highway, connecting the continents of North and South America has also led to the destruction of the forest. The government of Ecuador has relaxed environmental laws to allow companies to drill for oil, which could be used to produce energy for the country. Drilling for oil in the Amazon rainforest has led to deforestation, the release of heavy metals and dangerous toxins into ecosystems and the relocation of indigenous tribal communities. Drilling for oil in the Chocó rainforests could destroy millions of acres of the rainforest here too.

Figure 5.15 The location of the Chocó rainforests.

Local impacts of developments in the Chocó rainforest

Developments in the Chocó rainforests are having both local and global effects. **Monocultures**, such as oil palm plantations, have reduced the biodiversity of the area. Habitats have been lost, and species such as the jaguar and the spectacled bear are threatened. The vegetation found on plantations is less dense than the rainforest and so the risk of soil erosion is increased. Indigenous tribes, such as the Emberá have lost land to the developments. People from Curvaradó, a settlement beside the Atrato River, were forcibly removed from their land to make way for the plantations. The Atrato River is now polluted through the toxic **pesticides** that are used on the oil palm plantations.

Figure 5.16 Rainforests cleared for an oil palm plantation.

National and global impacts of developments in the Chocó rainforest

Many countries argue that the development of their rainforests is essential to their economic development. Colombia's 'Plan Pacífico' is a government-led scheme to use the Chocó forests to create wealth for the country. The plan encourages the development of plantations, mines and new highways. Other nations say that countries that contain areas of rainforest have a responsibility to protect them for the benefit of the planet. Rainforest plants take in carbon dioxide and store carbon. Carbon dioxide is a greenhouse gas and so the destruction of rainforests

💬 Discussion point

Why is road building an important part of any major development project in areas of tropical rainforest?

contributes towards global warming and climate change. The World Bank and the United Nations have promised funding to help Colombia, but only if the country carries out sustainable development within the Chocó rainforests.

The future

Rainforest Concern, WWF, Rainforest Rescue and the Woodland Trust are all working to protect the Chocó rainforests (Figure 5.17). Using charitable donations and expertise developed through previous conservation work, these groups have bought areas of the rainforest. They have set up protected reserves (Figure 5.18) and have created the Chocó Andean 'wildlife corridor' to allow animals to move between fragmented sections of forest. They also support local communities to help them to understand how they can use their surroundings more sustainably.

Assess to progress

1   Figure 5.16 shows rainforest that has been cleared for oil palm plantations. Describe how the plantation reduces biodiversity. `3 MARKS`

2   Explain the reasons for deforestation in areas of tropical rainforest. `6 MARKS`

3   Use Figures 5.17 and 5.18 and your own knowledge to explain how international organisations can protect tropical rainforest environments. `8 MARKS`

Tip

'Describe' means identify what you see in the image, rather than explain it.

It can be useful to use connectives to give depth to your answers, e.g. 'Large areas of tropical rainforest are cleared **in order to** provide space for palm oil plantations to be developed.'

Figure 5.18 Puro Fairtrade Coffee's reserve in Colombia helps the endangered golden poison arrow frog.

Key terms

biofuel: a fuel that is produced using living material, such as plants

monoculture: the farming of a single crop

pesticide: chemicals used on crops to kill pests and diseases

wildlife corridor: stretches of land that connect areas of native vegetation

Tip

The word 'international' means you should explain how organisations outside of Ecuador and Colombia are protecting the rainforest.

Figure 5.17 Rainforest Concern have nursery and rehabilitation facilities in Ecuador to rescue and take care of the golden-mantled howler monkey.

Did you know?

The golden poison arrow frog that lives in the Chocó rainforest is on the International Endangered Species list. It can poison up to 10 people.

5.6 Hydroelectric power generation in Madagascar's tropical rainforest

In this issue evaluation you will consider whether a hydroelectric power plant should be built in Madagascar's tropical rainforest.

Figure 5.19 A location map of Madagascar.

Figure 5.20 Lemurs in the Madagascan rainforest.

The island of Madagascar (Figure 5.19) is one of the world's most biodiverse places with a large number of plants and animals that are **endemic**. This means that 98 per cent of Madagascar's land mammals, 92 per cent of its reptiles, 68 per cent of its plants and 48 per cent of its bird species do not exist anywhere else in the world. Madagascar is home to every species of lemur (Figure 5.20), all of which are endangered, and two-thirds of all chameleon species. Plants that are found in Madagascar include orchids, pitcher plants and the Madagascan pink periwinkle (Figure 5.21) which thrive in the tropical climate.

Why has Madagascar been deforested?

The natural vegetation that grows in Madagascar is tropical rainforest. Deforestation occurred on a small and relatively sustainable scale up until 1950 where subsistence farmers engaged in shifting cultivation.

In 2014, Madagascar's national debt (Table 5.2) was $3697 million, an increase of $92 million since 2013. This is the equivalent to 34.65% of Madagascar's GDP.

Figure 5.21 Pink periwinkle, found in Madagascar.

Year	1980	1990	1995	2000	2004	2005	2006	2007	2008	2009	2010	2011	2012	2013	2014
Debt as a % of GDP	31	126.2	144	116.8	77.3	70.3	30.4	25.9	24.2	33.44	31.87	32.43	33.66	34.01	34.65

Table 5.2 Madagascar's debt burden

The population of Madagascar in 1960 was 5.1 million and this had grown to 23.6 million people in 2014, an increase of 362 per cent over the last 50 years (Figure 5.22). Its annual growth rate, which in 2014 was 2.62 per cent is amongst the highest in Africa. This combination of Madagascar's massive national debt and rapid population growth resulted in large-scale deforestation (Figure 5.23).

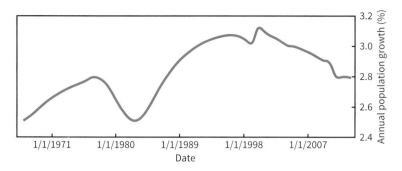

Figure 5.22 The population of Madagascar.

The government exploited its resources in order to pay off its debts and engaged in cash crop cultivation of coffee, rice and beef for export.

As a consequence, there is only 10 per cent of the original forest cover left and most of this is towards the east of the island where much of the coffee is grown as a cash crop.

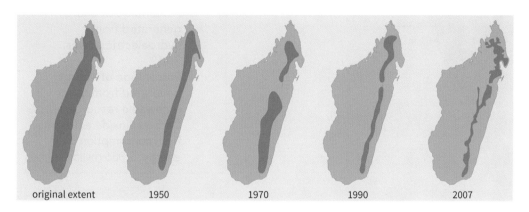

Figure 5.23 The loss of rainforest coverage in Madagascar, 1950–2007.

Slash and burn agriculture the main cause of deforestation in Madagascar

Despite the threat of large-scale deforestation to the forests of Madagascar, it is small-scale, but widespread clearance of vegetation for slash and burn agriculture and for firewood and charcoal which is the main cause of deforestation. As Madagascar's population has increased, this deforestation has also increased.

In the 1940s the death rate was 30 per 1000 per year in Antananarivo, the capital city. However, penicillin, antibiotics and chloroquine, which is used to fight malaria, were introduced and this lowered the death rate to 6.95 per 1000 per year by 2014. Consequently, as the birth rate remained high – 33.12 per 1000 per year in 2014 – Madagascar's population has increased more than fivefold from 4.1 million in 1950 to 23.6 million in 2014. The 2014 growth rate was 2.62%. Compare Madagascar's figures to other countries in the table below:

Country	Birth rate (per 1000) (2014)	Death rate (per 1000) (2014)	Population growth rate (%) (2014)
UK	12	9	0.54
Peru	18	6	0.97
Nepal	20	6	1.79
Sudan	29	7	1.72

This increase in population, coupled with the fact that these are people who are amongst the poorest in the world, has put pressure on the forests as the people try to survive by engaging in subsistence agriculture.

Is there a new threat?

However, there is perhaps a new threat to Madagascar's rainforests which has emerged from the country's need for energy generation.

A Pennsylvania-based engineering firm is currently working with the Madagascan Ministry for Energy and Mines to plan a hydroelectric power development project. The Volobe Amont site is located 30 km west/north-west of Madagascar's second largest city and largest shipping port, Toamasina (Figure 5.24 and 5.25).

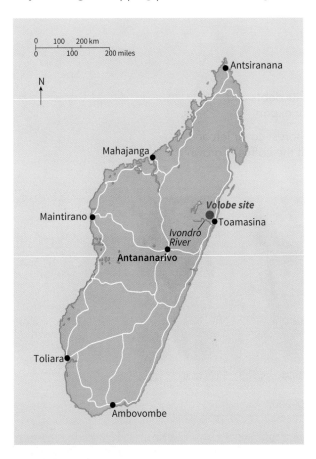

Figure 5.24 The location of the Volobe Amont development site.

Factfile

- **Electricity production:** 2.025 billion kWh
- **Electricity consumption:** 1.833 billion kWh
- **Madagascar energy** (2012 estimates)
  - **Reserves of oil:** None
  - **Reserves of natural gas:** 2.01 billion m³
  - **Reserves of recoverable coal:** None
  - **Imports of petroleum products:** 12.1 thousand barrels per day
  - **Percentage of energy generated from fossil fuels:** 69.6%
  - **Percentage of energy generated from hydroelectric power:** 30.1%
  - **Percentage of energy generated from other renewable resources:** 0.2%
  - **Carbon dioxide emissions from consumption of energy:** 2.886 million tonnes

Located on the Ivondro River in the middle of the tropical rainforest, it is hoped that the project will increase Madagascar's hydroelectric power capacity by 50 per cent. It will meet the energy needs of the population of Toamasina and also 20 surrounding rural villages.

'Our plan is to capture the River Ivondro's flow about 11 km upstream from the proposed hydroelectric plant. We are going to construct a wall less than 10 metres high, across the river. The water collected here will be transported through a large tunnel constructed by drilling and blasting the surrounding layers of rock between the two sites.' Engineer

One major disadvantage with the initial plan was that the debris and sedimentation build-up in the river mean that parts of the machinery would need to be replaced every five years due to erosion. To prevent this, a complex system of sluice tanks and tunnels will need to be built to remove the sediment.

Figure 5.25 Volobe Amont hydro-electric scheme site.

All of this development will have impacts on the economy and the environment (Figure 5.26) as well as the people that live in the area.

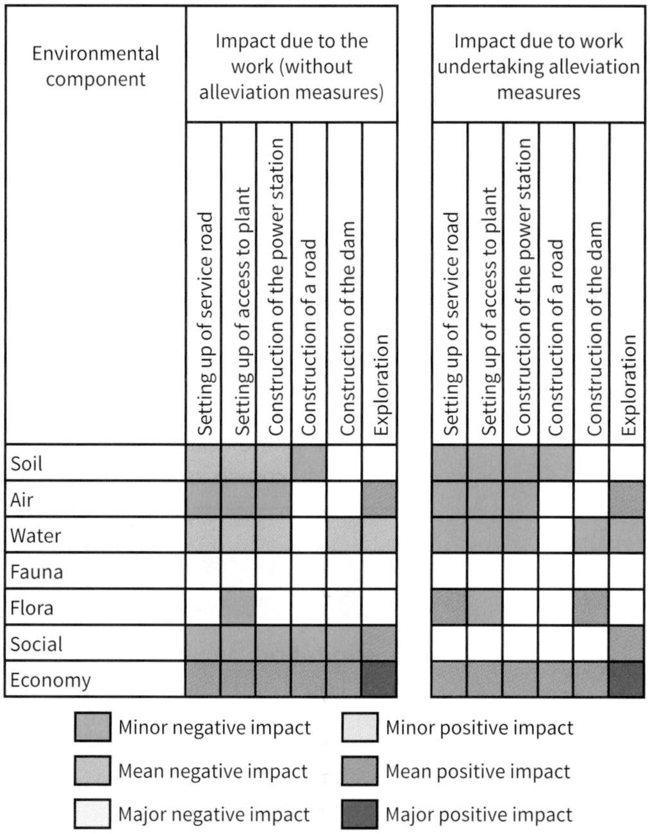

Environmental component	Impact due to the work (without alleviation measures)						Impact due to work undertaking alleviation measures					
	Setting up of service road	Setting up of access to plant	Construction of the power station	Construction of a road	Construction of the dam	Exploration	Setting up of service road	Setting up of access to plant	Construction of the power station	Construction of a road	Construction of the dam	Exploration
Soil	Minor neg	Mean neg	Mean neg	Minor neg			Mean neg	Mean neg	Mean neg			
Air	Mean neg	Minor neg	Minor neg		Major neg		Mean neg	Minor neg				Mean neg
Water			Mean neg	Mean neg	Mean neg				Mean neg	Major neg		
Fauna				Major neg						Major neg		
Flora	Major neg	Mean neg	Major neg	Major neg			Mean neg	Minor neg		Mean neg		
Social	Mean neg	Mean neg	Mean neg	Mean neg			Minor neg		Minor neg			
Economy	Mean pos	Mean pos	Mean pos	Mean pos	Mean pos	Major pos	Mean pos	Mean pos	Mean pos	Mean pos	Mean pos	Major pos

Legend:
- Minor negative impact
- Mean negative impact
- Major negative impact
- Minor positive impact
- Mean positive impact
- Major positive impact

Figure 5.26 The potential impacts without and with alleviation measures.

Many of the negative impacts will occur during the construction phase of the project as the surface soil is disturbed and there is increased dust, noise and emissions. There will also be deforestation so that infrastructure and access roads can be built.

✓ Assess to progress

1 Suggest two reasons why Madagascar has suffered from deforestation. **2 MARKS**

2 Use Figure 5.23 to describe how Madagascar's forest cover has changed from 1950 to 2007. **4 MARKS**

3 What is meant by the following terms:
- shifting cultivation
- cash crop **2 MARKS**

4 Should the hydroelectric power project go ahead at Toamasina? Justify your answer. **9 MARKS + 3 SPaG MARKS**

Tip

For question 2, you could describe the general pattern and then focus in on the specific detail. The map doesn't have any figures on it but you could estimate forest coverage.

Tip

For question 3, it could be helpful to make sure that you understand all of the terms that are used in the issue evaluation resource booklet. You could make an illustrated glossary of words with which you are not familiar.

Tip

For question 4, you first need to decide whether you think the project should go ahead or not. Write down all of the reasons for it going ahead and then write down the reasons why it shouldn't. Which list do you think is the most persuasive?

6 Hot deserts

In this chapter you will learn about...

- the distinctive characteristics of hot deserts
- how plants and animals have adapted to the harsh conditions found in hot deserts
- the economic opportunities that the hot desert environment provides
- why the exploitation of hot deserts is so challenging
- the causes and impacts of desertification
- the strategies used to reduce the risk of desertification.

6.1 What are the distinctive characteristics of hot desert ecosystems?

Climate, water and soils in hot deserts

A desert is defined as an area that receives less than 250 mm of rainfall a year, although some areas of hot desert have much less. Even though parts of large deserts such as the Atacama in South America and the Sahara in North Africa have no recorded **precipitation** for decades very few hot deserts get no precipitation at all. Average temperatures can be very high (Figure 6.1), however the **diurnal** range is often extreme – from over 50 °C during the day to around 0 °C at night.

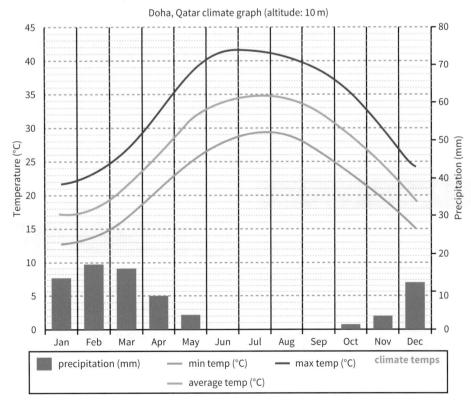

Figure 6.1 A climate graph for Doha, Qatar.

Did you know?

The Atacama is the driest desert on Earth but crops are still grown in some areas by trapping moisture in the coastal air on giant nets known as fog-catchers.

Key terms

precipitation: any form of water, both liquid and solid, which falls from the sky; rain, snow, sleet and hail are all examples of precipitation

diurnal: showing the altering conditions between day and night

Skills link

Accurate interpretation of climate graphs is an important skill for geographers. Describe the climate of Doha, Qatar (Figure 6.1), making precise observations about the precipitation and temperature.

Download Worksheet 6.1 from Cambridge Elevate for help with the Skills activity.

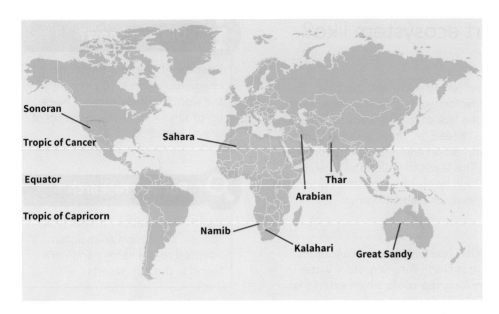

Figure 6.2 The distribution of the world's hot deserts.

The hot deserts of the world are mainly located within the tropics, in the centre or to the western side of the continents (Figure 6.2).

The traditional image of a hot desert is one of a dry, sandy landscape (Figure 6.3). While this description may offer a basic understanding of some of the characteristics of hot deserts, the reality is that they are much more complex. Rainfall can be very low but there is often just enough water to support different types of plants and animals. Furthermore, most deserts are not just sand but complex landscapes of rocks and gravel worn down by strong, dry winds (such as the Harmattan in the Sahara) and occasional flash floods.

Sometimes, after heavy rainfall in mountainous areas on the desert fringes, there can be flash floods which bring life to wilderness areas. Oases exist where the **water table** is close to the surface meaning palm trees and other vegetation can grow.

 Key term

water table: the level at which water is found under the Earth's surface

Figure 6.3 A sandy desert landscape.

6.2 What is the hot desert ecosystem like?

The hot desert ecosystem

Desert soils are generally not very fertile. Biodiversity is low and the biotic (living) and abiotic (non-living, e.g. soil, water light) factors are highly interdependent. There is plenty of sunlight in the desert but the lack of rainfall leads to limited plant growth. With a limited variety of plants to feed on, there is limited animal life. Both these factors lead to limited decomposition and a lack of the organic matter needed to give **nutrients** to the soil. This leads to infertile soils which, once again, limits plant growth. In addition, high levels of evaporation draw salts up through the soils making it even more difficult for plants to grow.

Life in the desert

Desert plants and animals need to find ways to cope with the dry conditions. Some plants, such as **succulents** like the cacti in North America, store water in their leaves, stems or roots. Others have very long **tap roots** which extend far down underground to find water. Some are largely **dormant** and may remain in the desert soils for many years before flowering after any kind of rainfall and living for a very short period of time (Figure 6.4).

Once plants have managed to find water, they need to avoid losing it. Many desert plants have few or no leaves, which reduces water loss by **transpiration**. Waxy coatings on leaves also help to reduce water loss.

Figure 6.4 The desert springs to life after rain.

Surviving in the host desert environment is also challenging for the animals which have had to adapt to the extreme conditions. The bat-eared fox, for example, has evolved large ears to allow it to cool down more easily (Figure 6.5). Its large ears provide a large surface area to volume ratio, which helps to maximise heat loss. Snakes such as the Peringuey's adder slide sideways to minimise contact with the hot sand (Figure 6.6). Some animals, such as the USA's cottontail rabbit, are entirely nocturnal and sleep under the ground in their burrows during the day (Figure 6.7). Others simply hibernate when it gets too hot.

Did you know?

Despite it being the traditional image, only about 10 per cent of the world's hot deserts are made up of sand – most are made up of gravel and rocks.

Further research

Use the internet to find out how specific plants and animals have adapted to the harsh conditions found in the hot deserts.

Key terms

nutrients: chemicals in the soil (often from decaying vegetation) that help plants to grow

succulents: plants with a thickened and fleshy structure which allows them to retain water

tap roots: long roots which extend far down in the soil to find water

dormant: when a plant's metabolism slows to a point where its growth and development are temporarily stopped

transpiration: evaporation of water from plant leaves

Figure 6.7 Cottontail rabbit.

Figure 6.5 Bat eared fox.

Figure 6.6 Peringuey's adder.

Camels are a good example of animals that are well adapted to the hot desert environment. They can go for long periods of time without water and can tolerate body temperatures of up to 40 °C. They also have a large surface area to volume ratio. Large flat feet help them to walk on sand while two rows of eyelashes keep sand out of their eyes.

ACTIVITY 6.1

1 Annotate a world map to show the distribution of the world's hot deserts.
2 Why do hot deserts experience such a high diurnal temperature range?
3 a Describe the climatic conditions which characterise hot deserts.
   b Explain how the animal and plant life of the hot desert ecosystem is adapted to the physical conditions.

Download Worksheet 6.2 from Cambridge Elevate for help with Activity 6.1, Question 1.

6.3 What development opportunities are found in hot deserts?

Despite the harsh conditions and the constant battle to find water and food, a surprising number of people live and work in hot desert environments. Estimates vary, but it is thought between 2.5 and 4 million people live in the Sahara desert while the Thar desert in India has a relatively high population density of 80 people per km². There are a number of ways in which people and communities have been able to use the resources found in hot deserts to develop the following economic opportunities.

Agriculture

Soils may be infertile but, around oases, there can be quite **intensive agriculture** using the water found in underground **aquifers** and distributed through irrigation systems. Dates, figs and fruits are the most commonly grown crops but wheat, barley and vegetables are also grown.

In some deserts, rivers flowing from other areas provide the water for large-scale **irrigation**. The Nile, flowing through arid areas in South Sudan, Sudan and Egypt, has provided water for agriculture for thousands of years (Figure 6.8). This has provided food for millions of people and sustained the growth of cities including Khartoum and Cairo.

In some areas of hot desert underground aquifers provide water for the development of large-scale commercial agriculture. The huge date farms of Saudi Arabia are an example of this.

Energy

The high levels of sunshine found in hot desert areas can provide an excellent opportunity for the generation of solar energy. Many countries with hot deserts, including Israel, Algeria and South Africa, have been experimenting with small-scale solar energy generation. Spain and the USA have developed large-scale solar arrays (Figure 6.9) and there is a lot of interest and investigation into whether this technology could be introduced into other areas including the Sahara desert in North Africa. Scientists believe it would be possible to supply the whole of the United States with energy from the Sun harnessed in the Sonoran, Mojave and Chihuahuan deserts found in the south-west of the country.

Mineral extraction

Many areas of hot deserts have significant reserves of natural resources. The most important resources from an economic perspective are oil and gas and some of the most oil and gas rich countries in the world are found in desert areas (Table 6.1).

Country	Oil reserves (in millions of barrels)
Saudi Arabia	270 000
Iran	150 000
Iraq	140 000
Kuwait	104 000
United Arab Emirates	98 000
Libya	48 000
Kazakhstan	35 000

Table 6.1 The hot desert countries with the largest oil reserves.

Figure 6.8 Irrigation-fed crops by the side of the Nile in Egypt.

Figure 6.9 A solar array in the hot desert.

Key terms

intensive agriculture: farming that requires large inputs of labour, chemicals, capital and so on, to produce as many crops or animals as possible on the available land

aquifer: a body of permeable rock that can store water and through which water can easily move

irrigation: taking water from a store such as an aquifer or river and distributing it across areas of landscape to make the land suitable for growing crops

Skills link

Although many graphs can now be produced quickly and easily using a computer, it is still a useful skill to be able to draw them by hand. Present the information in Table 6.1 using a bar graph. Include an appropriate scale, labelled axes and a title.

Download Worksheet 6.3 from Cambridge Elevate for help with the Skills activity.

Other economically important resources include precious stones and metals, the uranium used in nuclear power and other elements important in the chemical industries (Table 6.2).

Country	Minerals
Australia	Iron ore, Gold, Nickel, Uranium
Chile	Copper, Silver, Gold, Sodium Nitrate, Phosphorus
Morocco	Phosphorus
South Africa	Diamonds
USA	Silver

Table 6.2 Examples of resources found in hot desert areas.

Figure 6.10 The Burj Al Arab hotel in Dubai.

Desert tourism

Since the mid 1800s people have been attracted to the culture and ancient history that has developed in many countries with areas of hot desert. The pyramids of Egypt and the city of Petra in Jordan are just two examples of such sights that remain popular into the 21st century. Recently, many countries with desert landscapes have developed other tourist attractions in a bid to increase the foreign exchange they gain from tourism.

Adventure and extreme tourism has grown to include activities such as camel rides, four-wheel drive and dune buggy trips, and long-distance trekking. Cultural tours such as desert safaris and accommodation staying in desert camps with local nomadic groups are increasing in popularity. Tourists also visit to see the unique plant and animal life.

Some destinations, such as Dubai have built facilities as diverse as top quality golf courses and indoor ski centres and many wealthy visitors come to stay in luxury hotels including the 'seven-star' Burj Al Arab (Figure 6.10). Many companies also offer speciality holidays in desert areas including solar eclipse tours with the clear desert skies offering some of the best eclipse-viewing in the world.

ACTIVITY 6.2

1 'Hot desert environments provide a range of economic opportunities.' Discuss this statement.
2 Look at the brochure advert for Dubai.

> **Dubai, a land of opulence, wealth and mind-defying architecture that will never fail to impress.**
>
> Dubai is one of the world's most exciting destinations; a glittering city that guarantees fun, sun and endless tax-free shopping. In this ever-changing emirate, spectacular and innovative developments grace endless golden beaches and towering commercial centres, all the while contrasting starkly against a silent desert interior.
>
> *Source:* Zahid Travel

Why are increasing numbers of people visiting hot desert areas for their holidays?
3 Explain how the economic exploitation of hot desert environments may create problems for desert communities and environments.

 Further research

Another economic opportunity is the use of desert landscapes for films. Morocco is a favourite location and was used in the filming of *Star Wars* and *Gladiator*. Research examples of desert film locations and suggest how this type of activity contributes to the local economy.

6.4 What are the challenges of developing hot desert environments?

The main challenges associated with the development of hot desert areas are related to the extreme climate and the sheer scale of these areas. A further problem, especially for those involved in large-scale farming or mining operations, is the need to protect the fragile desert environment from over-use and destruction.

Managing water supply

In many hot desert areas water is taken from rivers and underground aquifers for agriculture and to supply growing populations. The water taken (abstracted) from these sources often comes from international river basins and this can cause conflict between different countries:

- Farmers and businesses in Ethiopia, Uganda, South Sudan and Sudan, for instance, take water from the River Nile, and its tributaries the Blue and White Nile (Figure 6.11), causing concern in Egypt about how much will be left for the country's own use.

- In the western United States the Colorado River is used for irrigation, drinking water and leisure facilities (including providing golf courses in the desert). So much water is now taken from the river that the delta where it enters the Sea of Cortez near Mexico has completely dried up.

- Water stored in underground aquifers is also under threat. Figure 6.12 shows the dramatic effect that pumping wells can have on **groundwater**. In some parts of the world underground stores hundreds of thousands of years old are being depleted (used up) at alarming rates. The graph (Figure 6.13) shows such a problem under the Central Valley of California. The development of towns and cities, golf courses, tourist resorts, farms and swimming pools in arid areas has added to water supply pressures.

Discussion point

What are the possible impacts on Egypt of countries upstream significantly increasing the amount of water they abstract from the River Nile?

Did you know?

The Nubian Sandstone Aquifer System found under Egypt, Libya, Sudan and Chad contains 375 000 cubic kilometres of water – enough to keep the whole River Nile flowing for 3750 years.

Key term

groundwater: water found underground in soil, sand or rock

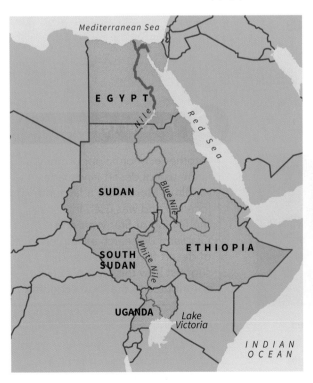

Figure 6.11 The course of the River Nile and its tributaries.

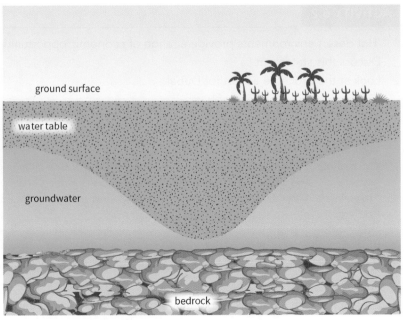

Figure 6.12 The effects of pumping water on an aquifer.

- Pollution of existing water supplies means even the water that is left is becoming unusable in many areas. Fertilisers and pesticides washed from fields in areas of intensive agriculture ultimately end up in the water supplies.

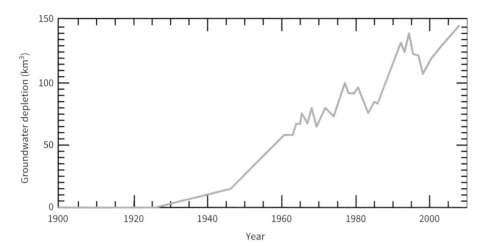

Figure 6.13 A graph showing the groundwater depletion in the Central Valley of California.

Extreme temperatures

The very high daytime temperatures in many hot deserts, often exceeding 50 °C, mean that people living and working in these areas face dehydration and possible death. At night temperatures can drop close to freezing. This high diurnal range makes it difficult for people to adapt.

Evaporation rates are also very high in the hot temperatures and **salinisation** occurs as salts are drawn up through the soil making the top layers infertile and unable to support the growth of crops.

Inaccessibility

The sheer scale of hot desert areas and the harsh nature of the environment – hundreds of miles of sand and gravel, rocky mountain ranges and **salt flats** – makes the building of roads and the transportation of people, goods and materials a significant challenge, both strategically and economically. In the Sahara desert a number of countries with areas of hot desert are landlocked making trade difficult.

The extreme temperatures mean vehicles can overheat quickly and specialist knowledge is needed to drive cars and trucks over sand and gravel. Breakdowns occur regularly and lack of support can mean inexperienced drivers are left stranded.

The challenge of developing hot desert environments

Desert environments are very fragile and easily damaged. Mining and mineral exploitation can be destructive because of the heavy machinery used (Figure 6.14). In some countries there is limited environmental protection and companies are not subject to laws designed to protect the environment. Although some governments and international environmental agencies such as the World Wildlife Fund try to put pressure on individual companies, the money that can be earned through the exploitation of these environments often outweighs any attempt to change their behaviour.

Key terms

salinisation: the accumulation of soluble salts in the soil, making the soil infertile

salt flats: areas of flat land covered with a layer of salt

Further research

Research hot desert countries that have built the infrastructure to overcome the challenges the environment poses for transportation.

Figure 6.14 Environmental damage at Chiquicamata copper mine in Chile.

ACTIVITY 6.3

1 Describe the challenges involved in the exploitation of hot desert environments.

2 Look at the environmental damage taking place in Figure 6.14.
  a What do you think is causing the damage?
  b In what other ways might mining put pressure on the environment?

3 What are the challenges of building roads and other infrastructure for transportation in desert areas?

6.5 What are the causes and impacts of desertification?

The term **desertification** was first used in 1949 to describe the way that semi-arid areas appeared to be turning into deserts in some parts of the world (Figure 6.15). It affects every continent except Antarctica but is most severe at the edges of hot deserts and particularly around the southern edge of the Sahara desert. In this area, known as the Sahel, severe droughts in the 1980s and 1990s caused the desert area to expand rapidly. With climate change affecting rainfall patterns and the prospect of further droughts in hot desert areas, estimates suggest that more than 30 per cent of the world could become desert in the future, affecting well over 1 billion people.

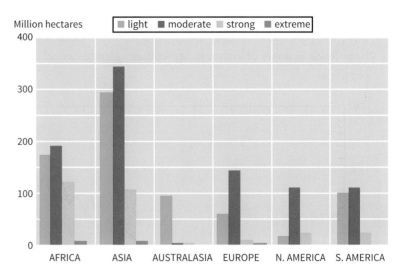

Figure 6.15 Areas affected by desertification.

Did you know?

The USA has already suffered the effects of desertification. In the 1930s poor farming techniques and years of low rainfall led to large areas of Texas and Oklahoma becoming known as the Dust Bowl. This resulted in 2.5 million people moving to areas as far apart as Detroit and California.

Causes of desertification

The causes of desertification vary, but it is increasingly considered to be the result of a combination of physical processes and human activity. It is often seen in those parts of the world where there is an increasing population putting pressure on land which is already at the limit of agricultural production.

In poor areas an increasing population has meant:

- trees have been removed for fuelwood (Figure 6.16) leaving areas open to soil erosion during periods of strong winds and rainstorms
- overgrazing when too many animals are farmed in too small an area leading to soil being compacted by the hooves of cattle and being unable to sustain plant life
- **overcultivation** as more people demand more food and too many crops are grown leading to a deterioration of soil quality
- people moving onto more **marginal land** where farming is very difficult and where the land is unable to support a large population.

In some countries, governments have also encouraged farmers to grow **cash crops** to increase exports and bring in money through foreign exchange. **Monoculture** has increased with this type of intensive farming taking nutrients from the soil and reducing soil quality. If the land is overworked it quickly becomes exhausted and crops begin to fail further adding to the possibility of desertification. Unreliable patterns of rainfall can add to this problem (Figure 6.17).

Together, all these different causes lead to the removal of vegetation which protects the fragile topsoil. Without this protection soils are very easily eroded by the wind and rain. As this happens the area gradually turns to desert and people are forced to move.

Key terms

desertification: a reduction in the biological productivity of the land which leads to desert-like conditions (as defined by the United Nations in 1977)

overcultivation: the excessive use of farmland to the point where productivity falls due to land degradation

marginal land: areas which can only be farmed when conditions (e.g. rainfall) are very good

cash crops: growing crops to make money, not for personal consumption

monoculture: the farming of a single crop

Figure 6.16 Fuelwood removal in Senegal.

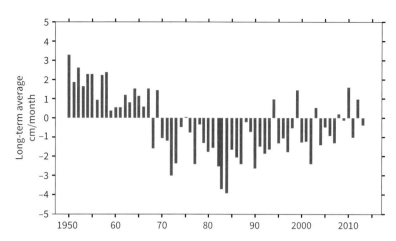

Figure 6.17 Rainfall patterns in the Sahel showing changes from the long-term average.

Impacts of desertification

The new environmental refugee crisis

The Sahel is one of the most vulnerable regions in the world to future climate fluctuations, and the effects are starting to take root.

Rain fall anomalies, increases in overall-temperature, persistent drought, and other issues like overgrazing have reaped turmoil in the Sahel. Sudden rainfall in 2013 led to flooding, destroying crops and houses for thousands of families. Following the disastrous flooding was a period of intense drought, which has continued for the past two years and is the source of food insecurity for over 18 million. Even worse is desertification, with researchers documenting a shift of sand 60 miles south into the Sahel.

While there is hope for renewed crops following a drought, desertification is almost impossible to revert. Some parts of the region have seen a temperature change of nearly 35 degrees in the past 50 years, and increasing pressure on water-supplies has helped fuel the cycle of desertification.

It's a deadly situation, which has left main residents of the Sahel with few options. With limited land-rights, no legal recourse and no economic ability to migrate, most are forced to abandon the agricultural life-style completely. Often their only destination is urban slums.

Source: The Huffington Post

Desertification has led to the loss of 650 000 square kilometres of farmland in the Sahel over the past 50 years. It needs to be remembered, however, that this is not just a problem for lower income countries (LICs). Large parts of Spain, Australia, the United States and other countries are also at risk.

 Watch an animation on Cambridge Elevate about soil erosion in the hot desert.

ACTIVITY 6.4

1 Explain the links between rainfall, drought and famine. Use the information in Figure 6.17 to help with your answer.
2 Using the newspaper article on the Sahel, examine the social, economic and environmental effects that desertification may cause.
3 Suggest how satellite photography has increased our understanding of desertification.

6.6 What strategies can be used to reduce the risk of desertification?

With large numbers of people affected by desertification, a lot of effort is being put into combating the problem. The solutions look at ways to stop wind and water erosion, to manage water and land resources and to reduce the population pressure on the land. It is important that appropriate technology is used to do this so that local communities are involved in the development of **sustainable** strategies.

In some hot desert areas, farmers are increasingly adopting techniques of dry farming. These methods are characterised by a lack of irrigation and instead use drought-resistant crops or moisture-enhancing techniques such as burying seeds deep in the ground or covering the soil with a layer of mulch to decrease evaporation. This type of cultivation can be very successful but does require constant monitoring by farmers to ensure sufficient moisture is available for crops to grow. It is an example of appropriate technology because the farmers can implement it themselves using the resources they already have.

Protecting the soil from erosion

Water management and the maintenance of moisture in the soil is crucial to protecting the land from desertification. Moisture allows vegetation to grow and vegetation cover reduces the risk of soil erosion. Methods used to retain moisture include:

- Cutting terraces into steep mountainsides means that heavy rainfall cannot wash the topsoil away (Figure 6.18).
- On more gentle slopes, earth **bunds** or small walls of stone will let water and soil build up behind them (Figure 6.19). The water will then filter into the soil maintaining moisture so that plants can grow.
- Building up stones in gullies or using steel cages filled with rocks known as gabions (Figure 6.20) traps rainfall by reducing runoff. The rainfall is then allowed to infiltrate through the soil.

Some of these methods involve whole communities making changes to the landscape. They cost very little apart from the labour involved and will protect the soil so can be considered to be sustainable. On a larger scale, irrigation systems can help to stop erosion by providing water to grow crops. The plant cover will then protect the landscape from erosion by wind and water. Some of these systems are only possible for LICs when they are set up by large companies with the resources to pay for them.

Because the soil is easily eroded by wind when vegetation is removed, replacing the vegetation is a fairly simple way of stopping desertification taking place. Ideas such as crop rotation keep nutrients in the soil by moving crops around fields over a large area on a yearly basis and keeping animals off certain areas of grassland will help to maintain the plant cover.

Afforestation programmes are also very important. Trees provide shelter and windbreaks which force winds up and over the cropland,

Key terms

sustainable: a method of using a resource so that it is not permanently damaged

bunds: low-lying rows of stones which reduce runoff and allow rainwater to infiltrate the soil

afforestation: planting trees

Figure 6.18 Terraces cut into a hillside to stop water erosion.

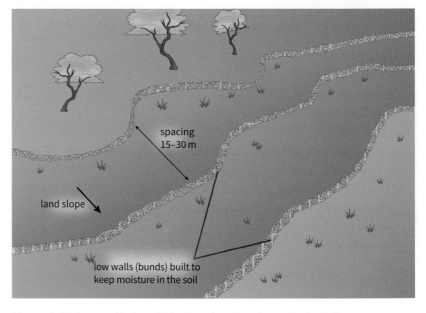

spacing 15–30 m

land slope

low walls (bunds) built to keep moisture in the soil

Figure 6.19 Low walls (bunds) built to keep moisture in the soil.

reducing wind erosion and allowing plants to grow. They can also provide fuelwood for local communities.

For any of these methods to work there is a need for education for the local people involved. Charities such as Excellent Development work with many farmers and communities in sub-Saharan Africa to help them understand the causes and consequences of desertification and how the risks can be reduced.

Combining strategies to tackle the problem

While all the methods outlined above work on a small scale none of them will solve the problem of desertification on their own. Organisations including many **NGOs** and the United Nations Convention to Combat Desertification (UNCCD) look at combining different techniques over bigger areas. By doing this local people can be involved in reducing the risk of soil erosion on their own land while at the same time being part of a larger strategy.

An example of this is the Great Green Wall of the Sahara. This is a plan to plant a wall of trees in order to stabilise the soils and prevent erosion in the countries across the southern edge of the Sahara desert. It will be combined with the other methods of tackling soil erosion outlined.

Figure 6.20 Gabions in a gully in Ethiopia.

Key term

NGOs: non-governmental organisations such as charities

Further research

The multi-dimensional approach used in the Great Green Wall of the Sahara is a very good example of how countries can work together to combat desertification. Find out more on the **Global Environment Facility website** (www.cambridge.org/links/gase40036).

ACTIVITY 6.5

1 Describe the methods used to protect soil from erosion and to stop desertification.
2 Explain how three of the methods to prevent soil erosion work and why they are appropriate for LICs.
3 a How might climate change increase the risk of desertification?
 b How might communities adapt to the increasing risks of desertification posed by climate change?

To find out more:

Visit the **BBC website** (www.cambridge.org/links/gase40037) for information about the characteristics of hot deserts

Visit the **Science Kids website** (www.cambridge.org/links/gase40038) for hot desert facts

Look at a slideshow about the landforms of hot deserts (www.cambridge.org/links/gase40039)

Visit the **US Geological Survey website** (www.cambridge.org/links/gase40040) for information about desertification

Visit the **United Nations Convention to Combat Desertification website** (www.cambridge.org/links/gase40041) for information about combating desertification

Visit the **Global Environment Facility website** (www.cambridge.org/links/gase40042) for information about the Great Green Wall

Visit **The Geological Society website** (www.cambridge.org/links/gase40043) for information about erosion in hot deserts

Visit the **CIA World Factbook** (www.cambridge.org/links/gase40044) for Information on hot desert countries

The hot desert environment of Qatar

Qatar – one of the world's richest countries – is an example of how economic development can take place in a desert environment. Located on the Arabian Peninsula, it lies to the east of Saudi Arabia in a very arid (dry) climate (Figure 6.21). During the early part of the 20th century, oil was discovered and the country has used the wealth generated by its exploitation to develop other parts of the economy (Figure 6.22).

The opportunities of the Qatari desert

Originally one of the poorest of the **Gulf States**, Qatar had an economy based almost entirely on pearl-fishing. Oil exports have been increasing year-on-year since the 1940s (Figure 6.23) yet it still has enough oil to last for at least another 50 years. The country is self-sufficient in producing its own energy and has used the oil and the money gained from its resources wisely. It has developed a wide range of industries including ammonia and fertilisers production, petrochemicals and ship repair. Qatar has also recently started to develop its tourism industry with the growth of luxury hotels and adventurous activities including four-wheel drive sand-dune trips and camel trekking.

Despite this diversification, oil and gas still account for 92 per cent of all export earnings. The development of the industry has led to the people of Qatar having high incomes and low unemployment. Socio-economic conditions have improved hugely for Qataris with, for example, life expectancy rising from 61 years in 1960 to 78 in 2015 and gross national income (GNI) per head increasing threefold over the last 30 years.

The challenges faced by Qatar's development

The huge amounts of money brought in by oil and gas have to compensate for a lack of other crucial resources. Qatar has no surface water but is one of the world's largest consumers at 500 litres of water per person per day. Very high temperatures (Figure 6.1) mean much of the scarce rainfall evaporates and is not available for use as a resource. Attempts to encourage people to use less water have failed so far so the government relies on expensive **desalinisation** plants to supply the population. Plans to build mega-reservoirs near Doha may help with future water supply problems.

Figure 6.21 Qatar.

Figure 6.22 The bright lights of Doha.

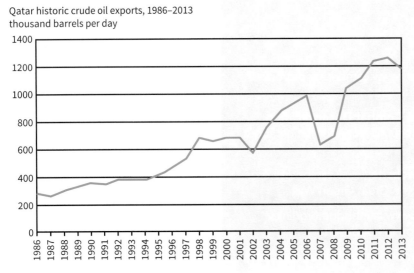

Figure 6.23 Oil exports from Qatar.

Key terms

Gulf States: the seven Arab states which border the Persian Gulf

desalinisation: industrial process to remove the salt from seawater

The growing population needs to be fed, but there is little indigenous farming because of the hot desert climate and the poor quality of the soils. Only 0.1 per cent of the nation's GNI is supplied through agriculture. Almost all food has to be imported but, with the vast majority of people living around the coastline and well connected by road, there are few of the accessibility problems found in other countries with hot deserts. Many areas away from the coast remain inaccessible except by camel and other traditional forms of desert transport and are, therefore, very sparsely populated. The few people who live in such areas do find it more difficult to obtain food. With rising global prices the Qatari government has decided to concentrate on becoming self-sufficient in food by 2023. Plans put forward to do this include:

- huge irrigation systems using desalinated seawater
- vertical farming with fields on different floors in huge skyscrapers
- hydroponics or soilless agriculture
- buying farmland overseas (including 7 300 km² in Australia) and shipping the food produced back to Qatar.

The future

Further development of the financial, industrial and business sectors shows Qatar is looking at ways to keep its huge wealth when the oil and gas run out. The falling price of oil at the end of 2015, perhaps suggests a greater urgency for these developments than the Qatari government's target of 2023. The challenge is in maintaining this while keeping its people fed and watered.

ACTIVITY 6.6

1 What are the challenges and opportunities which the hot desert environment of Qatar poses for the development of the country?
2 Describe the growth of oil exports shown in Figure 6.23.
3 Look at the picture of Doha in Figure 6.22. Use the picture to suggest ways in which Qatar has used the revenue from oil to develop.

Assess to progress

1 Describe the distribution of the world's hot deserts. `3 MARKS`
2 Which **one** of the following statements describes the climate of a hot desert? `1 MARK`
- A low diurnal temperature range with annual rainfall greater than 250 mm
- A low diurnal temperature range with annual rainfall less than 250 mm
- A high diurnal temperature range with annual rainfall less than 250 mm
- A high diurnal temperature range with annual rainfall greater than 250 mm
3 Describe **one** method used to protect the soil from erosion in desert areas. `2 MARKS`
4 Give **one** possible physical and **one** possible human impact of desertification. `4 MARKS`
5 Using a case study, explain how the environment in a hot desert provides both opportunities and challenges for development. `9 MARKS`

Factfile

- Co-ordinates: 25°30'N, 51°15'E
- Area: 11 586 km²
- Climate: Arid with mild, pleasant winters and very hot, humid summers
- Natural resources: Petroleum, natural gas, fish
- Population: 2 123 160 (2014)
- Natural hazards: Sandstorms
- Land use: Agricultural land 5.6%; forest: 0%
- Surface water: 0%
- GNI per head (PPP): £91 742 (2014) (highest in the world)
- gross domestic product (GDP) growth rate: 6.5% (2014)

Source: CIA World Factbook

Further progress

Qatar is investing heavily to promote its image as a dynamic, modern country. Plans to bring the Football World Cup and a Formula One Grand Prix to the country are examples of this. Find out about the challenges that the hot desert environment poses to the success of these plans.

Tip

When using a case study make sure you know the precise details of its location and learn all the appropriate facts.

7 Cold environments

7.1 What are the distinctive characteristics of cold environments?

Cold environments make up about 25 per cent of the Earth's surface (Figure 7.1) and are characterised by having temperatures below freezing. There are three main types of cold environment:

- **Polar environments**: these are the world's coldest places and are characterised by barren landscapes, ice sheets and glaciers. Polar environments occur at high latitudes and are defined by the Arctic Circle located at 66.5 °N, which includes the Greenland ice sheet (Figure 7.2), and the Antarctic Circle at 66 °S, which includes Antarctica.
- **Tundra environments**: the word 'tundra' comes from the Finnish word 'tunturia' which means 'treeless land'. It is land which is not permanently covered by ice but which experiences very cold temperatures (−34 °C) in winter. During the summer, the permafrost thaws just enough to let plants

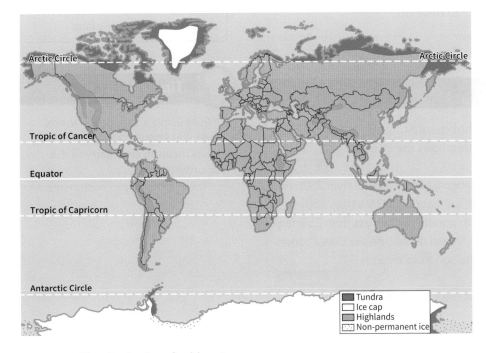

Figure 7.1 The distribution of cold environments.

grow and reproduce. The biodiversity of tundra is low although animals found in the Arctic tundra include caribou (reindeer), musk ox, arctic hare and polar bears. Tundra is located in high latitudes and covers parts of Alaska, northern Scandinavia, Siberia (Figure 7.3 and 7.4) and the edges of the Arctic and Antarctic.

- **High mountain or Alpine areas**: these include the Himalayas, Andes, Alps and Rocky mountains. Here, the type of environment is linked to the altitude as temperatures decrease 0.6 °C for every 100 m gained in height. Consequently, these areas have characteristics of tundra environments at lower altitudes and polar environments in higher and more exposed areas.

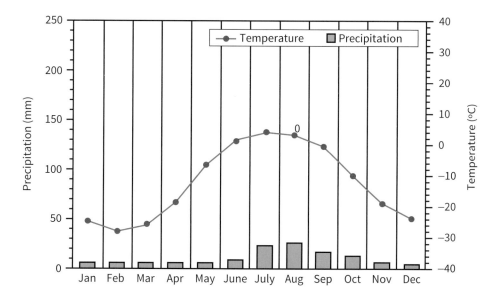

Figure 7.2 The polar climate: a climate graph for Greenland.

Figure 7.3 Tundra in Siberia.

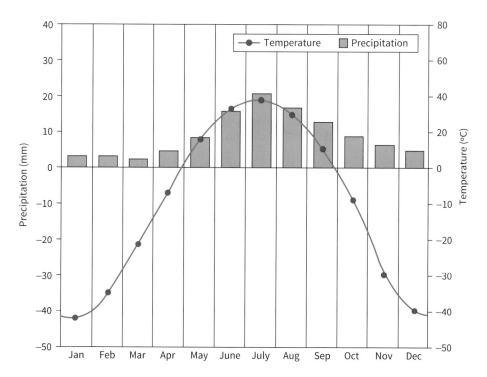

Figure 7.4 The tundra climate: a climate graph for Yatutsk, Russia.

7.2 How are plants and animals adapted to cold environments?

Plants and animals in polar environments

Cold environments are physically extreme; they are cold, dark and covered in permafrost. While temperatures vary throughout the year they rarely exceed 0 °C. Polar environments can also be considered as cold deserts because precipitation rates are so low. Precipitation rates in Antarctica are less than 200 mm per year on average. As a consequence, polar environments are barren landscapes. There is limited **biotic** life which can exist under such conditions.

Plants and animals in tundra environments

Tundra environments are characterised by **permafrost** or permanently frozen ground. The soil must be frozen (experience temperatures below 0 °C) for two or more years in order to be considered permafrost. However, tundra environments also experience seasonal variations in both temperature and precipitation. During the summer months, when temperatures rise above 0 °C, the surface layer (known as the active layer) of permafrost starts to melt. Soils may become boggy and waterlogged as the permafrost thaws and so plants have to adapt to these conditions. They also have adapted to the short summer growing season which is a period of rapid plant growth. Examples of plants that can be found in the tundra include arctic moss and the tufted saxifrage.

Arctic moss can grow in both freshwater arctic lakes and fens (wetlands that occur usually in the summer). It is very slow growing; when not growing, the moss stores nutrients. It grows close to the ground to protect it from the harsh winds, and can also grow underwater.

Tufted saxifrage has an extensive underground root system for storing nutrients. It grows in tight groups, which protects it from the wind and reduces transpiration.

Animals such as the caribou, snowy owl and arctic wolf have also had to adapt to the cold tundra climate. Caribou have thick coats and wide hooves to protect them from the cold temperatures and icy conditions. They can also smell and find lichens beneath the snow and manage to survive on this nutrient-poor food for a long time.

Snowy owls also have thick coats, which are made from layers of down with larger feathers on top. Even their legs and toes are covered with feathers to protect them from the low temperatures. They are white in colour, which means that they are well camouflaged in winter.

Arctic wolves are white and have two thick layers of fur. The outer layer grows thicker during the winter months and the inner layer forms a waterproof barrier. They also have padded paws which provide a good grip on the permafrost.

The interdependence of biotic and abiotic factors

It is clear from the nutrient cycle (Figure 7.5) that distinct interrelationships exist between both the biotic and **abiotic** aspects of the tundra ecosystem. The cold temperatures and minimal precipitation (both abiotic factors) mean that the decomposition of organic matter is limited. This in turn means that the soils are thin and infertile and there is limited **biomass** – only mosses, lichens and other hardy plants which have adapted to these conditions can survive. The environment is characterised by a lack of biodiversity and is therefore fragile. Plants in cold environments often have small leaves and grow close to the ground. This allows them to retain moisture more easily and to be protected from harsh winds.

Key terms

biotic: the living part of an ecosystem, e.g. plants and animals

permafrost: ground that has been frozen for two or more years; in the northern hemisphere over 19 million km² is covered in permafrost, most of which is found in Siberia, Alaska, Greenland, Canada and northern Scandinavia

abiotic: the non-living part of an ecosystem, e.g. climate and soils

biomass: organic material which comes from living or recently living organisms

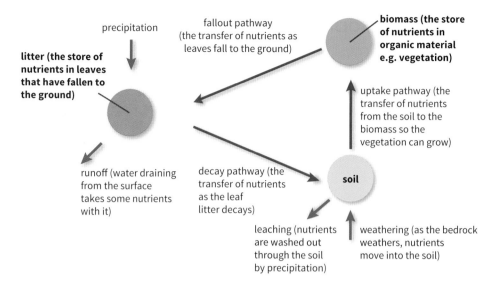

Figure 7.5 Nutrient cycling in the tundra. The red arrows indicate the influence of biotic factors, the blue arrows indicate the influence of abiotic factors.

They also tend to grow in shapes that help them to shed heavy snow more effectively. Deciduous trees lose their leaves in winter and go dormant. The leaves tend to evaporate water into the air and so they lose them in order to retain as much moisture as possible. Many species are evergreen which means that they keep their foliage. However, their leaves have a waxy coat to reduce moisture loss. Many animal species struggle to adapt to the cold winter conditions and lack of available food and therefore migrate to warmer environments in order to survive.

ACTIVITY 7.1

1 a On a blank world map locate the polar, tundra and alpine areas. You can use Figure 7.1 to help you.
  b Annotate your map with information about the climate and plant and animal species found in each environment. You may want to carry out some additional research to add detail.

 Download Worksheet 7.1 from Cambridge Elevate for help with Activity 7.1, question 1.

2 Carry out some research to explain why areas of higher latitude and altitude have lower temperatures.
3 Explain how plants and animals have adapted to the harsh conditions found in the tundra. The BBC series *The Frozen Planet* is a useful place to start.
4 Explain why the Arctic environment can be considered a 'fragile environment'.

 To find out more:

Take an in-depth look at the tundra biome on the **UC Berkeley website** (www.cambridge.org/links/gase40045), or on the **National Geography website** (www.cambridge.org/links/gase40046).

Visit the **Weather Underground website** (www.cambridge.org/links/gase40047) for information on permafrost.

 Discussion point

Cold environments cover about 25 per cent of the Earth's surface today. How and why might this coverage change over time?

 Did you know?

Although, on a global scale, the climate appears to be the same over a large area, climates do in fact show significant variation. For example, while Alaska is generally a tundra environment, its coast is warmer and its inland and more mountainous regions are much colder.

7.3 What are the challenges of developing cold environments?

Cold environments are physically extreme; they are cold, dark and covered in permafrost. The areas covered are also large and remote making it difficult to build communication links and supply services. Consequently, population densities in cold environments are typically very low and so the available workforce may be small. These physical and human reasons combined mean that attempting to develop cold environments can be a very expensive venture.

Extreme temperatures

The extreme temperatures found in cold environments make it difficult for living things to survive. Despite this, groups of **indigenous** people have lived in cold environments for centuries and have adapted to living in sub-zero temperatures. The Inupiat, who live in the North Slope region of Alaska, are hunter-gatherers. They hunt whale for its blubber, which provides them with energy to keep warm. Blubber is also rich in vitamins A and C; these vitamins are otherwise difficult to come across due to the lack of availability of fruit and vegetables, which cannot grow in the freezing temperatures. Some other indigenous people hunt reindeer, which provide them with fur for warm clothes and meat for energy. However, reindeer need grazing land to survive and so often these people are nomadic, travelling hundreds of kilometres to ensure the herds have enough to eat.

Permafrost

Permafrost occurs as a direct result of the extreme temperatures and developing areas of permafrost can be challenging. In the summer, when the active layer thaws, buildings can subside and transport routes become damaged (Figure 7.7). In the winter, when all the available water has turned to ice it can be difficult to find a reliable water supply. Also, when ice forms underground it expands causing frost heave which makes the ground shift. Consequently, people living in Alaska tend to construct their roads on gravel so that the effects of frost heave are less dramatic. They also build their houses on stilts. If they build their houses directly on the ground the residual heat melts the permafrost and causes subsidence. Therefore, one of the most effective methods to prevent this is to raise the base of the house above the ground so that the permafrost remains frozen.

Inaccessibility

Many cold environments are inaccessible and transportation costs to places that are inaccessible tend to be high. This means that the people living there cannot rely upon deliveries of resources such as food, building materials and fuel and instead have to be largely self-sufficient. It can also be difficult to provide the infrastructure to connect places to make them more accessible, which means that people struggle to reach help in an emergency situation. However, the remoteness of some cold environments has been considered an opportunity rather than a challenge by some governments who have located sensitive national security buildings in such areas for the reason that they are difficult to get to.

Provision of buildings and infrastructure

In addition to building houses on stilts in order to prevent permafrost from melting (Figure 7.8), there are other ways in which people have tackled the challenges of building in cold environments. For example, windows can contain triple glazing so that it is difficult for heat to escape. Roofs may have steep pitches so that it is difficult for snow and ice to accumulate on them. Accumulated snow is very heavy and can easily cause damage or present a danger to people and objects below. In areas where the outside temperature is below freezing, many buildings also use geothermal energy as a heating source. Underground pumps extract

Figure 7.6 Indigenous people with a herd of reindeer.

Figure 7.7 Melting permafrost causes damage to transport links.

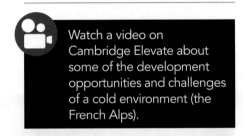
Watch a video on Cambridge Elevate about some of the development opportunities and challenges of a cold environment (the French Alps).

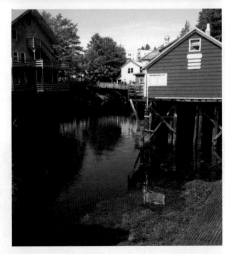
Figure 7.8 Houses on stilts in Alaska.

Figure 7.9 An ice dam forming on a snow-clad roof.

Figure 7.10 An Antarctic food web.

energy from the ground, which remains a constant temperature throughout the year, and transfer the energy for heating buildings.

Impacts on food chains and webs

Economic activity can have negative impacts on food chains. The challenge is how to manage this sustainably. Krill – small shrimp-like creatures – are vital to the food chain of cold ocean environments and support large animals such as penguins, whales and seals, which feed on them (Figure 7.10). These in turn sustain the indigenous people living in places such as Alaska. However, the numbers of krill are thought to be declining due to climate change and the overfishing of krill for fish food used in aquariums. This could have a significant impact on the ecosystem and the people living there.

Discussion point

How does the physical geography of cold environments makes development challenging? What about the human geography of cold environments?

Further research

Read the article 'Nepal Earthquake 2015: Aftershocks Devastate Nepal's Tourism Industry, Everest, Unesco Sites' in the *International Business Times* which discusses how the 2015 Nepal earthquakes may affect Mount Everest's tourism industry.

To find out more:

Look at the **Survival website** (www.cambridge.org/links/gase40048) or **BBC website** (www.cambridge.org/links/gase40049) to find out more about the Nenet tribe.

Look at the **WWF website** (www.cambridge.org/links/gase40050) for a more in-depth look at the impacts of Alpine tourism.

Read articles in **The Telegraph**, **'Is tourism turning Mount Everest into an open cesspit?'** (www.cambridge.org/links/gase40051) and the **Guardian**, **'Himalayas in danger of becoming a giant rubbish dump'** (www.cambridge.org/links/gase40052) to discover more about the negative impact that tourism is having on Mount Everest.

Visit the **Futurity website** (www.cambridge.org/links/gase40053) to investigate the impact that the reduction in krill may have on the polar food chain.

Key term

indigenous: native people who originate from a particular place

ACTIVITY 7.2

1 Why do cold environments have low population densities?

2 Carry out some research into the challenge of building in areas of permafrost. An article in *The Moscow Times*, 'Counting the Cost of Russia's Melting Permafrost' is a good place to start.

3 a Draw a food web of the Southern Ocean ecosystem. You could use Figure 7.10 to help you.

  b What would happen if the number of krill dramatically decreased?

  c What impact would this have on the ecosystem?

4 Search on the internet for clips about dangerous jobs in cold environments such as ice road truckers.

Development opportunities and challenges in Alaska

Cold environments are often **wilderness** areas and are extremely fragile ecosystems. However, they can also provide opportunities for economic activities such as mineral exploitation, energy, fishing and tourism. This creates opportunities for employment and increases local government spending on the improvement of local infrastructure. This improves the socio-economic conditions of the area and may encourage further development. This process is known as the **multiplier effect**.

Alaska is a US state located in the north-west of the North American continent, bordered by Canada, and the Arctic and Pacific oceans (Figure 7.11). Alaska is an example of a cold environment that has been developed.

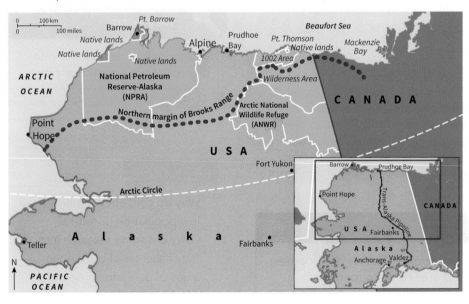

Figure 7.11 Map of North Slope, Alaska and the Trans-Alaska Pipeline.

Tourism

Tourism is one of the activities that has resulted in Alaska's economic growth. Around 1.9 million people visit each year to experience its unique wildlife and spectacular scenery. Most tourists visit between mid-May and mid-September when temperatures are warmer. Cruise companies operate their ships during this period and hikers can take advantage of the drier ground later in this season. Fewer tourists visit during the bitterly cold winter months. However, this is the best time to see the Northern Lights and to experience winter festivals such as the Yukon Quest dog sled race. One in every 13 jobs in the state is related to tourism, which adds $4 billion to the Alaskan economy each year.

As a result of increased tourism, conservation is also an important feature of spending and employment in the area.

Fishing

Commercial fishing takes place in the Arctic Ocean. The pollock fisheries in the Bering Sea are some of the largest and most significant in the world. The seafood industry provides 78 500 jobs in Alaska and contributes an estimated $5.8 billion to the Alaskan economy every year. Salmon is the most valuable commercial fishery.

Did you know?

At the turn of the nineteenth century, Alaska was one of the poorest US states. However, since the development of the oil and gas industry it is now one of the wealthiest.

Key terms

wilderness: a large area of land that has been relatively unaffected by human activity; these areas tend to have very low population densities and inhospitable environments

multiplier effect: the 'snowballing' of economic activity, for example, if new jobs are created this gives people more money to spend which means that more workers are needed to supply the goods and work in the shops

Mineral extraction and energy

Alaska has a wealth of mineral resources, including gold, copper, iron ore, lead, zinc, silver and others. Before oil was discovered in Alaska, gold was the main focus for mining.

The North Slope area of Alaska has significant petroleum resources which are mostly located in the Prudhoe Bay Oilfield which was discovered in 1968. The petroleum which is extracted from this region is transported 1 287 km south to Valdez via the Trans-Alaska Pipeline System (Figure 7.11, inset).

This process is further complicated by the fact that the Arctic National Wildlife Refuge (ANWR), a protected area, is located to the east of Prudhoe Bay. Those that are in favour of the exploitation highlight that the economy of the state would suffer without it. The oil and gas industry accounts for one-third of all jobs and 90 per cent of the revenue in Alaska. Those that are against exploitation focus on the environmental costs. The process of extraction emits harmful pollutants which degrades the clean air and water which animals such as walruses and whales depend upon for survival. There is also the threat of oil spills such as the Exxon Valdez oil spill which occurred in Prince William Sound, Alaska on 24 March 1989 (Figure 7.12).

The Alaskan pipeline reached its peak production in 1988 but has since been in decline. This is partly because the oil that is easiest to extract is being rapidly used up and the extreme conditions mean that developing new technologies to exploit reserves that are harder to reach can be prohibitively expensive. However, conservationists are concerned that melting sea ice may open up parts of Alaska and the Arctic so that these areas can be more easily exploited by oil and gas companies in the future.

Figure 7.12 The Exxon Valdez oil spill, Prince William Sound, Alaska.

Further research

Carry out some further research into the tourist attractions which can be found in Alaska.

Did you know?

More than one-third of Alaska's jobs are tied to the oil and gas industries.

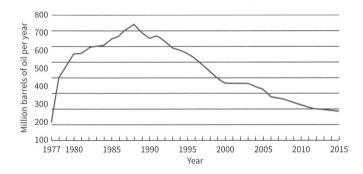

Figure 7.13 Oil extraction in Alaska from 1977 to 2015.

ACTIVITY 7.3

1 a What might be some of the challenges of living in a harsh environment such as Alaska?

   b Carry out some research about the problems of building the Trans-Alaska Pipeline and the problems of oil spills in the Arctic.

   c Should oil drilling be allowed in the ANWR?

2 a Describe and explain the pattern shown in the graph in Figure 7.13.

   b Is being reliant on natural resources, such as oil and gas, an advantage or a disadvantage?

The North Slope region, Alaska

The North Slope region of Alaska is located on the coastline of the Arctic Sea (Figure 7.14). It covers the area on the northern slopes of the Brooks Range mountains (Figure 7.11). The area has permafrost so only the active layer at the surface thaws each summer.

Development

Prudhoe Bay is the largest oilfield in North America. Oil is transported from Prudhoe Bay to Valdez, 1287 km away, by the Trans-Alaska Pipeline (Figure 7.15). The petroleum industry currently accounts for approximately one-third of Alaska's jobs, some 110 000. It also contributes over $14 billion to the state's economy.

Challenges

In addition to the challenges of extreme temperatures, inaccessibility and provision of buildings and infrastructure described in 7.3 What are the challenges of developing cold environments?, the development of the North Slope of Alaska has its own unique challenges. When the pipeline was initially proposed in the 1960s, native Alaskans were concerned that it would cross their traditional lands yet they would not benefit economically. Plans for the pipeline to be placed underground were also opposed by conservationists who were concerned that the seasonal

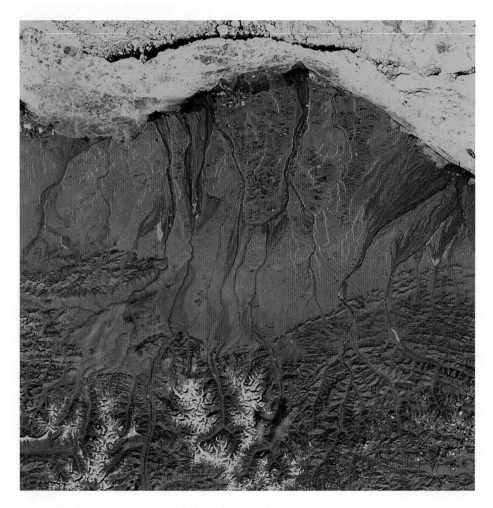

Figure 7.14 A satellite image of North Slope, Alaska.

melting and refreezing of the permafrost could damage the pipeline and lead to oil spills. Both groups launched legal campaigns and managed to postpone the building of the pipeline until 1973. However, after much legal wrangling an above-ground pipeline built on supports was given the go-ahead and construction commenced in 1974. Tens of thousands of workers moved to Alaska to work on the pipeline and it was finally completed in 1977 having cost $8 billion.

When it was built, the pipeline was a major feat of engineering. Maintaining it remains a challenge. It is especially difficult in winter when Alaska's bitter temperatures leave the pipeline and its pumping stations vulnerable to ice, wax build-up, and other operational problems. Despite this, 212 million barrels of oil are transported each year. While there have been spills caused by a combination of human error, maintenance issues and natural disasters such as the 2002 Denali earthquake, these have been relatively minor.

The future of the pipeline

The pipeline currently transports about 14 per cent of US domestic oil supplies and yet, since its peak in 1988 there has been an annual decline in production. While there are still substantial reserves in the Prudhoe Bay Oilfield these may soon be both unsafe and uneconomical to exploit. If the flow of oil drops then the flow slows and oil stays in the pipeline longer. This causes the oil to cool which can cause bacteria to grow which, in turn, can corrode the pipe increasing maintenance costs.

Oil companies instead believe the most profitable reserves are those where oil is relatively easy to reach. These areas include parts of the National Petroleum Reserve-Alaska (NPRA), the Arctic National Wildlife Refuge (ANWR), and the Chukchi and Beaufort seas just offshore of the North Slope. ANWR alone holds 10 billion barrels of untapped oil but drilling in this protected region could, in particular, damage the calving grounds of the porcupine caribou. In 2015 the US president, Barack Obama, put forward a proposal to increase the land designated as 'wilderness area' from 7.8 to 12.8 million acres. This would prohibit oil drilling and other development in this part of the refuge.

Figure 7.15 The Trans-Alaska Pipeline transports oil from the North Slope to Valdez in the south.

 Further research

Read the article 'How Much Life Is Left in the Trans-Alaska Pipeline? on the **Popular Mechanics website** (www.cambridge.org/links/gase40054) which highlights some of the problems the pipeline faces, and some possible solutions.

 Tip

- As the question asks for a case study, you must use one.
- Produce your case study at the beginning of the answer and build the answer around it by using facts and figures.
- It is better to use a small number of effectively developed points rather than a long list of basic ideas.

ACTIVITY 7.4

1 a Create a mind map which summarises the opportunities and challenges of development in Alaska. Try to include facts and figures to illustrate your points.

   b Colour code the mind map to show whether these opportunities and challenges are social, environmental or economic.

   c Do you think that the economic benefits are worth the environmental costs? Justify your answer.

 Assess to progress

1 Using a case study, explain how cold environments can provide both economic opportunities and development challenges.   9 MARKS

 Read the overview of the Trans-Alaska Pipeline that can be found on the **Alyeska pipeline service company website** (www.cambridge.org/links/gase40055) and **mindmap** about the environmental impacts of the Trans-Alaska pipeline (www.cambridge.org/links/gase40056).

7.4 How can cold environments be protected?

Cold environment ecosystems are complex and interconnected. However, because these ecosystems are also fragile, the equilibrium of the environment can be easily damaged if the numbers of a particular species decrease. For example, oil exploration has had a major impact on the number and health of polar bears in the Arctic. Contact with oil spills can reduce the insulating properties of the bear's fur and so they need to eat more calories in order to stay warm. Also, if the oil is ingested it can poison the bears causing liver and kidney damage. While numbers of polar bears are currently stable at around 20 000 to 25 000 these threats, combined with additional threats from climate change and overhunting, could cause population numbers to decrease which would have a knock-on effect to other species in the ecosystem. It is therefore important to balance the needs of economic development and the conservation of species under threat.

Managing Antarctica

Antarctica is an extremely fragile environment. The prolonged snow and ice cover, strong winds and associated wind chill and drought conditions mean that virtually no vegetation can survive. Antarctica is one of the most remote places on Earth and yet small-scale tourism began there in the 1950s. It now has around 50 000 visitors per year, approximately a third of whom are from the USA. Tourists are attracted by the wildlife, the scenery and the fact that this unique environment is very different from the one in which they live. This increase in visitor numbers has the potential to threaten penguin breeding patterns and passenger vessels could impact on the marine environment (Figure 7.16). In terms of economic development, the continent of Antarctica derives no economic benefits from tourism but can still suffer huge costs. Unlike other cold environments, there are no indigenous people to benefit from tourism in Antarctica and so tourism is not an alternative to local unsustainable economic activities. As a consequence, there is a need for governments (Figure 7.17), international organisations and conservation agencies to work together for the **sustainable management** of Antarctica.

The Antarctic Treaty

The Antarctic Treaty is an agreement which was signed in 1961 by 12 countries – Argentina, Australia, Belgium, Chile, France, Japan, New Zealand, Norway, South Africa, the Soviet Union, the UK and the United States – who were engaged in scientific research in Antarctica at the time. The purpose of the original treaty was to establish Antarctica as a peaceful region and to encourage scientists from different countries to work effectively together to research the continent. Today, there are three additional agreements which focus on the conservation of Antarctic seals and Antarctic marine living resources and a protocol on environmental protection.

Seven tour companies running tourist expeditions to Antarctica created the International Association of Antarctic Tour Operators (IAATO) in 1991. This built upon the treaty to provide practical guidelines for tour operators and tourists to minimise the impact. These include: do not feed wildlife or leave food scraps lying around and do not introduce any plants or animals into the Antarctic.

Figure 7.16 Tourist vessel approaching the South Pole.

 Key term

sustainable management: management that meets the needs of the present without compromising the ability of future generations to meet their own needs; it takes into account the environment, the needs of present and future generations and the economy

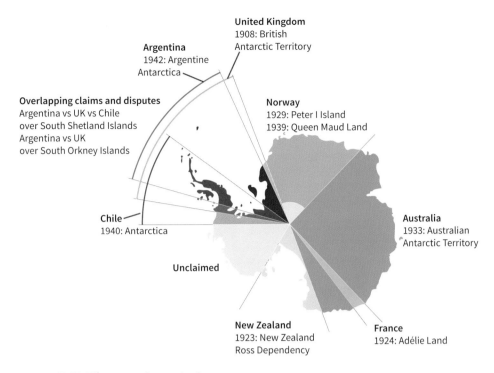

Figure 7.17 Who owns Antarctica?

Conservation agencies

Conservation agencies also work to protect and manage Antarctica. The World Wildlife Fund (WWF) is involved in creating a network of marine protected areas in the Southern Ocean to both protect the region's wildlife and its vulnerability to climate change. They are also developing a 'Polar Code' to ensure ships operate safely and sustainably in polar waters and are tackling the problem of illegal 'pirate' fishing.

Using technology to monitor the Antarctic

Since 1979, satellite data has been used to monitor sea ice coverage and thickness. In 2011, scientists installed fibre optic cables within the Antarctic ice sheet in order to measure ice sheet temperatures in and below the ice sheet. Scientists are particularly interested in observing changes at the base of the ice sheet, which is where melting begins.

ACTIVITY 7.5

1 Create a digital presentation that explains why cold environments are seen as fragile.
2 Look at Figure 7.17, which is an infographic that illustrates ownership claims on Antarctica. Create a similar infographic for 'Who owns the Arctic?'
3 Carry out some further research into management strategies in Antarctica. A good place to start is the Antarctic page of the **WWF website** (www.cambridge.org/links/gase40057). To what extent will management strategies help to ensure the sustainability of Antarctica?

Did you know?

Antarctica is the coldest, driest and windiest continent. It is also a cold desert. It has about 200 mm of precipitation along the coast and less than this inland and temperatures have reached −89 °C.

Discussion point

Does the fact that both the Arctic and Antarctica have complicated ownership/no ownership make them more or less vulnerable to development pressures?

Further research

Look at the guidelines for tourists to Antarctica on the **IAATO website** (www.cambridge.org/links/gase40058) and design an illustrated guide for tourists.

SECTION 3

Physical landscapes in the UK

You will learn about:
- the main features of the UK's landscape
- the physical processes that have influenced the UK's landscape

You will also study two of the following topics and will learn about:

8 Coastal landscapes in the UK

In this chapter you will cover:
- what happens when waves reach the coast
- how physical processes affect the coast
- landforms associated with erosion processes
- landforms resulting from deposition
- protecting coastlines from the effects of erosion using soft and hard engineering.

9 River landscapes in the UK

In this chapter you will cover:
- how river valleys change as rivers flow downstream
- how rivers erode the land and why they deposit sediment
- landforms arising from erosion
- landforms resulting from deposition
- why we need to protect river landscapes from the effects of flooding
- strategies used in flood protection and management

10 Glacial landscapes in the UK

In this chapter you will cover:
- glacial processes of weathering and erosion; movement, and transportation and deposition
- the characteristics and formation of landforms of upland and lowland erosion
- the characteristics and formation of landforms of deposition
- some of the economic activities that occur in glaciated areas
- that conflicts exist between different land uses and people
- how the Lake District National Park manages the pressures of tourism

Physical landscapes in the UK – an overview

The UK has a tremendous variety of natural landscapes, including high mountains, upland areas with fast-flowing rivers, broad lowland areas where rivers meander towards the sea and coastal areas with spectacular cliffs or areas of protected salt marsh. The variety of these landscapes is largely the result of the different underlying rocks and the processes that have operated on them.

The distinctive features of these landscapes have been created over millions of years by the action of water, wind and ice. Many of the distinctive landforms seen in the UK have been shaped by the processes of erosion, transportation and deposition. Understanding these processes allows us to see where they are active today and gives us the opportunity to consider how our relationship with the physical environment can be managed sustainably, especially in relation to river and coastal environments.

The United Kingdom

The United Kingdom consists of England, Scotland, Wales and Northern Ireland. It lies approximately between 50°N and 61°N and has a variety of natural landscapes which have developed from underlying rocks (Figures S3.2 to S3.4). The youngest rocks are to the south and east of a line drawn from the mouth of the River Tees to the mouth of the River Exe (Figure S3.5). Three main types of rock are found in the UK: sedimentary, igneous and metamorphic.

Figure S3.1 The Old Harry Rocks in Studland, Dorset, are a great example of landforms created by coastal erosion. Why do you think these landforms are not seen on every part of the UK coastline?

Sedimentary rocks

These are made from eroded material that has been compressed to form layers. Examples are:

- sandstone – deposited when the area had a hot desert climate
- chalk – fine deposits from the remains of marine animals
- carboniferous limestone – deposit of the remains of marine animals and calcium carbonate
- coal – remains of tropical swamps and forests
- clay – fine deposits of silt and mud

Igneous rocks

These are formed from molten material or magma resulting from volcanic activity. Examples are:

- basalt – formed on the Earth's surface from flows of lava
- granite – formed within the Earth and now exposed because of erosion

Metamorphic rocks

These are rocks that have been changed by heat or pressure. They are usually found near areas of volcanic activity.

Highlands, lowlands and rivers

Upland areas

Hills, moors and mountains form approximately 40 per cent of the UK land area. Upland areas are used for farming, forestry and recreational activities. Nearly 70 per cent of the UK's drinking water is stored in reservoirs in upland areas before being transported to urban populations in lowland areas. These areas are often made up of spectacular scenery and have unique plant and animal life. They also act as a store for billions of tonnes of carbon. The influence of glaciation can be seen in many parts of the UK (Figure S3.7). Glacial erosion created many of the spectacular mountain features while glaciation deposits are found throughout southern England.

Figure S3.2 Glen Coe in the Scottish highlands.

Figure S3.3 Soft alluvium cliffs on the Isle of Sheppey, Kent.

Figure S3.4 Chalk cliffs, south coast.

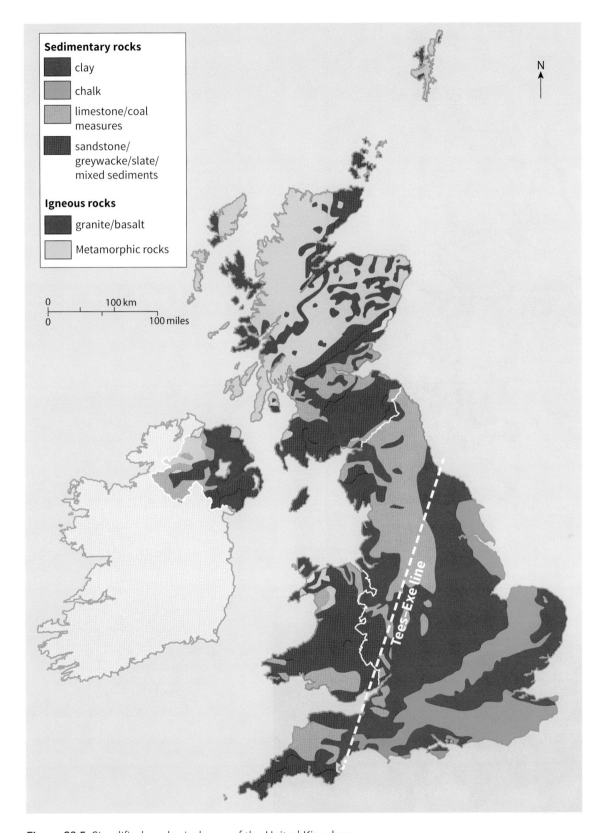

Sedimentary rocks
- clay
- chalk
- limestone/coal measures
- sandstone/greywacke/slate/mixed sediments

Igneous rocks
- granite/basalt
- Metamorphic rocks

N

0 ⎯ 100 km
0 ⎯ 100 miles

Tees–Exe line

Figure S3.5 Simplified geological map of the United Kingdom.

Figure S3.6 The major rivers, highland and lowland areas in the UK.

Figure S3.7 Mountains showing glacial features.

Lowland areas

Approximately 80 per cent of the population of the UK live in lowland areas and virtually all major towns and cities are found in these areas.

Rivers

The UK's rivers are a vital source of fresh water for people, industry, farming and wildlife. Chalk streams are unique physical features (Figure S3.8). There are only about 200 in the world, most of which are in southern England. Many of the UK's busiest ports are based around river estuaries and rivers provide important transport links for people and industrial goods.

The UK's longest rivers (km)		
1	Severn	354
2	Thames	346
3	Trent	298
4	Great Ouse	230
5	Wye	215

Figure S3.8 Chalk stream.

ACTIVITY

1 How does the geology of the United Kingdom show that:
   a the climate in the past has been very different?
   b the UK was once an active volcanic area?
2 Why do most people live in lowland areas?
3 a Construct a bar graph to show the length of the UK's five longest rivers.
   b Suggest why rivers are such a valuable resource.

Discussion point

'For a place the size of the UK, it has an enormous range of physical landscapes.'

Explain this statement.

Did you know?

The longest river in the world is the River Nile, which is estimated to be 6695 km in length.

8 Coastal landscapes in the UK

In this chapter you will learn about...

- what happens when waves reach the coast
- how physical processes shape the coastline
- the landforms resulting from erosion
- the landforms resulting from deposition
- how coastlines are protected from the effects of physical processes.

8.1 What happens when waves reach the coastline? (1)

How are waves formed?

When wind blows over the **sea** or **ocean**, friction occurs, causing the surface of the water to be pushed in the direction of the wind, creating waves (Figure 8.1). The stronger the wind, the greater the friction and the bigger the wave. The distance over which the wind is able to affect the ocean is called the fetch. The longer the fetch the greater potential wave energy.

Key terms

sea: a region of water within an ocean or partly enclosed by land

ocean: large body of salt water

Figure 8.1 Waves breaking on coastal defences in Eastbourne, East Sussex.

Why do waves break?

The energy within a wave moves in a circular motion. As a wave moves towards shallow, coastal waters, the base of the wave is in frictional contact with the seabed. This slows the base of the wave down, reducing the distance between waves (wavelength) and increasing the wave height. The wave breaks because the top of the wave is moving faster than its base (Figure 8.2).

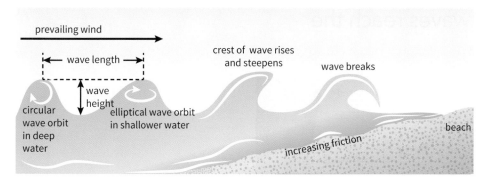

Figure 8.2 The formation of waves.

What happens when waves break?

Breaking waves force water and beach **sediment** up a beach as **swash**. When all of the energy from the breaking wave has been used gravity takes the water back down the beach as **backwash**.

Constructive and destructive waves

Waves are the most important force in helping to shape the coast. There are two main types of wave:

- **Constructive waves** (Figure 8.3) – which build up beaches by depositing sediment (Figure 8.4).
- **Destructive waves** (Figure 8.5) – which remove sediment from beaches (Figure 8.6).

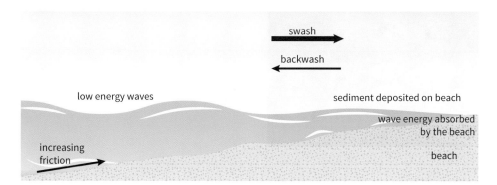

Figure 8.3 Constructive waves (also called spilling waves).

Figure 8.4 A constructive beach.

Did you know?

Coastal environments can be classified according to wave patterns:

- Storm wave environments are characterised by strong onshore winds which generate powerful waves, for example in the UK.
- Swell wave environments are characterised by less extreme winds but a large swell built up over a long fetch, for example in West Africa.
- Tropical cyclone environments are characterised by huge waves resulting from extreme winds, for example in Southeast Asia and the Caribbean.

Key terms

sediment: material moved and deposited in a different location

swash: movement of waves up a beach

backwash: movement of water down a beach (gravity)

Tip

Remember, a beach is an example of a **depositional** feature.

8.1 What happens when waves reach the coastline? (2)

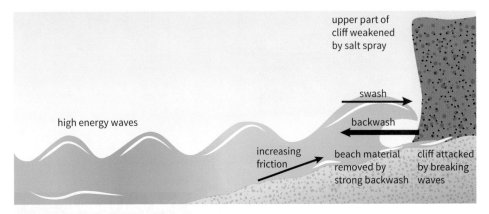

Figure 8.5 Destructive waves (also called plunging waves).

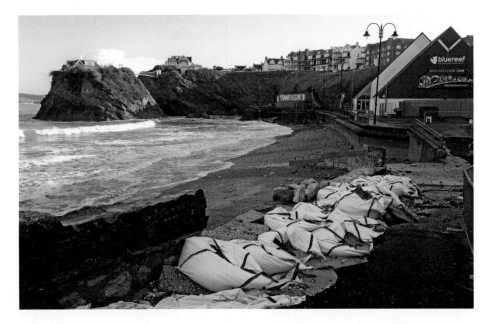

Figure 8.6 A beach affected by destructive waves.

Beaches created by constructive waves

As waves approach the coast they pick up sediment and when waves break the sediment is deposited. Constructive waves deposit more sediment than they remove, creating wide, gently sloping beaches (Figure 8.4). There are two main types of beach created by constructive waves: swash aligned beaches and drift aligned beaches.

Swash aligned beaches

Swash aligned beaches are formed when waves approach the coastline parallel to the beach. The swash and backwash moves sediment up and down the beach creating a wide beach with an even profile.

Drift aligned beaches

Drift aligned beaches are formed when waves approach the beach at an oblique angle to the coastline. The energy of the swash moves the sediment up the beach at the same angle, while the backwash moves the sediment back down the beach in a straight line, under the force of gravity (Figure 8.7). In this way sediment is moved along the beach by a process called longshore drift, creating a beach with an uneven profile. On some beaches groynes are built to slow down the movement of sediment along the beach.

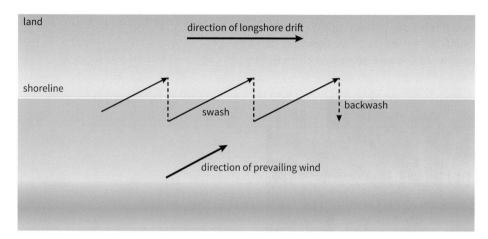

Figure 8.7 Longshore drift.

Figure 8.8 A constructive, drift aligned beach.

ACTIVITY 8.1

1 Explain how waves break.
2 Examine the relationship between wind speed and wave energy.
3 Use annotated diagrams to describe:
   a constructive waves
   b destructive waves.
4 Use Figure 8.7 and Figure 8.8 to describe the characteristics of beaches created by constructive waves.

8.2 How do physical processes affect the coast?

The coastline is the frontier between land and sea (Figure 8.9) and is being constantly reshaped by the action of waves and the weather. Where coastlines are made of more resistant rocks, or are sheltered from the prevailing wind and storm waves, changes occur slowly. Where coastlines are made of weaker rock or open to the full force of storm waves and heavy rainfall, whole areas can change in minutes as a result of landslides or rock falls.

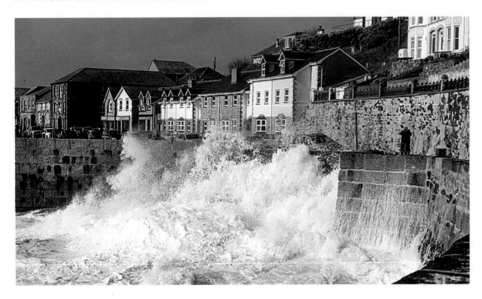

Figure 8.9 Waves battering the coast at Porthleven, Cornwall.

The coastal system

The shoreline acts like a giant conveyer belt (Figure 8.10): rocks are broken down and worn away in some places. The resulting sediment is transported by waves and wind and deposited in other places.

Figure 8.10 The coastal conveyer belt.

Weathering in coastal areas

Coastal areas are mainly affected by two types of **weathering**.

Mechanical weathering

Rocks are broken down without any change in their chemical composition. Examples found in coastal areas are:

1 **Wetting and drying** – softer rocks such as clays and shales are affected by water. These rocks expand and contract as they become wet and then dry out. As they dry out, cracks develop so rainfall and sea spray can more easily penetrate the rock, making them unstable. This can lead to landslides.
2 **Freeze-thaw** – moisture in rock surfaces freezes and expands at night and thaws during the day. In cold coastal areas the constant expansion and contraction weakens the rock surface until it begins to crumble.

Key term

weathering: the breaking up of rocks that occurs *in situ* (the same place) with no major movement taking place

Did you know?

The increasing use of coastal areas for recreation can put pressure on the coastal environment. The over-use of clifftop footpaths can weaken the rock structure and the removal of vegetation can leave clifftops more vulnerable to weathering and erosion.

Chemical weathering

Water reacts with the minerals in the rock to change its structure. The main type of chemical weathering found in coastal areas is solution. Carbon dioxide in the air dissolves in water, making the water slightly acidic. This weak acidic solution is able to dissolve some types of rock, especially limestone.

Marine erosion

The sea is a powerful force, especially during storm conditions. The coastline is affected by three main types of marine (coastal) **erosion**.

1 **Abrasion (also called corrasion)** – during storm conditions waves pick up sand and pebbles. When waves break the sediment that they have been carrying is hurled at the cliff face, creating a sandblasting effect. It is suggested that abrasion is the most powerful type of erosion affecting coastal areas in the United Kingdom.

2 **Attrition** – sand grains and pebbles are constantly being moved as waves break on a beach. This constant movement wears away the beach material, making it increasingly smaller and more rounded.

3 **Hydraulic power** – the sheer force of waves breaking against a cliff will cause parts of the cliff to break away. As waves hit a cliff face air is compressed in cracks in the rocks, 'blasting' away small fragments of material. During storms, hydraulic power can remove enough rock at the base of a cliff to make the cliff face unstable, resulting in a rock fall (Figure 8.11).

Figure 8.11 Rock fall on a chalk cliff.

Key term

erosion: the breaking up of rocks that is the result of movement

Tip

Remember that coastlines are affected by both **weathering** and **erosion**. Weathering weakens and breaks rocks down which means they are then more easily attacked by the forces of erosion.

Did you know?

During storms, breaking waves can exert force of up to 50 tonnes per square metre on a cliff face.

ACTIVITY 8.2

1 Explain what is meant by the 'coastal system'.
2 Construct a table with the heading, 'How physical processes affect the coast'.
   a List the processes of weathering and erosion down the left hand side of your table.
   b Write a brief definition of each type of weathering and erosion.
3 Explain how weathering and erosion helps to 'reshape' the coast.
4 Suggest three factors that might affect the rate of erosion.
5 When walking along a shingle beach in front of a cliff, what evidence might you look for to identify the different physical processes taking place?

Download Worksheet 8.1 from Cambridge Elevate for help with Activity 8.2, question 2.

8.3 What landforms are associated with coastal erosion?

Key term

headland: a narrow piece of land extending out into the sea, sometimes called a promontory

The influence of geology on coastal landforms

Rock type and structure can have a significant influence on coastal landforms. Rocks that have a stronger structure, such as chalk and limestone, erode more slowly and often produce spectacular cliff and **headland** features. When chalk cliffs are undercut by erosion, caves are formed but the cliff does not easily collapse because of the strength and structure of the rock. Weaker rocks, such as clays and sands, have less structural strength and are more easily affected by weathering and erosion, resulting in slumping and landslides.

Landforms associated with headland erosion

Many coastlines are made up of headlands and bays. Because headlands are made of more resistant rock they erode more slowly and often form spectacular scenery. This can be clearly seen on chalk headlands where wave action produces a number of distinctive landforms. Figure 8.12 shows how these features are formed.

Tip

When describing the formation of coastal landforms, remember the sequence of their formation. Use geographical terms to identify processes and individual features.

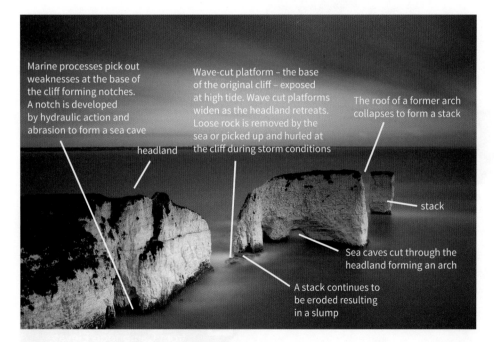

Figure 8.12 Annotated photograph showing chalk headland features.

Mass movement in coastal areas

Mass movement is the downhill movement of material caused by gravity. Weathering and erosion weakens cliffs and they become increasingly unstable, resulting in a mass movement event. Examples of types of mass movement include:

- **Landslides** – where a mass of unconsolidated material moves down a slope, often after a period of rainfall.
- **Slumping** – where a section of cliff drops down along a line of weakness (Figure 8.13).
- **Rock fall** – where material falls from a cliff face and lands at the base of a cliff. This is often seen on chalk cliffs.

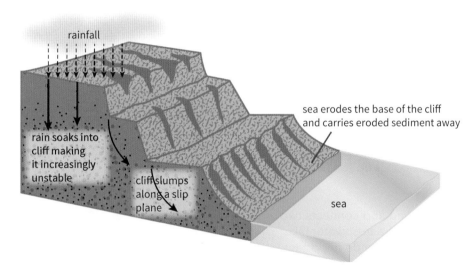

Figure 8.13 Diagram showing rotational slumping.

 Watch an animation on Cambridge Elevate about how erosion affects coastal headlands.

ACTIVITY 8.3

1 How does rock type affect rates of erosion?
2 Explain the processes that lead to rotational slumping.

8.4 What landforms are created by coastal deposition?

Beaches

As waves approach the coast they pick up sediment and carry it towards the beach. In some coastal areas, especially where there are sheltered bays or the slope of the seabed is gentle, wave energy is reduced. In these areas more sediment is deposited than is transported away from the coast, so wide, gently sloping beaches are created (refer back to Figure 8.4).

Sand dunes

Sand dunes are formed when strong onshore winds blow sand from the beach inland. The sand forms into mounds, held together by long rooted grasses such as marram (Figure 8.14). Where no vegetation is present the sand will continue to be blown inland. In some places grasses are planted or fences built in an attempt to stabilise the dunes.

Spits and bars

Spits and bars are ridges of sand or shingle that has been transported along the coast by longshore drift (refer back to Figure 8.7). Where the coastline changes direction sediment continues to be transported in the same direction of the original coastline, eventually being deposited to form a ridge of sediment sticking out into the sea. The seaward end of the spit is usually shaped into a curve by wave action and ocean currents. The area behind the spit is sheltered from wave action so mud and **silt** deposited by rivers builds up to form **mudflats** and **salt marshes** (Figure 8.15).

A bar is formed when a spit extends across an opening and connects two areas of coastline. This often results in the formation of a freshwater lake or lagoon behind the bar. Areas behind spits and bars provide important habitats for plants and animals and are often considered to have high environmental value. They are often protected by being designated as Nature Reserves or Areas of Outstanding Natural Beauty (AONB). Many environmentally protected coastal areas are used by the public for recreational and educational activities.

 Did you know?

Millions of tonnes of sand and gravel are dredged each year from the seabed for the construction industry.

Figure 8.14 Marram grass helps stabilise sand dunes.

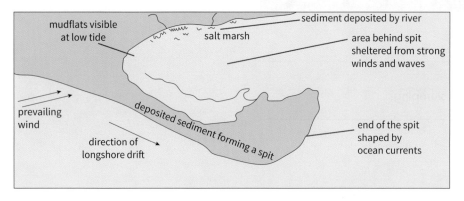

Figure 8.15 The formation and features of a spit.

 Key terms

silt: very fine material deposited by water

mudflats: a stretch of muddy land which is uncovered at low tide

salt marsh: an area of coastal grassland regularly flooded by seawater

Example

The Holderness coast, Yorkshire

The Holderness area on the Yorkshire coast has a mixture of hard and soft rocks, resulting in a range of spectacular coastal landforms (Figure 8.16).

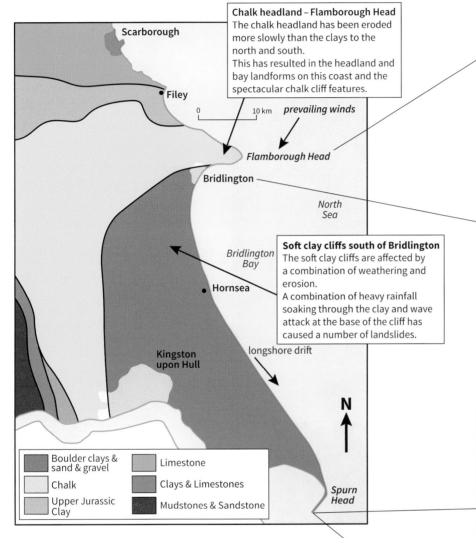

Chalk headland – Flamborough Head
The chalk headland has been eroded more slowly than the clays to the north and south.
This has resulted in the headland and bay landforms on this coast and the spectacular chalk cliff features.

prevailing winds

Soft clay cliffs south of Bridlington
The soft clay cliffs are affected by a combination of weathering and erosion.
A combination of heavy rainfall soaking through the clay and wave attack at the base of the cliff has caused a number of landslides.

longshore drift

N

Scarborough

Filey

Flamborough Head

Bridlington

North Sea

Bridlington Bay

Hornsea

Kingston upon Hull

Spurn Head

Legend:
- Boulder clays & sand & gravel
- Chalk
- Upper Jurassic Clay
- Limestone
- Clays & Limestones
- Mudstones & Sandstone

Figure 8.16 Simplified geological map of the Holderness coast, Yorkshire

Figure 8.17 Flamborough Head.

Figure 8.18 Slumping at Bridlington.

Figure 8.19 Spurn Point.

Spurn Point is a curved spit created by deposited sediment transported south along the Holderness coast. It is approximately 6 km in length but only 50 m wide in places. It is made of sand and shingle, held together by marram grass. Mudflats and salt marshes have developed behind the spit, creating an ideal habitat for wildlife.

The Holderness coast, Yorkshire (continued)

The following example shows a landslide that occurred to the south of Scarborough.

The vanishing coastline

Guests at the Holbeck Hall hotel in Scarborough woke up to find cracks in their bedroom walls and part of the hotel garden missing. Guests were quickly evacuated as it became clear that a landslide was likely to occur. Over the next 48 hours about a million tonnes of clay and gravel slid into the sea, taking much of the hotel with it.

The landslide followed a period of heavy rainfall which meant that the cliff had become increasingly unstable.

A local geologist said, 'After a long dry period there were a number of deep cracks in the clay. Heavy rainfall then penetrated the cracks, lubricating the clay and making it very unstable. Eventually the pressure of weight and gravity did the rest!'

Scarborough is on part of one of Europe's fastest eroding coastlines – in some places the soft cliffs are retreating at over 10 metres a year.

Figure 8.20 Holbeck Hall landslide – 3 June 1993.

Find out more about hard and soft coastlines at the **Channel Coastal Observatory website** (www.cambridge.org/links/gase40062)

Find out more about Spurn Head at:

- **BBC – Seven Wonders – Spurn Head** (www.cambridge.org/links/gase40059)
- **The Yorkshire Wildlife Trust** (www.cambridge.org/links/gase40060)
- **Spurn Bird** (www.cambridge.org/links/gase40061)

Skills link

See 19.9 Cartographic skills – coordinates, to remind yourself about grid references and other map skills.

Spurn Head spit

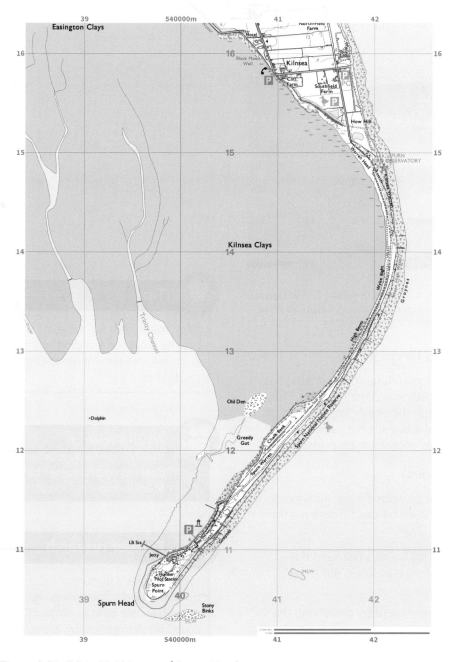

Figure 8.21 OS 1 : 25 000 map of Spurn Head.

ACTIVITY 8.4

Using Figure 8.21:

1 Give the six-figure grid reference for:
   a Spurn Bird Observatory
   b the southernmost point on Spurn Head spit
2 What is the direct distance, to the nearest kilometre, from Spurn Bird Observatory to the southernmost point on Spurn Head spit?
3 What is the direction of longshore drift in grid squares 4115/4116?
4 Why have mudflats and salt marshes formed behind the spit?
5 Why have sand dunes formed on the spit?
6 Using evidence from Figure 8.19 to help you, explain why coastal spits are valuable environmental and recreational areas.

8.5 Protecting coastlines from the effects of physical processes

Why is there a need to protect coastal areas from the effects of physical processes?

The increasing use of coastal areas, both as places for industrial development and places to live, has meant that a growing number of people are at risk from coastal erosion and flooding. The cost, both in terms of risk to life and property loss, of coastal flooding and mass movement events can be significant. There has always been a need to protect vulnerable coastal areas from the elements, but this need is likely to increase in the future as climate change brings about rising sea levels and an increasing number of winter storms (Figure 8.22).

Figure 8.22 Railway workers inspecting the main Exeter to Plymouth railway line at Dawlish which was closed due to parts of it being washed away by the sea on 5 February 2014.

How are coastlines managed in the United Kingdom?

In the UK the Department for Environment, Food, and Rural Affairs (Defra) has overall responsibility for the protection of the coastline against flooding. They work with local councils and landowners to manage the coast. Not all parts of the coastline need to be protected. Some coastal areas are not vulnerable to erosion or flooding or are not considered valuable enough to be protected.

What methods are used to protect coastlines?

If it is decided that an area of coastline should be protected a decision has to be made about the management strategy that should be used. There are three main strategies of coastal protection.

1   **Hard engineering** – controls the sea by building barriers between the sea and the land, or forces waves to break before they reach the coast by building offshore breakwaters.
2   **Soft engineering** – works with the natural environment by preserving the beach. A wide, gently sloping beach absorbs wave energy, reducing the threat of erosion and flooding.
3   **Managed retreat (coastal realignment)** – involves allowing the sea to flood land up to a new line of defence further inland. Salt marshes are then developed on the newly flooded land, providing a natural barrier against storm tides.

Tip

Remember: The main threats to coastal areas are flooding, erosion and mass movement (landslides/ mudslides).

Watch a video on Cambridge Elevate about coastline management at Studland Bay in Dorset.

Further research

In many areas a mixture of both hard and soft engineering techniques are used to protect the coast. An example of this is the West Bay Coastal Defence Scheme in Dorset. Find out more at the **Dorset County Council website** (www.cambridge.org/links/ gase40063)

Hard engineering

Hard engineering attempts to control the power of the sea by using artificial structures. These structures are designed to reduce wave energy or simply create a barrier between the land and the sea so that the storm waves cannot reach cliffs or flood low-lying coastal areas (Figures 8.23 and 8.24).

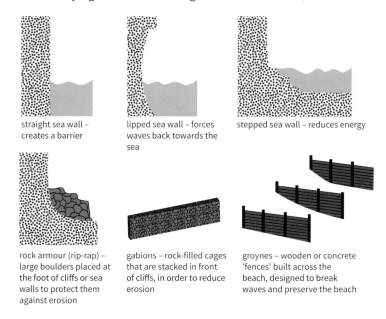

straight sea wall – creates a barrier

lipped sea wall – forces waves back towards the sea

stepped sea wall – reduces energy

rock armour (rip-rap) – large boulders placed at the foot of cliffs or sea walls to protect them against erosion

gabions – rock-filled cages that are stacked in front of cliffs, in order to reduce erosion

groynes – wooden or concrete 'fences' built across the beach, designed to break waves and preserve the beach

Figure 8.23 Examples of hard engineering techniques.

ACTIVITY 8.5

1 What factors might be taken into account when a decision is made about whether a coastal area should be protected or not?
2 Why are the terms 'hard' and 'soft' engineering used to describe those two methods of coastal protection?
3 Explain how hard engineering techniques protect coastal areas from physical processes.
4 Using Figures 8.24 and 8.25:
  a Describe the hard engineering methods used at Cleveleys.
  b Suggest why hard engineering was used at Cleveleys.

Figure 8.24 Sea defences at Cleveleys, Lancashire.

Figure 8.25 A storm at Cleveleys.

Costs and benefits of hard engineering

Costs	Benefits
Very expensive	Effective against the threat of flooding
Disruptive during construction	Reduction of wave energy
Need for constant maintenance	Can provide amenity valve (promenades)
Not always very attractive	Reduces risk of damage, making residents and local business feel more secure
May disturb/damage wildlife habitats	

Table 8.1 The costs and benefits of hard engineering.

8.6 Protecting coastlines from the effects of physical processes – soft engineering

What is beach nourishment?

Beach nourishment is a method of adding sediment to a beach so that it acts as a natural defence against storm waves (Figure 8.26). A wide and gently sloping beach acts as a natural defence because it is able to absorb wave energy and prevent waves from moving inland. It is used to reduce the flood risk in low-lying areas.

Beach nourishment is an example of 'soft engineering' because it provides a natural defence against the sea without having to build large, artificial structures. Consequently it is seen as environmentally friendly and more suitable in sensitive environments. The following example, at Pevensey Bay in East Sussex, describes how soft engineering techniques have been used to reduce the risk of flooding.

Why was coastal management needed at Pevensey Bay?

Pevensey Bay is a low-lying coastal area which has a history of flooding and faces increasing flood risk as sea levels rise. Just inland are the Pevensey levels, a low-lying freshwater environment which would be significantly damaged by saltwater flooding. The coastal area is home to about 1000 people and there are a number of local tourism related businesses. The main road and rail links are close to the coastline.

There are four main coastal management techniques being used at Pevensey Bay:

- **Beach nourishment** – replacing beach material lost during storms and as a result of longshore drift. Sand and gravel is taken from the seabed and sprayed back onto the beach using a specially adapted ship (Figure 8.26).
- **Beach recycling** – longshore drift carries beach material from west to east (25 000 m³ a year). Lorries are used to move the material back to its original position, making sure that the width of the beach remains the same along its length (Figure 8.27).
- **Beach reprofiling** – during storms beach material is moved towards the sea by the strong backwash, creating a beach with an uneven profile. This can leave the upper parts of the beach very low. Bulldozers push the material back up the beach, creating an even profile (Figure 8.28).
- **Technology** – global positioning systems (GPS) are used to track the movement of the beach material and identify areas where replenishment is required.

Dune regeneration

Dune regeneration is another type of soft engineering. It is used in areas where there are sandy beaches and strong on-shore winds which can blow the fine sand inland. In order to preserve the beach and stabilise the sand dunes, the area is planted with marram grass and fencing is used to trap the sand, creating a natural barrier between the land and the sea. In some areas boardwalks or coconut matting is used to protect the sand dunes from erosion caused by walkers and bike riders.

Managed retreat (coastal realignment)

Managed retreat is where low-lying areas are allowed to flood up to a new line of defence. This creates an area where salt marshes can be developed. When the salt marshes are fully developed they absorb wave energy during storms and therefore act as a natural defence. The area of salt marsh also allows rising

Key term

soft engineering: working with the environment in order to reduce the risks of flooding and erosion

Figure 8.26 Beach nourishment.

Figure 8.27 Beach recycling.

Figure 8.28 Beach after reprofiling.

tides to spread out, reducing the flood risk on surrounding coastal areas. The newly formed salt marshes provide an ideal habitat for wildlife and are often designated as **nature reserves**.

Managed retreat is often called coastal realignment because a new coastline is formed inland, so the coastline is 'realigned'.

Costs and benefits of soft engineering

Costs	Benefits
Requires regular maintenance	Has a more natural appearance
Not very effective against storm waves	A wide, gently sloping beach reduces wave energy
Removes sand from elsewhere	Uses natural materials
Can be expensive to set up	Creates a wide beach with high amenity value

Table 8.2 The costs and benefits of soft engineering.

Costs and benefits of managed retreat (coastal realignment)

Costs	Benefits
Can be very expensive to set up	Creates a natural barrier against storm waves, reducing flood risks
Causes the loss of land and amenities	Develops an **intertidal habitat** (salt marsh)
Cost of relocating infrastructure	New habitat encourages wildlife
Conflict with local landowners	Nature reserves create environmental and social opportunities

Table 8.3 The costs and benefits of managed retreat (coastal realignment).

Find out more on the **RSPB website** (see www.cambridge.org/links/gase40065) or the **SCOPAC website** (see www.cambridge.org/links/gase40066)

Assess to progress

1 Name and describe two types of coastal erosion. `4 MARKS`

2 Suggest how geology influences coastal landforms. `4 MARKS`

3 Explain the formation of a coastal spit. You may use a diagram. `6 MARKS`

4 Describe the features of erosion and deposition in a section of UK coastline you have studied. `6 MARKS`

5 Using an example you have studied, describe the features of a coastal management scheme and explain why management was required. `9 MARKS`

Further research

Find out more at the **Pevensey Bay website** (see www.cambridge.org/links/gase40064)

Think about:
- why soft engineering was used in this area
- how successful the scheme has been.

Key terms

nature reserve: a conservation area which is set aside to preserve plants and animals

intertidal habitat: land exposed at low title and covered at high tide

ACTIVITY 8.6

1 Explain why a beach is a good defence against storm waves.

2 Explain why Pevensey Bay required three types of beach management techniques (nourishment/recycling/reprofiling).

3 Why would hard engineering not be appropriate at Pevensey Bay?

4 Why was Medmerry a suitable area for a managed retreat scheme?

5 Why is the Medmerry managed retreat scheme described as a 'sustainable way to manage this coastal area'?

Coastal management scheme, Ventnor to Bonchurch, Isle of Wight

The coastline between St Catherine's Point and Bonchurch, on the south coast of the Isle of Wight lies within an area known as the Undercliff. The **geology** of this area is a complicated mixture of clay, sands and chalk and it is one of the most unstable coastal areas in the whole of Europe. Because of this, the area has a long history of landslides and rock falls, especially after heavy rainfall and major storms (Figure 8.29).

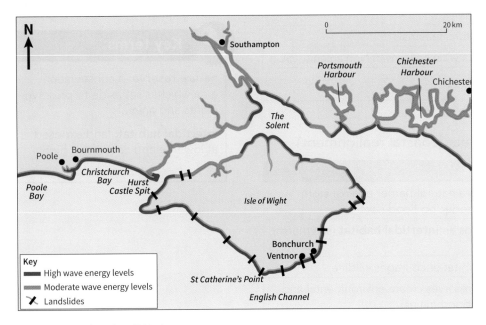

Figure 8.29 The Isle of Wight.

Did you know?

If there was no coastal management in this area it is estimated that the **cliff retreat** would be 48–160 m over the next 100 years.

Key terms

geology: study of the Earth, especially rocks

cliff retreat: cliff eroding away and the position of the coastline moving back

sub-aerial processes: processes that affect the face and top of cliffs

Why is coastal management needed in this area?

There are a number of physical and economic reasons why this area needed coastal protection. The following statement describes some of these reasons.

'This is a high energy coast which is affected by powerful winter storms. Along most of the coast there is no real beach to break the energy of the waves. The cliffs are very unstable and the area has a long history of landslides, some of which have caused significant property damage. Increasingly wet winters have increased the risk of ground movement. The area is both a holiday resort and a residential area with millions of pounds' worth of property. The main east–west road link runs close to the coast and all the main services run underground. A major landslide event in the area would have serious economic and social consequences.'

What type of hard engineering techniques have been used in this area?

This area has been protected from the effects of erosion and weathering for many years but historically coastal defences were built along short parts of the coast as a response to a particular storm or landslide event. However, over the last 40 years there has been a coordinated plan to use hard engineering techniques to protect the whole of the coastline between Ventnor and Bonchurch (Figures 8.31 and 8.32).

Further research

Investigate the recent landslips in the Undercliff area to the west of Ventnor to see the impacts of major ground movement. Search the web with terms such as, recent landslide, Undercliff, Isle of Wight.

Figure 8.30 Coastal management between Ventnor and Bonchurch.

Figure 8.31 Hard engineering techniques – Ventnor to Bonchurch. The photo on the left shows tetrapods; the photo on the right shows rock armour (rip-rap).

The plan aims to:

- protect the cliffs from the effects of storm waves and **sub-aerial processes**
- stabilise the areas where there are risks of landslides.

Has coastal management been successful in this area?

The two key aims of coastal management were to protect the area from the effects of erosion and to reduce the risks from landslides. In terms of these aims the scheme has been successful. There are no areas where the sea can now directly attack the cliff and the area has seen no major landslides for the last 20 years. Sections of coast to the east and west of this area, which are not protected, have had major landslide events. The cliffs are still affected by weathering and there are minor rock falls occasionally, but these are not significant. Despite the success of coastal management in this area, there are conflicting views about it. Some of these are shown below.

'It has reduced the risk of landslides and included a wonderful promenade where you can walk right along the coast.'

'All the concrete has spoilt the look of the area and it needs to be repaired after every winter.'

'It may not look nice but it is very effective and was really the only option for the area.'

'It may have affected wildlife habitats and has made the area less attractive for visitors.'

'Without the scheme there would be a massive risk of erosion and landslides, with a loss of homes and business. Now it is stable people are more likely to set up new businesses in the area.'

ACTIVITY 8.7

1 What is meant by a 'high energy coast'?
2 Explain why coastal management was needed in this area.
3 a Describe the hard engineering techniques used on the Ventnor–Bonchurch coast.
  b Explain how the techniques reduce the risks of erosion and landslides.
4 Why do some people think that hard engineering is not a very environmentally friendly way to manage the coast?
5 a Do you think that coastal management has been successful in this area?
  b What other evidence might you collect in order to make a more detailed judgement about the effectiveness of coastal management in this area?

9 River landscapes in the UK

In this chapter you will learn about...

- how river valleys change from source to mouth
- the processes that affect river valleys
- landforms resulting from river erosion
- landforms resulting from deposition by rivers
- assessing and managing flood risk.

9.1 How do river valleys change as rivers flow downstream?

The long profile of a river

The **long profile** shows the gradient of a river from its **source** to its **mouth**. It is not always a smooth line and can have:

- steep slopes in upland areas
- shallow areas where lakes may form
- breaks in the slope where waterfalls can be seen.

The following examples shows the characteristics and long profile of the River Tees, in northern England (Figures 9.1, 9.2 and 9.3).

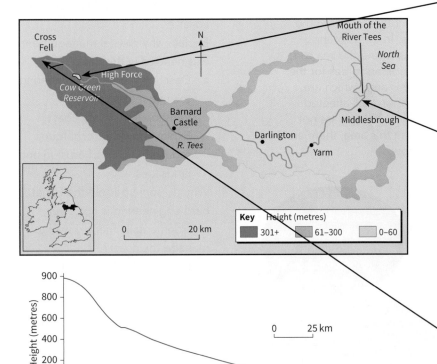

Figure 9.4 Long profile of the River Tees.

Figure 9.1 Cow Green reservoir, northern England.

Figure 9.2 River Tees, northern England.

Figure 9.3 Cross Fell – the source of the River Tees.

The cross profile of a river valley

The **cross profile** of a river valley shows the shape of the valley from one side to the other (Figure 9.5). As a river flows downstream, its cross profile changes. In upland areas the cross profile often has steep sides and a narrow valley floor. Downstream the cross profile becomes flatter, with lower valley sides and a broader valley floor (Figure 9.6).

Key term

cross profile: a cross-section drawn across the river valley

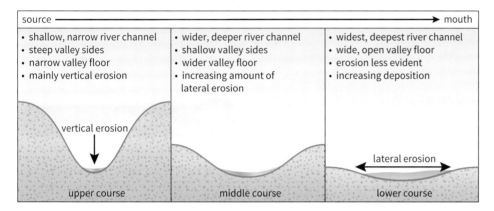

Figure 9.5 The cross profile of a river.

Figure 9.6 The estuary of the River Tees.

Tip

You could use GIS to complete Activity 9.1, for example ArcGIS Online.

ACTIVITY 9.1

1   Describe the long profile of the River Tees.
2   Suggest why there is more vertical erosion near the source of the river.
3   Using Figure 9.6 to help you, describe the characteristics of the River Tees at its source.
4   Suggest why river estuaries provide opportunities for the development of industry.

9.2 What are the processes that affect river valleys?

Rivers and their valleys are affected by these main processes (Figure 9.7):

Figure 9.7 River valley processes.

Weathering

Weathering breaks down rocks and makes it easier for the processes of **erosion** to operate. In upland areas **freeze-thaw** weathering breaks down rocks on valley sides. Rock fragments fall into river channels and are transported and eroded by moving water.

Erosion

There are four main processes of erosion.

- **Abrasion** (also known as **corrasion**): **sediment** carried by a river can act like an abrasive, wearing away the bed and banks of a river. When a river is in flood it can carry tons of sediment, so the effect of abrasion will be greater. Abrasion is the major force for deepening valleys (**vertical erosion**) in upland areas.
- **Attrition**: flowing water pushes pebbles and rocks against each other, making them smoother and decreasing their size.
- **Hydraulic action:** the power of moving water hitting riverbanks can weaken rocks and lead to collapse (Figure 9.8). The loose material is then carried away by the river.
- **Solution**: chemical action occurs when slightly acidic water comes into contact with chalk and limestone. The calcium in the rock dissolves, causing the rock to weaken.

What factors affect the rate of erosion?

The more energy that the river has the greater the rate of erosion. The energy will vary according to local and seasonal conditions. After heavy rainfall the volume of water in the river and the speed of flow will increase, sometimes causing flash flooding. During these periods the river will have the energy to carry huge quantities of material which has the power to completely change a river valley (Figure 9.9). During periods of low rainfall a river may be reduced to little more than a trickle or even dry up completely. Consequently the power of the river to transport and erode material is reduced. Other factors affecting rates of erosion include:

- **Gradient** – if a river has a steep gradient the amount of energy will be greater, increasing the rate of erosion. The gradient can be seen by looking at the long profile of the river.
- **Amount of bedload** – if there are a lot of rocks, pebbles and sediment in the river the rate of abrasion will increase.
- **Human factors** – changing the landscape can affect the rate of erosion. For example, removing vegetation may decrease the time it takes for rainfall to get into a river. This will increase the volume of water in the river and the potential energy of the river.
- **Building concrete riverbanks** will reduce **lateral erosion**.

Key terms

weathering: the breaking up of rocks that occurs *in situ* (the same place) with no major movement taking place

erosion: the breaking up of rocks that is the result of movement

freeze-thaw: weathering of rocks by continued freezing and thawing of moisture in cracks

sediment: material moved and deposited in a different location

vertical erosion: downward erosion of a riverbed

bedload: larger particles moved along a riverbed

lateral erosion: erosion of the sides of a valley

Watch a video on Cambridge Elevate about river erosion.

Figure 9.8 An eroded riverbank.

Figure 9.9 The River Irwell in flood, with debris being swept along.

Transportation

Rivers transport material downstream in four main ways (Figure 9.10):

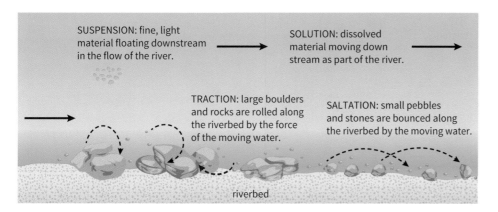

SUSPENSION: fine, light material floating downstream in the flow of the river.

SOLUTION: dissolved material moving down stream as part of the river.

TRACTION: large boulders and rocks are rolled along the riverbed by the force of the moving water.

SALTATION: small pebbles and stones are bounced along the riverbed by the moving water.

riverbed

Figure 9.10 Processes of transportation.

Figure 9.11 The mouth of a river.

Deposition

Deposition occurs where a river does not have enough energy to carry the load any further. Larger boulders are deposited first because they are heavier. Finer material can be transported further, often being deposited near the mouth of a river as mudflats (Figure 9.11).

ACTIVITY 9.2

1 If you were completing a river investigation what evidence would suggest that abrasion, attrition and hydraulic action were taking place?
2 a Draw a sketch of the photograph in Figure 9.8.
  b Annotate your sketch to explain what is happening.
3 Explain how the seasonal pattern of rainfall is likely to affect rates of erosion in a river.
4 Why does the bedload of a river usually become smaller and smoother from its source to its mouth?

Download Worksheet 9.1 from Cambridge Elevate for help with Activity 9.2, question 1.

9.3 Distinctive landforms resulting from different physical processes (1)

Near the source of a river, in upland areas, vertical erosion is the dominant force, creating steep sided valleys and in some areas **rapids**, waterfalls and gorges. When a river reaches lower land the valley is often wider and flatter and the river begins to **meander** across the valley. Here lateral erosion and deposition are often the most important processes.

Nearer the mouth of the river, as it flows across the flood plain, deposition is the major process and levées can be formed. Most rivers flow into tidal estuaries as they reach the sea. Here, much of the fine silt and mud being carried by the river is deposited, forming mudflats. These are visible at low tide and provide an opportunity for the development of salt marshes, an important habitat for wildlife.

Figure 9.12 shows some of the features along the river in the River Tees **drainage basin**.

Key terms

rapids: fast flowing river over an uneven riverbed

meander: a large bend in a river

drainage basin: the area of land drained by a river system

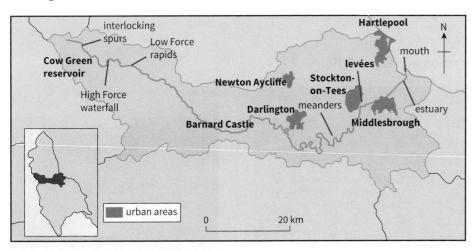

Figure 9.12 The features along the river in the River Tees drainage basin.

Landforms resulting from erosion

Interlocking spurs

In upland areas vertical erosion is a powerful force, creating a V-shaped valley. As the river erodes downwards it winds and bends around areas of more resistant rock, so when looking back up the valley these projecting ridges, which are called 'spurs' appear to overlap or 'interlock' (Figure 9.13).

Figure 9.13 Interlocking spurs in the Upper Tees.

Waterfalls and gorges

The gradient of a river becomes less steep as it flows downstream from its source. A **waterfall** is a break in the profile of a river and is usually the result of the river flowing over an area where harder rock (called a 'caprock') overlays softer rock. Erosion, especially by hydraulic action, undercuts the softer rock until the caprock collapses, leaving rocks and boulders in the valley below (Figure 9.14). The continued headward (backward) erosion of the waterfall leaves a steep-sided feature called a gorge (Figure 9.15).

Key term

waterfall: a steep fall of water along the course of a river

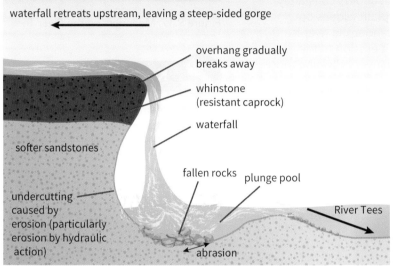

Figure 9.14 The formation of a waterfall.

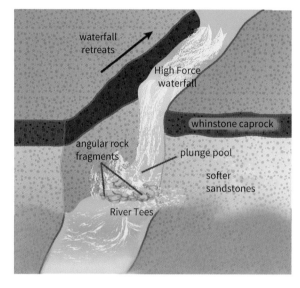

Figure 9.15 The formation of a gorge.

High Force waterfall – River Tees

In upper Teesdale an outcrop of a hard igneous rock called whinstone has created the rapids at Low Force and a spectacular waterfall at High Force (Figure 9.16). The whinstone is the caprock and softer sandstones and limestones are found below it. At 22 m, High Force is one of the tallest waterfalls in England. The gradual retreat of the waterfall has created a steep sided gorge.

ACTIVITY 9.3

1 Explain the difference between vertical and lateral erosion.
2 Why are interlocking spurs found in upland river valleys?
3 Use an annotated sketch to explain the formation of a waterfall and a gorge.

Further research

Investigate why the area around High Force is a popular recreation and tourist destination.

Figure 9.16 High Force waterfall and gorge.

9.3 Distinctive landforms resulting from different physical processes (2)

Landforms resulting from erosion and deposition

Meanders and ox-bow lakes

Meanders are bends in the course of a river (Figures 9.17 and 9.18). They begin to form as the valley floor becomes wider and flatter and lateral erosion becomes a more significant force than vertical erosion. Meanders constantly change their shape, forming broader loops as the **flood plain** widens out towards the mouth of the river. On the outside of the meander the river is flowing faster and is able to carry more sediment, causing erosion to the riverbed and riverbank. On the inside of the meander the river is flowing more slowly and sediment is deposited (Figure 9.18).

Ox-bow lakes are formed when continued erosion of the neck of the meander causes the river to break through (Figure 9.19). It will then take the most direct route, avoiding the meander. Deposition will block off the meander, creating an ox-bow lake. Gradually the ox-bow lake will dry up, forming a meander scar.

Figure 9.17 Meanders on the River Tees.

Key term

flood plain: area of flat land which is prone to flooding

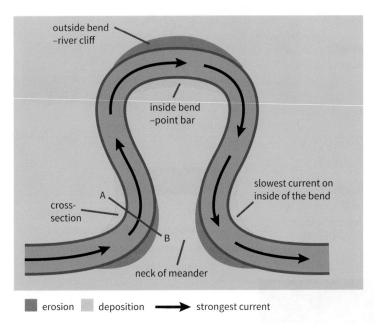

Figure 9.18 Features of a meander.

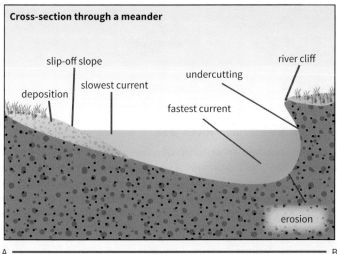

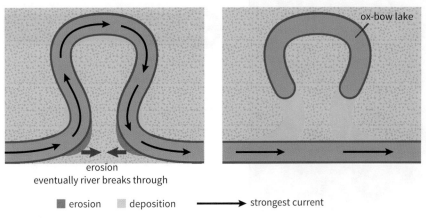

Figure 9.19 Formation of an ox-bow lake

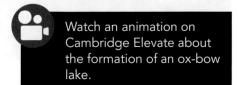

Watch an animation on Cambridge Elevate about the formation of an ox-bow lake.

Levées and flood plains

Levées and flood plains are the result of deposition left by repeated flooding. Flood plains develop where sediment is deposited across a wide area of the valley floor. Levées are elongated ridges of sediment deposited alongside the river (Figure 9.20). In lowland areas the natural flow of a river pushes sediment towards the sides of a river where it is deposited, gradually building up so that it stands up above the general level of the flood plain. When river levels are high levées form a natural defence against flooding. They are often artificially increased in height in order to reduce flood risks in built-up areas.

Estuaries

Most rivers flow into a tidal **estuary** when they reach the sea. At this point the land is at sea level so the river uses its energy to widen the river channel. Because of the volume and **velocity** of the river it is able to carry vast amounts of fine sediment. Much of this is deposited as silt and mud and in some places salt marshes develop, forming an important wildlife habitat. Estuaries are often adapted for human use by draining the flat land either side of the river and deepening the river channel. This provides an opportunity for the development of port facilities and heavy industry. This can be seen in the River Tees estuary, which is a major UK port and an important area for heavy industry (Figure 9.21).

Key terms

estuary: wide part of a river where it nears the sea

velocity (of river): speed of flow, usually measured in metres per second

Figure 9.20 Levées on the River Tees.

Figure 9.21 Aerial photograph of the Tees estuary.

ACTIVITY 9.4

1 Use an annotated diagram to explain:
   a the features of a meander
   b the formation of an ox-bow lake.
2 Explain how levées provide a natural defence against flooding.
3 Using Figure 9.21:
   a Describe the physical characteristics of the river.
   b Identify the land uses shown in the photograph.
   c Suggest why estuaries:
      i are important areas for industrial development
      ii provide a valuable area for wildlife.

9.4 Assessing and managing flood risk

When it rains, water will either:

- be lost through evaporation
- be held in storage, for example, in lakes
- make its way into rivers, either on the surface or by soaking through the ground.

What is a storm hydrograph?

A storm (or flood) hydrograph shows how a river responds to heavy rainfall and can be used to help predict floods and plan for future flood events. Rainfall is shown on the left-hand vertical and **discharge** on the right-hand axis (Figure 9.22).

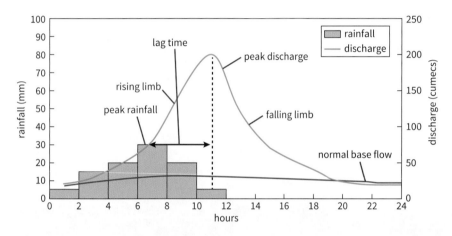

Figure 9.22 Features of a storm hydrograph.

Key terms

discharge: the volume of water at a given point in a river (measured in cumecs)

precipitation: any form of water, both liquid and solid, which falls from the sky; rain, snow, sleet and hail are all examples of precipitation

flash floods: rapidly rising river levels leading to a rapidly developing flood situation

runoff: all precipitation that reaches a river

Measuring river discharge

River discharge is a calculated at a particular point on a river by measuring:

- the velocity (speed) of the river
- the cross-sectional area of the river.

Discharge = velocity × cross-sectional area, expressed in cumecs (cubic metres of water per second).

What can a storm hydrograph show?

A storm hydrograph can show:

- a steep rising limb means that **precipitation** is reaching a river quickly, increasing flood risks
- A short lag time means that river levels may rise quickly, adding to the risk of **flash floods**
- a gently sloping falling limb means that the river takes a longer time to get back to normal flow. If there is more heavy rainfall during this time flood risks will increase.

River authorities will know how high the peak discharge can rise before a river floods. If it is predicted that the peak discharge will go above this figure people can be warned about the risk of flooding or even evacuated.

What factors affect river discharge?

There are a number of factors that can affect river discharge, including:

- **Precipitation** – heavy rainfall can increase river levels.
- **Geology** – some rocks allow water to soak through them because they are porous/permeable. This can increase the lag time. Some rocks are impermeable so rainfall flows on the surface and gets into rivers quickly.
- **Relief** – on steeper slopes surface **runoff** will be quicker so river levels will rise faster.
- **Land use** – removing vegetation can decrease lag times so that rainfall reaches rivers more quickly. Building hard surfaces and putting in artificial drains decreases the time it takes for rainfall to reach a river, reducing the lag time.

Managing flood risk

Flooding is a significant hazard which can affect people and property. Consequently flood management strategies are used in order to reduce the flood risk.

Hard engineering strategies

These attempt to control rivers by using technological solutions which often include large-scale building projects. Hard engineering strategies include building dams and reservoirs which can be used to control river flow (Figure 9.23) or creating flood relief channels to direct water away from flood risk areas. Other strategies include building artificial riverbanks (levées) or embankments to prevent a river flooding across a flood plain. Also, rivers can be deepened and straightened to enable the water to flow more quickly.

Soft engineering strategies

These attempt to work with natural processes in order to reduce flood risks. A major soft engineering strategy is flood plain zoning where building is restricted in areas where the flood risk is high. Other strategies include replanting trees (afforestation) to increase lag times and consequently reduce discharge, and increasing the amount of green space and permeable surfaces in urban areas (Figure 9.24). Issuing flood warnings and educating people about flood awareness are also important strategies.

Further research

Investigate other methods of managing river flooding risks at the **Environment Agency website** (www.cambridge.org/links/gase40067)

ACTIVITY 9.5

1 Look at Figure 9.22.
   a Describe the pattern of precipitation and discharge shown on the hydrograph.
   b Suggest what might have happened to the discharge had there been heavy rainfall between 12 and 14 hours.
2 Explain how hydrographs are helpful when trying to manage the flood risk.
3 Write a paragraph explaining how physical and human factors can affect river discharge.

Further research

Investigate hard and soft flood management strategies and consider the advantages and disadvantages of each type of strategy.

Figure 9.23 The Derwent reservoir.

Figure 9.24 Urban greening.

Responding to a flood event – the Boscastle Flood 2004

The Boscastle flood of 2004, on the North Cornwall coast was an example of a flash flood. This type of flood is unusual in the United Kingdom, where slow or rapid onset floods are generally more common.

What were the causes of the Boscastle flood?

The physical geography of the area made the village of Boscastle especially vulnerable to flooding. The reasons for this are:

- a small river catchment which includes the relatively impermeable upland area of Bodmin Moor
- steep sided valleys converging as they run towards the sea, funnelling water towards Boscastle
- the area is in the path of rain-bearing westerly winds.

Added to this, a number of unusual factors occurred during the summer of 2004.

- It was an exceptionally wet summer so the ground was already saturated.
- The area had been affected by a series of summer storms.
- On 16 August the remains of hurricane Alex moved across the Atlantic Ocean. The storm clouds brought unprecedented amounts of rain, estimated at over 1400 million litres in under 2 hours (Figure 9.25).

The flood risk in Boscastle had been increased by the amount of building alongside the river and the construction of a number of small bridges. The bridges trapped boulders and trees which had been washed down the river, and acted like dams eventually breaking and sending a torrent of water down the valley.

Factfile

There are three main types of flood:

- **slow onset floods** – gradually develop over a number of days and often last a week or more
- **rapid onset floods** – occur more quickly, often in upland areas where lag times are shorter. They do not often last long but can cause a lot of damage
- **flash floods** – an immediate response to heavy rainfall. River levels rise quickly, causing devastation and a danger to life

Further research

Find out more about the Boscastle flood – build up a factfile, describing the effects of the flood.

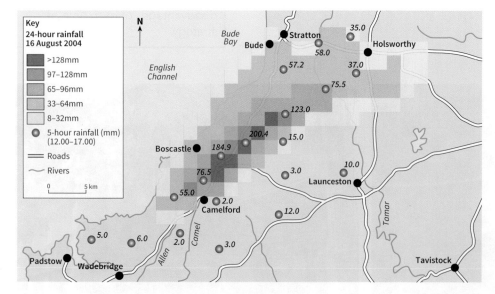

Figure 9.25 Rainfall map of Boscastle, 16 August 2004.

The effects of the flood

The article, Figures 9.26 and 9.27 describe what it was like on the day of the flood.

Village devastated by floods

The morning of 16 August started quietly with holidaymakers enjoying the early morning sunshine while admiring the beautiful scenery or shopping for souvenirs in local gift shops. Within hours heavy rain began to fall and by late afternoon the village was a disaster zone. A 3-metre wall of water flooded through the village at over 60 kph, flooding the whole area. The water contained huge boulders and trees, acting like a battering ram against the buildings in the village. Cars were swept away and buildings were destroyed. A major disaster incident was declared at 5.10 pm and local RAF search and rescue teams were alerted. By the time the flood subsided over 20 business properties had been destroyed and four bridges had been washed away. A number of people were injured but fortunately there were no fatalities.

Responding to the flood

When the floodwaters subsided thousands of tons of mud and debris were left throughout the village. The local economy was devastated because the area relies on tourism for 90 per cent of its income.

Figure 9.26 The scene during the Boscastle floods on 16 August 2004.

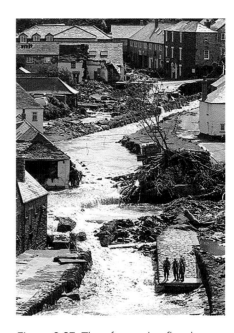

Figure 9.27 The aftermath – flood devastation in Boscastle 2004.

The Environment Agency investigated the causes of the flood and found that a number of human factors had added to the flood risk:

- artificially narrowing the river as it flowed through the village
- building low bridges which trapped boulders and trees
- allowing trees to grow alongside the river; during the flood these were washed into the river channel, blocking the flow of water
- building alongside the river and not leaving space for floodwater to flow into the river
- removing vegetation further back up the valley, which reduced the time it took for rainfall to reach the river.

In order to reduce the risks of future flooding a £5 million flood defence scheme was put in place in 2008. The main features of the scheme can be seen in Figure 9.28.

How effective have the management strategies been?

After the flood the Environment Agency, the National Trust and local people worked together to develop a plan to reduce future flood risks (Figure 9.28). This helped to bring the community together and the flood was seen as a significant driver for change in the area. People are now more involved in the local community and prepared to help each other.

The Environment Agency estimated that the flood management strategy has reduced the risk of flooding from a one in 10 year event to a one in 75 year event. Since 2004 there have been several periods of heavy rainfall but the defences have contained the rising rivers, even in 2012 when other parts of Cornwall were flooded.

The rebuilding of Boscastle and the subsequent improvements have increased investment in the area and the number of tourists has increased. Environmentally,

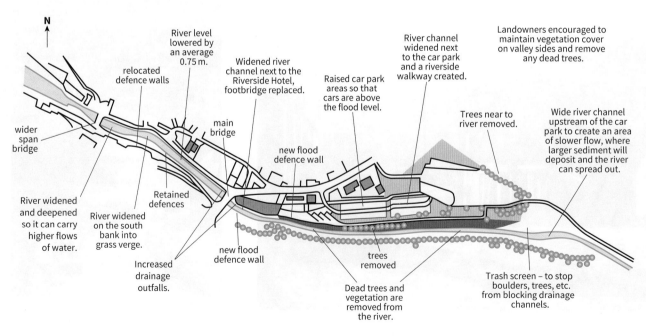

Figure 9.28 The Boscastle flood management strategy.

the flood management plan did not please everyone. Some people felt that it changed the character of the village, especially replacing the old Elizabethan stone bridge at the harbour mouth with a metal structure (Figure 9.29).

Figure 9.29 The new bridge at Boscastle.

ACTIVITY 9.6

1 Explain how physical and human factors contributed to the flood risk in Boscastle.
2 Describe the effects of the flood on the village of Boscastle.
3 a Write a paragraph explaining why a flood management scheme was needed in Boscastle.
   b Describe the main features of the flood management scheme.
   c Explain the social and economic advantages of the scheme.
   d Why might some people be concerned about the environmental effects of the scheme?

 Assess to progress

1 Name and describe two ways that sediment is transported by a river. **4 MARKS**

2 How do hydrographs explain the relationship between rainfall and the discharge of a river? **4 MARKS**

3 Explain the formation of ox-bow lakes. You may use a diagram. **6 MARKS**

4 Describe the major landforms of erosion and deposition in a UK river valley that you have studied. **6 MARKS**

5 Using an example you have studied, describe the features of a flood management scheme and explain why the scheme was required. **9 MARKS**

10 Glacial landscapes in the UK

In this chapter you will learn about...

- the glacial processes of weathering and erosion, movement and transportation and deposition
- the characteristics and formation of landforms of upland and lowland erosion
- the characteristics and formation of landforms of deposition
- some of the economic activities which occur in glaciated areas
- conflicts that exist between different land uses and people
- how the Lake District National Park manages the pressures of tourism.

Key term

Quaternary period: a period in geologic time which stretches from about 2.5 million years ago to the present; it is divided into two main sections: the Pleistocene (2.5 million years ago to 11.7 thousand years ago) and the Holocene (11.7 thousand years ago to the present)

10.1 Why do glacial landforms exist in the UK?

Today, approximately 10 per cent of the Earth is covered in ice. Some 90 per cent of this ice is found in Antarctica in ice sheets that are nearly 5000 metres thick. Ice is also found in other high latitude regions such as the Arctic and areas of high altitude such as the Alps. However, this has not always been the case. Over the last 2.5 million years (the **Quaternary period**), the global average temperature and therefore the volume of ice has fluctuated (Figure 10.1).

Glaciers are slow moving rivers of ice that are formed by the accumulation, or build-up of compacted snow. **Ice sheets** are vast domes of glacier ice that cover more than 50 000 km². Significant ice sheets include the Greenland and Antarctic ice sheets. **Ice caps** are smaller than ice sheets and cover less than 50 000 km². The Vatnajökull ice cap in Iceland is the largest in Europe at 8100 km² and 400 m thick.

Interglacial and glacial periods

Currently we are in a warm, **interglacial** period where the global average temperature is approximately 15 °C. However, in colder, **glacial** periods this figure decreased to around 11 °C and ice spread out from the poles and mountain ranges to cover around a third of the Earth's surface. The last glacial period, or ice age, occurred between 110 000 and 120 000 years ago. Glaciers have advanced and retreated several times during this period, reaching their maximum advance about 22 000 years ago (Figure 10.2).

The distribution of ice in the last ice age

Massive ice sheets extended southwards from the Arctic and covered Canada, the northern part of the United States and most of the UK around 22 000 years ago. Ice sheets also extended northwards from Antarctica but, due to the location of landmasses, these had a limited effect. In the UK, the northern half of the country was covered with an ice sheet hundreds of metres thick (Figure 10.3). The south of the UK would have been

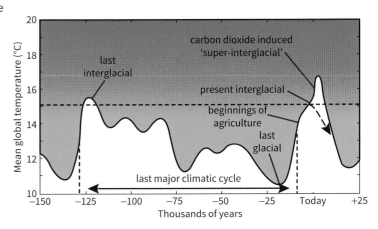

Figure 10.1 Temperature changes over the last 15 000 years.

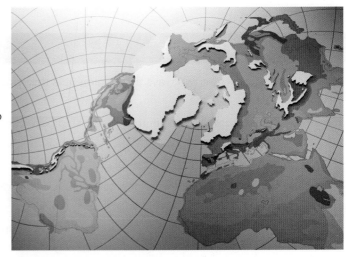

Figure 10.2 The maximum extent of the ice cover during the Pleistocene 22 000 years ago.

Figure 10.3 The maximum extent of the ice cover in the UK during the Pleistocene 22000 years ago.

permanently frozen ground (**permafrost**) with icy winds blowing off the ice sheet and **tundra** vegetation; conditions similar to those found in Siberia today.

As the ice advanced it changed the landscape by eroding it. As temperatures began to increase the ice then began to retreat so that, by 10000 years ago, the ice covering the UK had completely melted. While we don't have any permanent glaciers currently shaping the UK landscape, the last ice age has left dramatic scenery, particularly in the north of the UK and in upland areas such as the Lake District and Snowdonia.

ACTIVITY 10.1

1 Using Figure 10.1, describe changes to the global average temperature over the last 15000 years. Use the information to:
   a describe the general trend
   b quote specific figures to illustrate the general trend
   c identify any anomalies.
2 Create a guide to the different types of glacier. The **National Snow & Ice Data Centre (NSIDC) website** (www.cambridge.org/links/gase40068) is a good place to start. Include an image and an example for each glacier type.
3 Read the information on the **Live Science website** (www.cambridge.org/links/gase40069) which explains the Pleistocene in detail. Create an illustrated factsheet about the Pleistocene using this information.

Key terms

glaciers: slow moving rivers of ice formed from compacted snow

ice sheets: large masses of glacial ice which are over 50000 km²

ice caps: smaller mass of glacial ice which are less than 50000 km²

interglacial: a warm period where average global temperatures increase to around 15 °C

glacial: a cold period where average global temperatures decrease to around 11 °C

permafrost: ground that has been frozen for two or more years; in the northern hemisphere over 19 million km² is covered in permafrost, most of which is found in Siberia, Alaska, Greenland, Canada and northern Scandinavia

tundra: vegetation found in cold environments which is mainly composed of shrubs, grasses, mosses and lichens – the low temperatures mean that it is difficult for trees to grow

Did you know?

70 per cent of the world's fresh water is frozen in the Antarctic ice sheets; that's 25.4 million km³ of ice at an average of 1829 m thick.

Discussion point

What evidence is there to suggest that these changes in climate happened?

Watch a video on Cambridge Elevate about the evidence for how the UK's climate has changed over time.

10.2 What are glaciers?

Glaciers do not simply appear overnight. They start out as a layers of snow which fall in winter but, due to cooler summer temperatures, do not melt. Over time these layers of snow, which become compressed to form ice, build up to form glaciers (Figure 10.4). They build up in high altitudes (zone of accumulation) where temperatures are lower and move down the valley due to gravity. The higher temperatures in the valley cause the glacier to melt (zone of ablation) (Figure 10.5). The end of the glacier is known as the glacier snout.

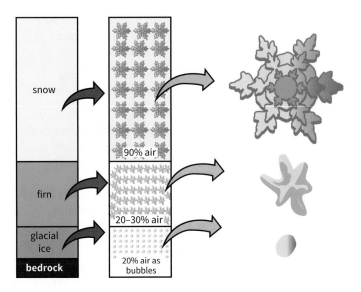

Figure 10.4 The formation of glacial ice.

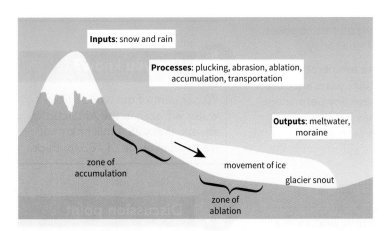

Figure 10.5 The glacier as a system.

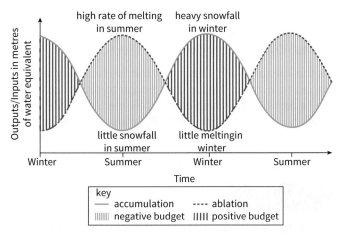

Figure 10.6 Seasonal variation in accumulation and ablation.

The glacial budget is the balance between accumulation and ablation over a year. In winter there is more accumulation than ablation and in summer there is more ablation than accumulation (Figure 10.6). If the summer and winter budgets cancel each other out then the glacier is described as stationary and the snout stays in the same position. If there is more ablation in summer than accumulation in winter then the glacier begins to retreat and the snout moves up the valley. However, if there is more accumulation in winter than ablation in summer then the glacier advances and the snout moves down the valley. When a glacier retreats it looks as though the glacier is moving uphill but it is not! Ice is still moving downhill and being transferred from the zone of accumulation to the zone of ablation.

10.3 How do glacial processes shape the landscape? (1)

Ice shapes the landscape in two main ways, through **weathering** and through **erosion**.

Freeze-thaw weathering

The main type of weathering which occurs in glacial environments is freeze-thaw weathering. Water (from rainfall or meltwater) runs into naturally occurring cracks in the rock. When the temperature of the area falls below 0°C the water freezes; expanding by approximately 10 per cent as it does so. When temperatures rise above 0°C the ice subsequently thaws leaving a deeper and wider crack. This process is repeated until the rock shatters (Figure 10.7). The fragments of rock are usually angular and, if they become embedded in the ice, can help the process of erosion known as abrasion as the ice moves down the valley. **Scree** is a product of freeze-thaw weathering and can usually be found at the bottom of mountains that have been glaciated (Figure 10.8).

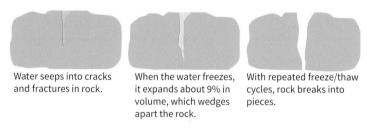

Water seeps into cracks and fractures in rock.

When the water freezes, it expands about 9% in volume, which wedges apart the rock.

With repeated freeze/thaw cycles, rock breaks into pieces.

Figure 10.7 Freeze-thaw weathering.

 Watch some of the clips from *Earth: the Power of the Planet* on the **BBC website** (www.cambridge.org/links/gase40070).

Glacial erosion – plucking and abrasion

There are two main processes of glacial erosion which occur as a result of glacial movement: abrasion and plucking (Figure 10.9). A glacier is not simply made up of ice; it also has angular fragments of rock embedded within it as a result of freeze-thaw weathering. When the glacier moves, these fragments scratch the **bedrock** leaving deep grooves known as striations. This is abrasion. Striations can be found on boulders in Central Park, New York, suggesting evidence for the presence of glaciers at a latitude of 40°N during the last glacial period. When a glacier meets an obstacle, such as a piece of rock sticking out from the bedrock on the valley floor, it partially melts due to the pressure and friction which builds up behind the obstacle. As the glacier moves around the obstacle and the pressure is released, the glacier refreezes (regelation) but continues to move. The ice which is frozen and stuck to the rock then pulls or plucks it apart as it moves leaving an angular, jagged surface.

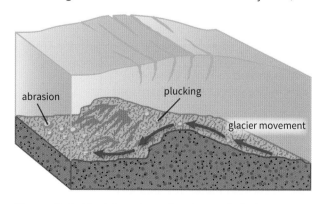

Figure 10.9 Glacial erosion: abrasion and plucking.

Figure 10.8 Scree slopes lead to the towering cliffs of the north face of Crib Goch, Snowdonia National Park, Wales, UK.

10.3 How do glacial processes shape the landscape? (2)

Glacier movement

A glacier is able to erode because it flows slowly downhill due to gravity (Figure 10.10). The speed of movement depends on the volume of ice, the gradient of the slope and the volume of **meltwater** at the base of the glacier. The weight of the glacier puts a huge amount of pressure on the base which causes the bottom few centimetres of ice to melt. This reduces friction between the glacier and the bedrock and allows the glacier to move; this process is called basal slippage. If the rock beneath the glacier is curved, such as in a corrie, then the ice moves in a semicircular motion called rotational slip. A second process of movement is called internal deformation or internal flow. This is where the layers which make up the glacier flow at slightly different speeds. The layers at the surface flow fastest as they have the least weight and pressure on them. Also, the middle of the glacier moves faster than the sides as these are in contact with the valley sides and friction slows it down.

Key term

meltwater: water which is formed by the melting of snow and ice

Watch the animation on Cambridge Elevate to see how glaciers move.

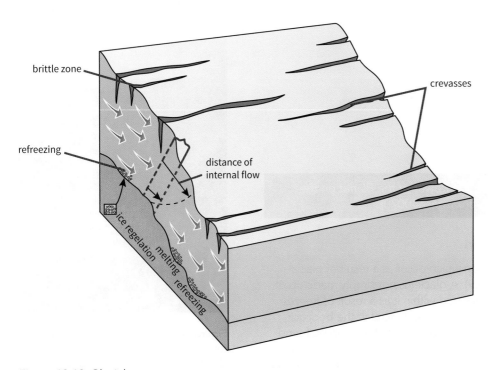

Figure 10.10 Glacial movement.

Did you know?

Glaciers tend to move at speeds of close to zero and up to half a kilometre per year, although the fastest moving glacier is in Greenland and is estimated to move at 12.6 km per year!

Glacier transportation and deposition

As the glacier moves downhill it transports fragments of rock that have been eroded or weathered. These fragments can be transported either at the base of the glacier, frozen within the glacier or carried on top of the glacier. The rocks that are frozen into the ice cannot collide with each other (like they would in a river). As a consequence, they remain angular and increase the amount of abrasion that the glacier is capable of as they can freely move and erode the valley sides. Due to the fact that glaciers have so much energy, they are able to transport both tiny particles and massive boulders.

A glacier melts when temperatures are warmer, for example when the glacier reaches lowland areas, or where there is friction between the glacier and the valley sides. Where the glacier melts it begins to deposit its load (Figure 10.11). Most deposition occurs at the snout of the glacier because the majority of melting takes place there. The glacier will deposit **unsorted** material of all different sizes. This is known as glacial till. If the glacier advances then the material will be pushed forwards by the glacier snout. This is known as bulldozing which is the power of the glacier to move debris. Meltwater running from the snout of the glacier will also carry deposited material further down the valley which is known as outwash.

Key term

unsorted: deposits where all different sizes of rock – from tiny particles to massive boulders – are mixed together

Figure 10.11 Glacially deposited rocks on the Munro Beinn Sgulaird above Glen Creran between Oban and Glen Coe, Scotland, UK.

ACTIVITY 10.2

1 Design and carry out an experiment with different sized ice cubes and a piece of wood or plastic tray to act as a slope to see how the different factors – volume of ice, gradient of the slope and volume of meltwater – affect the movement of ice.

2 Read the information on **The National Snow & Ice Data Centre (NSIDC) website** (www.cambridge.org/links/gase40071) which explains why glaciers move in more detail. Use an annotated diagram to explain the movement of a glacier.

3 Create an illustrated dictionary of all of the key words that you have met so far in this chapter. You can add to this as the chapter progresses and add other words that you think might be useful.

10.4 What landforms result from glacial erosion in upland areas?

Ice is a very powerful erosive force and can consequently form spectacular features in upland areas. Although the UK no longer has glaciers, much of the scenery in Snowdonia and the Lake District was shaped during the last ice age (Figure 10.12).

In upland areas glaciers erode to create four distinctive landforms: corries, arêtes, pyramidal peaks and truncated spurs.

Figure 10.12 Panorama taken from the summit of Helvellyn – the corrie and tarn (circular corrie lake) can be seen.

Corrie

A corrie is a bowl-shaped hollow which has a steep backwall and is sometimes filled with a circular lake, known as a tarn. A corrie starts to form where there are small, sheltered hollows on the side of a mountain. Snow collects in the hollows and does not melt. In the northern hemisphere this tends to happen on north facing slopes which are generally cooler than those facing south. The following year another layer of snow falls and again, does not melt. This process repeats year after year until a small glacier fills the hollow (Figure 10.13). The glacier moves slowly within the hollow and erodes the bottom making it deeper, and the sides making them steeper. The movement of the glacier happens in a circular motion called rotational slip (it's a bit like the movement when you slip off the front of your chair). It causes the glacier to pull away from the backwall creating a deep crevasse known as a bergschrund.

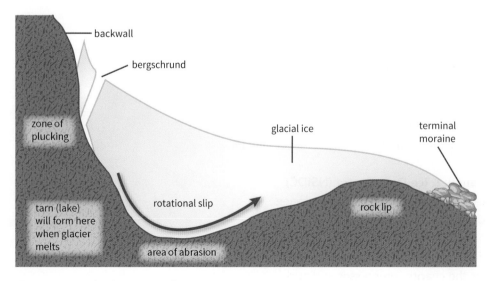

Figure 10.13 The formation of a corrie.

Different processes of erosion work more effectively in different places within the corrie; abrasion happens mainly at the bottom of the hollow while plucking and freeze-thaw weathering mainly steepen the sides. As the glacier moves towards the front of the corrie the ice begins to thin and doesn't erode as much; this forms a rock lip. After glaciation, scree slopes – angular fragments of rock – build up at the base of the backwall. The corrie can also be filled with glacial meltwater and rainwater to form a small circular lake called a tarn.

Arête

When two corries form back to back, their backwalls erode to create a steep, knife-edged ridge known as an arête (Figure 10.14). An example of an arête is Striding Edge on the side of Helvellyn in the Lake District.

Pyramidal peak

Where three or more corries form back to back, this can form a pyramidal peak, a pointed mountain top (Figure 10.15), such as that found at the summit of the Matterhorn in Switzerland.

Truncated spur

Truncated spurs are formed when glaciers move through the main valley and cut off spurs (Figure 10.16).

Figure 10.14 The arête Tryfan from Bristly Ridge on Glyder Fach, Snowdonia, Wales.

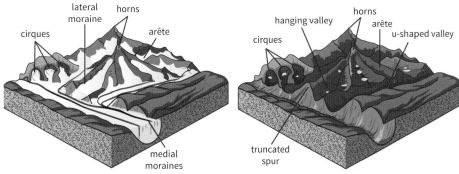

Figure 10.16 Landforms of upland and lowland glacial erosion during and after glaciation.

ACTIVITY 10.3

1 Find a photograph of a corrie on the internet. Either print it off or use an app such as Skitch to annotate the photograph. Make sure your annotations include:
   a a description of the corrie
   b key processes that have formed the corrie
   c any other useful features or characteristics.
2 Which direction is the corrie in your photograph facing and how do you know?
3 Cwm Idwal in Snowdonia, North Wales has a backwall height of 420 m. Generally, the length to height ratio of a corrie is 3 : 1. What is the approximate length of Cwm Idwal?

Figure 10.15 The peak of Stob Dearg, Buachaille Etive Mor in Glencoe, Scottish Highlands – an example of a pyramidal peak.

10.5 What landforms result from glacial erosion in lowland areas?

Glacial troughs

As corrie glaciers move downhill they become valley glaciers. The moving glacier erodes the valley sides, through plucking and abrasion, making it wider and deeper. The cross-profile of the valley becomes a steep sided trough with a broad base and steep valley sides. These glacial troughs are sometimes called U-shaped valleys (Figure 10.17).

There are several features of erosion which can be found in glacial troughs. Smaller, tributary glaciers branch off from the main one. These tributary glaciers do not erode as deeply as the main valley and so, when the ice melts, these smaller valleys are left hanging above the main valley floor. Unsurprisingly, they are called hanging valleys and often contain small streams which flow as waterfalls into the main glacial trough below.

Often, long, narrow streams and lakes are found in the bottom of these glacial troughs. The streams are known as misfit streams because they look a little out of place and could not have eroded the dramatic and massive glacial trough through which they flow. Long, narrow ribbon lakes are also found in the bottom of glacial troughs and are usually fed by misfit streams. Some of the most dramatic ribbon lakes in the UK are Windermere and Wastwater (Figure 10.18); lakes which give the Lake District its name.

Figure 10.17 Martindale valley, Lake District National Park, Cumbria, England, UK – a U-shaped glacial trough.

Figure 10.18 A view of Wastwater (a ribbon lake) from Scafell Pike in the Lake District.

Visit the **Cool Geography website** (www.cambridge.org/links/gase40072) or the **online geography website** (www.cambridge.org/links/gase40073) for detailed summaries of the formation of upland and lowland glacial landforms.

Look at the **animation** of how a corrie is formed on the **e-chalk website** (www.cambridge.org/links/gase40074).

10.6 What landforms result from glacial deposition? (1)

Landforms of glacial deposition are usually found in lowland areas where the ice melts due to an increase in temperature or friction between the ice and the valley sides and bottom. These landforms are usually made up of glacial till – fragments of angular and unsorted rocks – which are dropped by the ice when it melts.

Moraine

Moraine is the term given to piles of glacial till which are transported and then deposited by a glacier. There are several different types of moraine and their names and characteristics are determined by the location of the deposited material in the valley (Figure 10.19). For example, lateral moraine are ridges of glacial till which are deposited at the sides of the valley. Medial moraine is a ridge of moraine found in the middle of the valley floor. It occurs where two glaciers meet and the lateral moraine on each side of the valley joins. The Kaskawulsh glacier in Canada has a ridge of medial moraine that is 1 km wide.

Recessional and terminal moraines are both found near the glacial snout (end of the ice). Terminal moraine marks the furthest extent of the ice. The size and shape of the terminal moraine depends upon the shape of the glacier at its snout and also the time that the snout has stayed in the same place. Recessional moraine are small ridges of debris which tend to be deposited at a temporary pause when the glacier is retreating.

Ground moraine is, as the name suggests, found on the valley floor. Till accumulates at the base of the ice and, as it melts, is deposited to form a blanket of silt and clay. Ground moraines can be shaped by glaciers as they advance and retreat and eventually can form drumlins.

Drumlins

Drumlins are another example of glacial deposition. They tend to be found in swarms (groups) (Figure 10.20). They are small hills which have a shape like an egg; round at one end and tapered at the other. The pointed, tapered end points in the direction of ice flow. While there is no one accepted theory for drumlin formation it is generally believed that they are formed as a result of the glacier advancing and then retreating. As it does so it shapes deposits of ground moraine into these elongated hills.

 Did you know?

A corrie is also known as a cirque in France and a cwm in Wales.

 Discussion point

In the northern hemisphere, corries usually form on north-facing slopes.

a Why do you think this might be?

b How would you investigate this to see whether or not it was the case in the Lake District?

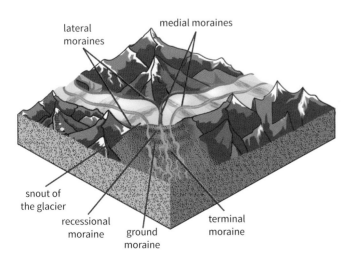

Figure 10.19 Diagram of moraine location in a glacial valley.

Figure 10.20 A swarm of drumlins on the Llwyn y Brain estate, Snowdonia.

10.6 What landforms result from glacial deposition? (2)

Erratics

Erratics are pieces of rock which differ from the bedrock in the area in which they are found (Figure 10.21). Due to their size and power, glaciers are able to transport large boulders, which can weigh several thousand tonnes, over hundreds of kilometres. The Indian Rock in Montebello, New York is an example of an erratic. It is a boulder of granite gneiss which was transported around 30 km from the Hudson Highlands to its current location of sedimentary sandstone and shale rock. It is estimated to weigh around 15 700 tonnes and was transported during the last glacial maximum, around 21 000 years ago.

Discussion point

How might features created by glaciation in the UK have been changed by physical processes since the last ice age?

Further research

To find out more about different types of moraine go to the **National Geography education website** (www.cambridge.org/links/gase40075).

Figure 10.21 Large hilltop erratic boulder over looking Traigh Allt Chailgeag, near Durness, Sutherland, Highland, Scotland, UK.

ACTIVITY 10.4

1   Carry out some research into other examples of erratics – you could use the information on **The National Snow & Ice Data Centre (NSIDC) website** (www.cambridge.org/links/gase40076) to help you.
2   Describe what a drumlin looks like and explain how it is formed.

10.7 What economic opportunities do glacial landscapes provide?

Glaciated landscapes provide opportunities for a range of economic activities. The beautiful, dramatic landscapes formed by glaciation make glaciated areas a perfect place for tourist activities. In areas that used to be glaciated, such as the Lake District, mountain walking is popular (Figure 10.22) and the large lakes can be used for water sports and fishing.

Figure 10.22 Walking on Catbells in the Lake District.

In addition to tourism, agriculture is an economic opportunity which glaciated areas provide. The steep sided fells which have thin soils and cool temperatures are only really suitable for hill sheep farming. However, in lowland areas where there are warmer temperatures and the ground moraine makes the soils more fertile, crops and commercial forests can be grown. Commercial forests are forests grown to provide timber and fuel.

Quarrying is also a major economic activity found in many glaciated areas. For example, the Meadowside Quarry at Kincraig in the Cairngorms extracts around 55 000 tonnes of minerals such as bauxite and iron ore per year.

10.8 Why do we need strategies for managing glacial landscapes?

Glacial landscapes provide opportunities for a wide variety of activities for many different groups of people – both visitors and residents. There are advantages and disadvantages to large numbers of tourists visiting the area. Tourists create income and employment but may require different services to those needed by the local population. For example, tourists are likely to want cafes and hotels while residents are likely to need services such as supermarkets and schools. The residents may then have to seek out these services elsewhere. Tourists are attracted to the natural scenery created by glacial landscapes and, as a consequence, this encourages the development of facilities and services to support tourism. However, if tourism is developed at the expense of conserving the natural glacial environment, then this can lead to conflict between the two. Tourist activities may conflict with local agriculture and wildlife. Other problems include litter and footpath erosion. All of this can lead to conflict between different groups and, if left unchecked, may destroy the very environment which attracted the tourists in the first place. For all these reasons, glaciated environments need to be carefully managed in order to remain sustainable.

 Discussion point

What economic opportunities are offered by glaciated landscapes? Why do these landscapes create these particular opportunities? How might these different activities lead to conflict both with each other and with the needs of conservation?

The Lake District

About 2 million years ago the Lake District was a mountainous area broken up by rivers flowing from the highest point. During the last ice age glaciers built up in the higher areas and moved downhill following the path of the existing river valleys. This created the classic glaciated landscape that we see today. It is the ribbon lakes and tarns which have given the Lake District its name. Many of the rocks in the Lake District are volcanic which means that water does not seep into them. The high levels of rainfall (the Lake District is one of the wettest places in England with over 2000 mm per year) and the deep glacial troughs means that the lakes contain vast amounts of water. The largest lake in England, at 14.8 km² is Lake Windermere (Figure 10.23), while the deepest is Wastwater at 74 m.

Figure 10.23 Lake Windermere, Lake District.

Did you know?

Red Tarn is about 25 m deep and contains brown trout and schelly (a type of whitefish which can only be found in the Lake District lakes of Haweswater, Brothers Water, Ullswater and Red Tarn).

Further research

Look at one of Alfred Wainwright's guides to the Lakeland Fells which describe in detail, with maps and sketches, what it is like to climb the fells in the Lake District.

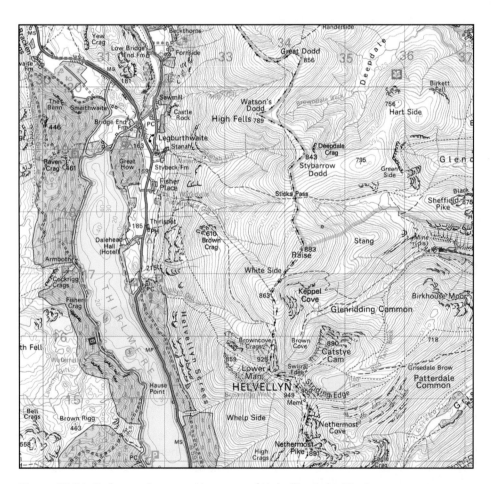

Figure 10.24 Ordnance Survey map extract of Helvellyn, Lake District.

The Lake District also has some of the highest peaks in the UK. Helvellyn is the third highest peak in the area at 950 m and is formed by a number of corries and arêtes (Figure 10.24). Red Tarn on the eastern side of Helvellyn is an example of a corrie and Striding Edge and Swirral Edge (Figure 10.25) are examples of arêtes.

Figure 10.25 Swirral Edge, an arête on the flanks of Helvellyn in the Lake District.

ACTIVITY 10.5

1 Find photographs of glacial features in the Lake District. Locate these on a suitable map.

 Download Worksheet 10.1 from Cambridge Elevate for help with Activity 10.5, question 1.

2 Look at the map extract of the Lake District (Figure 10.24). Identify the glacial landform at each of the following grid references:
   a 3116
   b 347 152
   c 309 158

3 Figure 10.25 is a photograph of Swirral Edge (grid reference 3 415 on Figure 10.24).
   a What is the name of the lake shown in Figure 10.25?
   b What are two features of Swirral Edge?

 Discussion point

Look at the OS map of the Lake District. What area could be used to investigate glacial landscapes? Justify your choice.

The Lake District National Park

The Lake District National Park is England's largest and most visited National Park. Its aim is to conserve and enhance the natural beauty, wildlife and cultural heritage of the area and promote opportunities for the public to understand and enjoy its special qualities. It is also concerned with the economic and social well-being of local communities within the National Park.

Physical and human characteristics of the area

The dramatic high mountains and deep valleys shaped by glaciers make the Lake District a unique landscape. At 978 m, Scafell Pike is the highest mountain in England. At lower altitudes there is moorland, deciduous woodland and evidence of glacial deposition. For example, drumlin swarms are found in the Vale of Eden. Due to the high levels of rainfall in the Lake District, species such as Atlantic moss and liverwort have the chance to thrive. The large glacial lakes are also unique physical features of the area and provide opportunities for tourism and recreation (Figure 10.27).

The Lake District is one of the most highly populated National Parks in the UK. The area is associated with several writers and artists such as William Wordsworth and Beatrix Potter who have been inspired by the landscape. Every year around 70000 Japanese tourists visit Hill Top, Beatrix Potter's family home.

Impacts

Social impacts

Around 89 per cent of tourists travel to the area by car and so there are severe problems with traffic congestion, particularly in the summer months (Figure 10.26).

In addition, 20 per cent of the properties in the National Park are second homes. These are used at weekends or during holidays but not throughout the year, which can mean local services, such as schools, shops and public transport, become unviable.

Factfile

- The Lake District National Park contains many honeypot sites – that is places which attract large numbers of tourists. This can put pressure on the local environment and people.
- The area is used by tourists for many activities including water sports, fell walking and fishing.
- Around 16.4 million tourists visit the Lake District per year spending 23 million visitor days.
- Over 50 per cent of the Lake District economy comes either directly or indirectly from tourism making the local people dependent upon it for an income.
- Tourism contributes £1 146 million to the local economy per year and provides over 16000 jobs within the National Park.

Figure 10.26 Traffic congestion in Keswick in the Lake District.

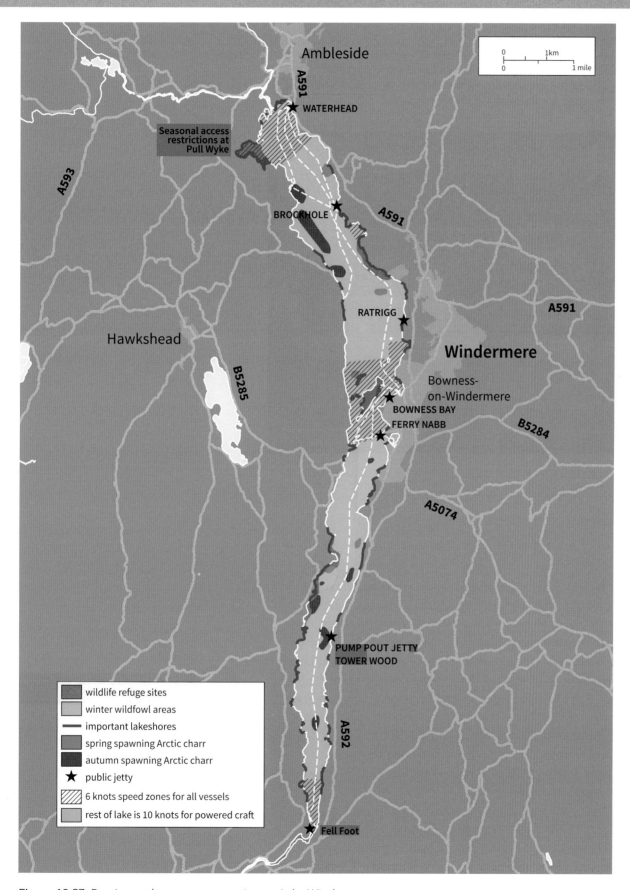

Figure 10.27 Boating and nature conservation on Lake Windermere.

Economic impacts

The price of housing is kept high meaning that local people who want to buy a house in the area often cannot afford it (Figure 10.28). There is a narrow range of jobs on offer for locals in the area and most are focused upon tourism which typically provides only seasonal employment and low wages.

Figure 10.28 The market for holiday homes in the Lake District pushes up property prices.

Strategies used to manage the impact of tourism

There are a number of initiatives that aim to manage the impact of tourism on the Lake District National Park:

- The Windermere Waterfront Programme is a regeneration project which aims to provide a sustainable experience for visitors to this area. Improved transport options are being introduced and a 10 knots limit has been in force on Lake Windermere since March 2005 in an attempt to halt the growing numbers of speedboats using the lake and to reduce noise and disturbance for residents and wildlife. A survey in 2014 suggested that, despite the limit, there is still conflict between open water swimmers, large sailing boats, and water skiers.
- 'Fix the Fells' is a conservation plan to repair and maintain mountain paths in the Lake District. The Lake District has many rare plant species such as the blanket mire. By repairing the footpaths, walkers are less likely to stray onto these rare habitats and destroy them.
- The Lake District National Park Authority is also working towards making the area more accessible for those who might otherwise find the physical environment a challenge. They have introduced guided walks which welcome those with mobility or visual problems and have developed a 'Miles without stiles' guide that highlights footpaths suitable for wheelchairs and pushchairs.

 Further research

Find out more by looking at the The Windermere Management Strategy 2011 produced by the Lake District National Park and the South Lakeland District Council.

Key facts about tourism in the Lake District can be found on the Tourism page of the Lake District National Park.

Environmental impacts

Other pressures and challenges focused specifically on the lakes include water levels and quality which affect the visitor experience. In 2010, The Great North Swim (a mass participation swimming event in Windermere) was cancelled due to blue-green algae in the lake. Also, flooding around Windermere, following heavy rainfall in October 2008 and November 2009, as well as the drought conditions which were experienced in 2010, demonstrated the impacts of extreme weather events on local businesses and recreational activities.

Conflicts exist among those who want to use the lakes for recreation. Windermere, Derwentwater, Coniston Water and Ullswater are particularly popular places for recreation. People can fish, sail, canoe, water-ski, swim, powerboat or take the ferry. Often these activities conflict with each other. For example, people using powerboats may struggle to see swimmers in the water and may collide with them causing injury or death. There is also conflict between those who use the lakes and those concerned with the preservation of wildlife in and around the lakes.

ACTIVITY 10.6

1 Watch the film, *ShowMeBritain:* Episode 7: Lake District, on YouTube which introduces a number of different activities that tourists can do in the Lake District. Write a commentary for the film which describes the activities on offer.

2 Do you think the costs of tourism in the Lake District outweigh the benefits?

Download Worksheet 10.2 from Cambridge Elevate for help with Activity 10.6, question 2.

3 A honeypot site is a place where tourists flock to, particularly in the summer months.

   a Which specific places in the Lake District do you think could be described as honeypot sites?

   b What might be the specific opportunities and challenges for these places?

Assess to progress

1 Explain how tourism might put pressure on the physical environment of a glacial environment such as the Lake District National Park.   6 MARKS

2 Using a diagram, explain the formation of a corrie.   6 MARKS

3 With the help of Figure 10.19, describe and explain the location and formation of one type of moraine.   6 MARKS

Tip

- Question 2 asks you to include a diagram so it is very important that you include one and that it adds something to your answer.
- Question 3 asks you to link the activities of tourists to pressures on the physical environment.
- It also mentions the Lake District National Park and so it is a good opportunity to add named example detail here.

Urban issues and challenges

11 Changing urban areas

In this section you will cover:

- why a growing proportion of the world's population live in urban areas
- the factors affecting the rate of urbanisation
- the growth of megacities
- how urban growth creates opportunities and challenges in Low Income Countries (LICs) and Newly Emerging Economies (NEEs)
- the UK's urban landscape
- the location and importance of one city in the UK
- how urban change has created opportunities and challenges in one city in the UK
- how urban transport strategies can reduce traffic congestion
- how urban areas can be made more sustainable.

Urban issues and challenges – an overview

In the early part of the 21st century the world became an 'urban world', with more people living in urban areas than rural areas for the first time in history. As lower income countries (LICs) and newly emerging economies (NEEs) continue to develop, this trend is going to continue as more people migrate to urban areas in the hope of improving their quality of life.

In 1950, New York was the world's only megacity, with a population of 10 million people. In 2015, it was estimated that there were 35 megacities, the largest of which were home to more than 20 million people.

Urban areas are dynamic places that create opportunities for people. However, with so many people living in relatively small areas they are not always easy to manage. All cities face the challenge of providing adequate facilities for the population while ensuring that pressures on the environment are kept to an acceptable level. Cities in LIC/NEE countries face the added pressure of managing rapid population growth and the challenge of large numbers of people living in slum conditions.

The challenge for urban planners is to ensure that people have access to basic facilities and that cities are increasingly sustainable.

Figure S4.1 Urban areas create a wide range of social and economic opportunities but also a number of management challenges. What opportunities and challenges are suggested by this photograph?

11 Changing urban areas

In this chapter you will learn about...

- why a growing proportion of the world's population live in urban areas
- the factors affecting the rate of urbanisation
- the growth of megacities
- how urban growth creates opportunities and challenges in Low Income Countries (LICs) and Newly Emerging Economies (NEEs)
- the UK's urban landscape
- the location and importance of one city in the UK
- how urban change has created opportunities and challenges in one city in the UK
- how urban transport strategies can reduce traffic congestion
- how urban areas can be made more sustainable.

11.1 Living in an increasingly urban world

In 1950 fewer than one in three people in the world lived in a town or city. In 2006 the United Nations estimated that the world's population was evenly split, with 3.2 billion people living in urban areas and the same number living in rural areas. In 2015 it was estimated that 55 per cent of the world's population lived in urban areas and this figure is expected to increase to approximately 70 per cent by 2050 (Figure 11.1).

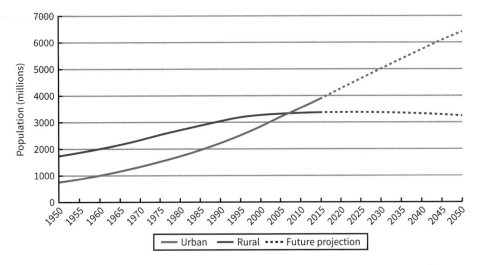

Figure 11.1 Urban and rural population of the world, 1950–2050 (*Source*: United Nations).

Rates of urbanisation

Rates of urbanisation vary across the world (Figure 11.2).

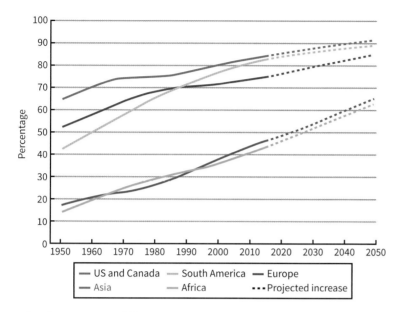

Figure 11.2 **Urbanisation** in different regions, 1950–2050 (*Source*: The Green City Index).

Key terms

rate of urbanisation: the percentage increase in the urban population

urbanisation: an increase in the proportion of people living in towns and cities

migration: the movement of people from one place to another

Did you know?

Just three countries – India, China and Nigeria – together are expected to account for 37 per cent of the projected growth of the world's urban population. Between 2014 and 2050, India is projected to add 404 million urban dwellers, China 292 million and Nigeria 212 million.

Source: United Nations, World Urbanisation Report (2014)

As countries change from largely rural, agricultural economies to more industrial, commercial economies the population becomes increasingly concentrated in towns and cities.

- Countries that have reached high levels of economic development generally have urban populations of 75 per cent plus.
- Countries at very low levels of economic development generally have an urban population of less than 50 per cent.
- Countries going through rapid economic development have high rates (%) of urbanisation.
- Asia and Africa are urbanising more rapidly than other parts of the world.

Rapid Growth of Chinese Cities

Over the last 20 years an average of 10 million rural Chinese have moved into the cities each year, making it the largest **migration** in history. In Shanghai and Beijing, along with many other cities, the Chinese economic boom has created millions of jobs and attracted migrants from hundreds of kilometres away. In 1980 Shanghai had only 121 buildings that were over 8 storeys high. By 2015 this figure had risen to over 5000.

11.2 What are the causes of urbanisation in LICs and NEEs?

There can only be two reasons for the growth of population in urban areas:

1 **Net migration** – where more people are moving into an urban area than are leaving.
2 **Natural increase** – where there are more births than deaths. The greater the difference between the birth rate and death rate, the faster the population grows.

The major cause of urbanisation in developing countries is rural–urban migration. Poor social and economic conditions in rural areas encourage people to move to urban areas where there may be better opportunities for people to improve their quality of life (Figures 11.3 and 11.4). Natural disasters in rural areas and civil conflict can also force people to leave, as can a lack of land ownership.

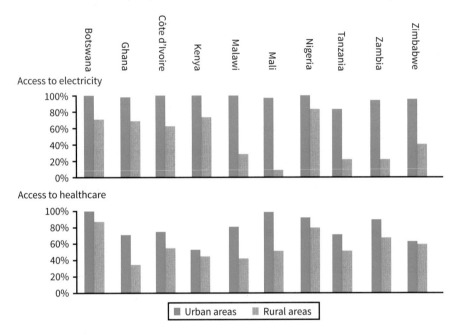

Did you know?

The 'push-pull' theory of migration is often used to explain the causes of migration. Push factors are factors that encourage people to move away from a place. Pull factors are factors that attract people to a place.

Figure 11.3 Access to services in Africa for selected countries. (*Source*: Afrobarometer, 2014–15 data).

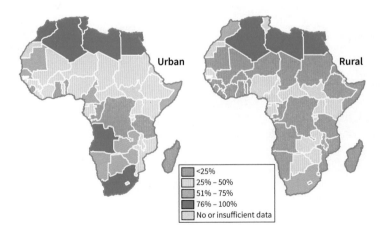

Figure 11.4 Use of improved sanitation facilities in Sub-Saharan Africa in 2010. One map shows usage in urban areas within each country and one shows usage in rural areas within each country. (*Source*: UN.)

Download Worksheet 11.1 from Cambridge Elevate for some questions about urbanisation.

Urban sprawl

Urban areas do not only grow in terms of population numbers, they also grow in area. This is called **urban sprawl**. As the population of urban areas increases, there is more demand for land. As land in the centre of urban areas is limited there is increasing pressure for development on the edge of the urban area. This continued expansion can create huge urban areas; for example, Mexico City has expanded so much that it is now over 50 kilometres from one side to the other (Figure 11.5).

Key term

urban sprawl: the expansion of an urban area into surrounding, less populated areas

Figure 11.5 Mexico City showing massive urban sprawl.

 Did you know?

Urbanisation has become more rapid as industry has spread across the world. It took London 130 years to grow to 8 million people, while Bangkok took 45 years and Seoul only 25 years to reach this number.

Source: World Health Organization

ACTIVITY 11.1

1 Explain the link between economic development and urbanisation.
2 Explain how push and pull factors encourage rural–urban migration.
3 Explain how migration and natural increase are linked factors in urban growth.
4 Estimates for the population of Mexico City are given as 18, 23 and 35 million in three different sources.
   a Suggest reasons for the differences.
   b Why might estimates of future urban populations in LICs not be accurate?

 Tip

Make sure that you can describe and explain the causes of rural–urban migration in lower income countries (LICs) and newly emerging economies (NEEs). Use examples to develop your ideas.

11.3 The growth of megacities

What is a megacity?

A megacity is defined as a city that has 10 million or more people. In 1950 New York was the only global megacity. In 2015 it was estimated that there were 35 megacities, most of them in LICs and NEEs; the majority in Asia, the continent with the largest number of people living in urban areas.

The growth of megacities is likely to continue as more people move towards urban areas in LICs and NEEs (Figure 11.6). Recent estimates suggest that the number of megacities will increase beyond 50 by 2050.

It is not so easy to calculate when a city becomes a 'megacity'. In many countries cities are growing so rapidly it is impossible to calculate the number of people living in them accurately, especially since many people live in illegal **squatter settlements** and unplanned slum areas.

Key term

squatter settlements: illegal settlements where people have no legal rights over the land on which they live

Note: The circles indicate population sizes ranging from ○ (10 million) to ◯ (39 million). The circles do not reflect the physical extents of the cities and any overlap between them merely reflects their relative population sizes and not any official acceptance or endorsement of any geographical sovereignty.

Source: UN (2012).

Figure 11.6 The rise of megacities.

'China's Pearl River Delta overtakes Tokyo as world's largest megacity'

Several hundred million more people are expected to move to cities in East Asia over the next 20 years as economies shift from agriculture to manufacturing and services, according to a World Bank report.

China's Pearl River Delta has overtaken Tokyo to become the world's largest urban area in both size and population, according to a report from the World Bank. The megacity – which covers a significant part of China's manufacturing heartland and includes the cities of Shenzhen, Guangzhou, Foshan and Dongguan – is now home to more people than the countries of Canada, Argentina or Australia.

Urbanisation which took place over a period of several decades in Europe and North America is happening in just a few years in East Asia, which already contains eight megacities (with population above 10 million) and 123 cities with between one and 10 million people. With almost two-thirds of the region's population (64%) still non-urban at present, several hundred million are expected to move to cities over the next 20 years as economies shift from agriculture to manufacturing and services…

Nick Mead, *Guardian*, 28 January 2015

Did you know?

- Urban growth in many LICs is occurring so quickly that urban planners do not know the actual population of many cities.
- There are a billion urban dwellers who have no access to basic services. That number is expected to double by 2030.
- Unplanned urbanisation makes it difficult to manage basic services such as water, sanitation and electricity.

Source: UN Habitat (2015)

ACTIVITY 11.2

1 Using Figure 11.6, describe the expected changes to the number and distribution of megacities.

Mumbai – a city of national and international importance

Mumbai is the wealthiest city in India. It is located in the Maharashtra state on the western coast of India, facing the Arabian Sea. The city was originally a number of separate islands. They were joined together by a **land reclamation** project over a hundred years ago.

Key terms

land reclamation: gaining land from the sea

peninsula: a piece of land almost surrounded by water

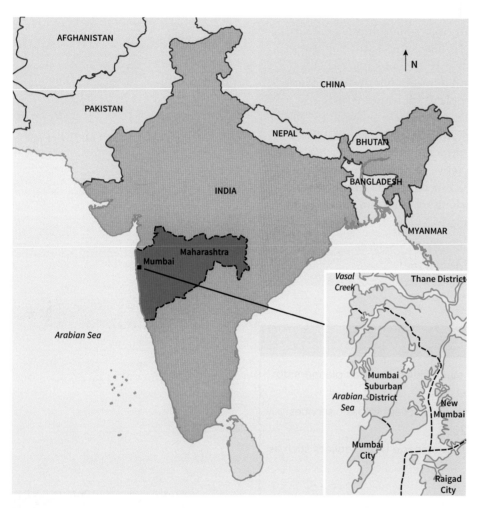

Figure 11.7 The location of Mumbai.

The growth of Mumbai

Mumbai has seen rapid growth over the last 50 years, both in terms of population numbers and area. In the 1950s the population was concentrated around the city centre and port district to the south of the **peninsula**. Since then the city has spread to the north and east and is now a sprawling metropolitan area of over 23 million people (Figure 11.7).

The growth of Mumbai has been driven by rural–urban migration. A recent study found that:

- Migration into Mumbai was averaging nearly 1 person per minute!
- 70 per cent of all migrants were from the state of Maharashtra.
- The average age of migrants was 20–21 years and 64 per cent were male.

In virtually all cases the reason for migration was described as 'economic'.

Natural increase is also a significant contributor to the growth of Mumbai, despite a decreasing birth rate resulting from improved family planning measures.

 Factfile

The economic importance of Mumbai

Mumbai is the commercial and financial capital of India and handles approximately 60 per cent of all of India's sea trade. Over 3 million people commute each day to the city for work. There is a growing hi-tec industry which includes call centres, online banking and software development. The city contributes 40 per cent of the total income to the whole state of Maharashtra. Nationally, Mumbai contributes 33 per cent of all income tax and 60 per cent of all custom duty from trade.

- In 2014 Mumbai was the most globalised city in S.E. Asia and had:
  - the largest number of transnational corporation (TNC) headquarters
  - a large number of international companies, including:
    - Bank of America/Citigroup
    - GlaxoSmithKline
    - Johnson & Johnson
    - Volkswagen
    - Walt Disney
  - the main Indian stock market
  - the largest amount of foreign investment
  - the busiest port and airport in India.
- Mumbai is seen as a hub for smaller businesses, including design, fashion, jewellery and tourism.
- It is home to the biggest cultural industry in Asia – Bollywood.
- It is a **transport hub** with transport links to all of the major industrial cities in India and air transport links to many world cities.

Source: Mumbai: India's global city, a case study for the Global Cities Initiative: a joint project of Brookings and JPMorgan Chase. Written by Greg Clark and Tim Moonen, December 2014.

Figure 11.8 Mumbai's city centre.

 Did you know?

Mumbai has the highest percentage of internet access of any Indian city (12 million people in 2013).

 Key term

transport hub: connecting point for transport links

Opportunities and challenges of urban growth in Mumbai

Cities in LICs are often seen as overcrowded and polluted, with high proportions of their population living in slums where conditions are very poor. However, that is only one side of the story. These cities are often the centre of economic activity where there are opportunities for development and a regular income. Also, because a lot of people live in a small area it can be easier to supply basic services such as water, sanitation systems and electricity. Consequently, cities can be a positive force for development.

Opportunities of urban growth in Mumbai

The growth of Mumbai has created many socio-economic opportunities. This is because:

- There are more employment opportunities. In Mumbai urban industrial areas are stimulus for economic development and the city has a flourishing manufacturing industry producing electronic items, jewellery and textiles.
- Incomes in the city are higher and more reliable than in rural areas.
- Access to education and healthcare is easier in Mumbai than in surrounding areas.
- In poorer residential areas there is often considerable community spirit and support.
- The urban poor provide a massive labour force who carry out essential jobs and keep the city running.

The following example describes the social and economic opportunities found in Dharavi.

Dharavi – a thriving community

To many people, Dharavi is just an urban slum near Mumbai city centre where nearly a million people live in squalid conditions. A closer look reveals a different side to the area. A local resident explained:

'Many people have lived here for forty or more years and their homes are made of bricks and have electricity and running water. There are schools and local community groups where people help each other. There are thousands of small businesses, some in tiny workshops, others, like the clothing factories, with computerised machinery which make clothes and bags which are exported all over the world. There are many shops, you can buy just about anything here and many small family businesses producing a wide variety of foods. Dharavi produces billions of dollars a year for the local economy and is an important part of the city of Mumbai. All it lacks is decent housing and **infrastructure**.'

Source: Mumbai: India's global city, a case study for the Global Cities Initiative: a joint project of Brookings and JPMorgan Chase. Written by Greg Clark and Tim Moonen, December 2014.

Recycling in Dharavi

Among the shacks and open sewers are an estimated 16 000 small factories employing over a quarter of a million people. Amongst all of this is what has been described as a 'recycling miracle' (Figure 11.9). Virtually everything here is recycled, even small pieces of soap from hotel bathrooms! Thousands of people work recycling huge quantities of industrial waste, much of which is imported

Did you know?

According to a 2014 United Nations Habitat report, in poor countries urban populations are generally better off than rural populations. They tend to have greater access to education and health services, literacy rates are higher and life expectancy is longer.

Figure 11.9 Commercial activity in Dharavi.

Further research

Do a web search for:

- Dharavi
- Industry in Dharavi
- Community development in Dharavi.

Key term

infrastructure: the framework of transport and energy networks, including roads, railways, ports and airports, plus energy distribution

Further research

Investigate the range of products recycled in Dharavi.

from other parts of the world. It is estimated that 80 per cent of Mumbai's own waste is recycled – a figure the west would be proud of.

Challenges of urban growth in Mumbai

While the growth of Mumbai has created many social and economic opportunities, it has also created a number of challenges for urban planners.

Managing the growth of squatter settlements

It is estimated that around 40 per cent of the population live in poor quality housing or on the streets (Figure 11.10). Many of the poorest inhabitants of the city cannot afford to rent housing so end up living in illegal squatter settlements. It is very difficult to supply services to these unplanned, overcrowded areas.

There are strategies in place to manage the growth of squatter settlements and the needs of the people that live in them. Mumbai's Slum Rehabilitation Authority is a planning authority that attempts to develop better housing on squatter sites or to relocate residents to alternative accommodation.

Providing clean water, sanitation systems and energy

In many parts of Mumbai access to basic services is limited. A study of one slum found that 95 per cent of households had lower access to clean water than the minimum amount recommended by the World Health Organisation (WHO). Lack of access to basic services increases the risk of illness and disease.

Providing access to services – health and education

Satisfying the growing demand for health services and education is a real challenge. The youthful population means that there is a constantly rising demand for school places and maternity services. The poor environmental conditions in squatter areas means that the risk of disease is high, putting increasing pressure on health services.

Reducing unemployment and crime

The growing demand for skilled labour has meant that opportunities for unskilled workers are increasingly limited and wages are low. This can lead to rising crime rates as people turn to illegal activities in order to survive.

Managing environmental issues

Dealing with increasing volumes of human and industrial waste is difficult and expensive. The lack of sanitation systems in some parts of the city means that streams and rivers are used to dispose of sewage. The growth of unregulated industry and increasing vehicle numbers add to the problems of air and water pollution.

The overall challenge for Mumbai is being able to manage growth so that the population has adequate access to basic services and can earn a reasonable living, leading to an acceptable quality of life. A significant part of this challenge is to prevent the continued growth of urban slums and to make sure that the urban poor have suitable places to live.

Figure 11.10 Slum area in Mumbai.

ACTIVITY 11.3

1 Explain the statement: 'The reason for people moving to Mumbai was mainly economic.'

2 What evidence suggests that Mumbai is a city of national and international importance?

3 Suggest three challenges that Mumbai might face as the population continues to grow.

4 Why are poor areas often important to the development of cities?

5 Dharavi has been described as a 'slum of hope'. Explain this statement.

6 'Urban growth puts pressure on things like housing, water and electricity supply, sanitation systems and road networks.' Explain this statement.

7 Explain how the conditions shown in Figure 11.10 are a risk to health.

8 Why is urban growth a challenge for urban planners?

9 Suggest three ways that life for the urban poor in developing cities could be improved.

Urban planning in Mumbai

Improving living conditions and opportunities for the urban poor is a major challenge in LICs and NEEs. Rapid urban growth is leading to the growth of **urban slums** and according to UN-Habitat, this is likely to continue in the future (Figure 11.11).

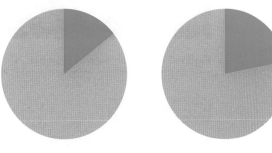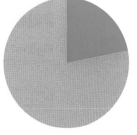

2003
1 billion slum dwellers
out of 6 billion world
population

2030
2 billion slum dwellers
out of 8.3 billion world
population

2050
3.5 billion slum dwellers
out of 9.1 billion world
population

Figure 11.11 Projected growth in the number of slum dwellers.

People living in urban slums often suffer from overcrowding, inadequate housing, limited access to basic services and poor environmental quality. Often poor migrants can only afford to live in makeshift slums or squatter settlements. While these areas are often unregulated and have many problems, they can also be strong communities where people work together and support each other. The challenge for urban planners is to improve the conditions for the urban poor while at the same time making sure that social communities are protected. The following examples describe some of the projects taking place in Mumbai, India.

The Mumbai slum resettlement scheme

As part of the Mumbai Urban Transport Project (MUTP) a slum area alongside the railway line was cleared and residents moved to a new housing area in a different part of the city. One slum dweller, Jyothi Pujari, recalls what life was like before the move:

'Home was a tin shack right next to the railway line (Figure 11.12). There were open drains and a constant battle to get fresh water. The stench of the open toilets was terrible and the fear of accidents from the trains was there all the time.'

Now, sitting in her new home Jyothi recounts how her life has changed.

'For the first time we own our own home and don't have to worry about being homeless. The new apartment is made of solid material and has a water supply and proper drains. Now we have plumbing the risk of typhoid, stomach problems and other infections has fallen and the children are not ill as much so don't miss as much schooling. There are lots of small shops and businesses so it is easier to earn money. We now have a television and a fridge and sleep on beds instead of the floor.'

A United Nations spokesperson said that, 'schemes like this remind us that people need homes, not just shelters'.

The Mumbai slum sanitation project

The rapid growth of slum areas in Mumbai has put enormous pressure on sewage systems throughout the city. In some areas hundreds of people share a

Key term

urban slums: poor areas lacking in services; they are often called 'shanty towns' but also have localised names such as 'favelas' in Brazil, 'barriadas' in Peru and 'bustees' in India

Factfile

The United Nations Sustainable Development Goals

By 2030, all areas should have:

- Number 4 (Quality Education)
  - free and accessible primary/secondary education for all
- Number 6 (Clean water and sanitation)
  - safe and accessible water for all
  - adequate sanitation and hygiene for all
- Number 11 (Sustainable cities)
  - safe and affordable housing for all
  - safe and affordable public transport
  - a reduction in the number of people affected by natural disasters
  - improved environmental quality (air pollution/waste management)
  - an increased number of green areas

Figure 11.12 Slum by a railway in Mumbai.

Figure 11.13 Community toilet blocks, Mumbai.

single toilet and it is estimated that 1 in 20 people are forced to use the street as a toilet. The slum sanitation project aims to improve sanitation facilities for up to a million slum dwellers across the city. So far over 300 community toilet blocks have been built (Figure 11.13), housing more than 5100 individual toilets, with separate facilities for men and women.

Incremental housing strategies

Incremental housing strategies are a way of developing informal slums into permanent residential areas by making gradual improvements (Figure 11.14). Families are given the right to the land on which their home is built and a grant which can be used for improvements. They work with an architect to design their home or even plan a new house. The local community is involved in the design and layout of the area and individual residents make decisions about their homes, including the colour of outside walls. As one planner said, 'this is a way of giving people what they want and keeping communities together rather than knocking their houses down and splitting up the community by moving people to different parts of the city.'

Figure 11.14 Incremental housing strategy, Mumbai.

The Mumbai slum electrification project

Many slum areas do not have access to electricity and rely on bottled gas for cooking and heating. This is expensive and dangerous. Also the fumes created can cause health problems. The Global Partnership on Output Based Aid (GPOBA) have recently completed a project which is providing 10 000 slum dwellers with new or upgraded electricity connections.

Further research

Investigate the Dharavi redevelopment plan. Describe the main changes suggested in the plan and consider the advantages and disadvantages for the local community.

ACTIVITY 11.4

1 Suggest why the number of people living in slums is likely to increase.
2 Explain how access to clean water can improve both health and wealth.
3 What does the statement 'people need homes, not just shelters' mean?
4 How might incremental housing strategies encourage people to:
   a take pride in their homes
   b develop a sense of community spirit?
5 How might the examples shown on these two pages help to achieve the UN Sustainable Development Goals outlined in the Factfile?

11.4 The UK's urban landscape

How is 'urban' defined in the UK?

In the UK, urban areas are defined as 'built-up areas that have resident populations above 10 000 people' (Gov.uk). In England, the 2011 **population census** showed that 43.7 million people (82.4 per cent of the population) lived in urban areas, but urban areas only accounted for 15 per cent of the land area.

In the 2011 Census urban areas were divided into four (Table 11.1):

	Major conurbation	**Minor conurbation**	**City/town**	**Sparse city/town (urban with significant rural)**
% total population	35.4	3.6	43.2	0.2
Population	18 800 000	1 900 000	23 000 000	89 000
(numbers approximate)				

Table 11.1 Classification of urban areas in the UK (data for England 2011).

The most populous urban areas of the UK

Figure 11.15 The 25 most populous areas in the UK.

Key terms

population census: an official count of the population. In the UK there is a census every 10 years: the last census was in 2011

conurbation: extensive urban areas resulting from the expansion of several towns or cities so that they merge together but maintain their separate identities. For example, the West Midlands conurbation includes the cities of Birmingham, Coventry and Wolverhampton, as well as many large towns, including Sutton Coldfield, Dudley, Walsall and West Bromwich

Did you know?

It is important to be clear when using the term 'United Kingdom', as different terms mean different things.

The United Kingdom: England, Scotland, Wales and Northern Ireland

Great Britain: England, Scotland and Wales

The British Isles: England, Scotland, Wales and the whole of Ireland (note that this is a geographical description, not a political one)

The ten most populous urban areas in the UK (Figure 11.15) all:

- have a population greater than half a million people
- are home to approximately 40 per cent of the urban population in the UK
- have a **population density** of over 3000 people per square kilometre
- cover approximately 20 per cent of all urban land area
- have grown by over 600 000 people and 60 km² in the last ten years.

Comparing the ten most populous urban areas in the UK

In order to compare different urban areas, the census looks at a number of population characteristics, including:

- **age** – with a particular emphasis on under-16 and post-retirement age
- **sex** – the balance between males and females
- **recent migration** – people moving into the area from another part of the UK or another country in the last 12 months
- **ethnicity** – belonging to a social group that has common cultural traditions

The information in Table 11.2, taken from the 2011 Census, identifies differences between the ten most populous urban areas in the UK.

All data in %	Age		Sex			Migrants	
	Under-16	Pension age	Male	Female	Ethnic minorities*	From other parts of the UK	From outside the UK
Greater London	20.0	15.0	48.4	51.6	25.9	2.0	1.6
West Midlands	22.0	18.4	48.6	51.4	20.4	1.6	0.47
Greater Manchester	21.2	17.6	48.6	51.4	9.6	2.1	0.52
West Yorkshire	21.6	17.2	48.3	51.7	14.1	2.8	0.58
Greater Glasgow	19.2	19.2	47.3	52.7	3.8	3.1	0.54
Tyneside	19.3	19.4	48.3	51.7	3.4	2.5	0.55
Liverpool	20.9	18.3	47.7	52.3	3.9	2.1	0.37
Nottingham	19.8	17.3	49.2	50.8	7.9	3.7	0.71
Sheffield	19.5	19.0	48.8	51.2	7.8	3.2	0.63
Bristol	19.4	17.3	48.9	51.1	6.3	4.0	0.86
UK average	**20.2**	**18.4**	**48.6**	**51.4**	**14.0**	**N/A**	**N/A**

Table 11.2 Population characteristics for the ten most populous urban areas in the UK (2011).
*Ethnic minority defined as people who identify with an ethnic group other than white (ONS.gov.uk).

Factfile

Urban areas are often identified by their characteristics and generally have:

- high levels of population density
- a mixture of land uses (commercial, industrial, recreational, transport and residential)
- a complex transport and communications infrastructure, often with transport hubs
- an influence beyond the immediate boundary of the urban area

Find out more at the **Office of National Statistics website** (see www.cambridge.org/links/gase40077)

Key term

population density: number of people per square kilometre (a measure of 'crowdedness')

ACTIVITY 11.5

1. a Why is it difficult to define an 'urban area'?
   b Suggest which three functions you might expect to find in a city but not a small town.
   c Explain your choices.
2. Why do conurbations develop?
3. Using Table 11.2:
   a Identify any significant differences between the urban areas.
   b Suggest reasons for those differences.

Birmingham – national and international links

Birmingham lies near the geographic centre of England and is at the centre of the West Midlands conurbation. It is a city of both national and international importance. It is the United Kingdom's second largest city and has five major universities, with over 60 000 students from across the world. The article gives some background information about the importance of Birmingham in the wider world.

University of Birmingham – Make it in the Midlands

According to the … European Cities Monitor 2009, Birmingham is rated among the top 15 best cities in Europe in which to locate a business. It is home to around 31 000 companies and provides space for more than 300 organisations based outside the UK who have operations here. More than £13 billion is to be invested in its infrastructure development as a city of global importance over the next decade. Birmingham also has around 200 law firms, many top accountancy firms, 50 major property services and banking specialists and one of Europe's largest insurance markets. As well as large employers such as Arup, Atkins, BBC, Deutsche Bank, Dollond & Aitchison, E.ON, Fujitsu, Jaguar Landrover, Muller and Severn Trent Water, to name but a few large employers, the West Midlands is home to numerous small and medium sized enterprises (SMEs), some at the cutting edge of new technological developments.

Source: University of Birmingham website

Did you know?

- Nearly 35 million people visited Birmingham in 2015, bringing in over £35 billion.
- Over 60 000 people work in the leisure and tourism industry in the area.
- The number of visitors from Southeast Asia has risen rapidly in the last 10 years.

Find out more at the **Visit Birmingham website** (www.cambridge.org/links/gase40079)

Birmingham as a transport hub

Railway links

Birmingham has direct links to many cities in the UK. New Street station is the busiest station outside of London. In 2014 over 34 million people used the station. Planned redevelopment of the area around the station will create 14 000 new jobs and include 2 000 new city centre homes, bringing in £1.3 billion a year.

Birmingham Airport

Birmingham Airport had over 9 million customers in 2014 – it has been redesigned to cope with planned growth of up to 18 million customers. The airport is a global communications hub offering flights to a wide range of business and leisure destinations. The completion of the runway extension in 2014 has enabled Birmingham Airport to take larger aircraft and offer direct flights to China, South America, Africa and the west coast of the United States.

The impacts of national and international migration on Birmingham

Birmingham is one of the most **culturally diverse** cities in the United Kingdom. In 2015, the city had a population of 1.1 million drawn from 187 nations (Figure 11.16). Birmingham has:

- an age profile which makes it Europe's 'youngest city'
- one of the highest proportions of migrants from other parts of the United Kingdom

Key term

culturally diverse: having a variety of cultural/ethnic groups within a society

- a significant proportion of foreign migrants from a wide range of countries
- a wide range of socio-cultural and recreational/entertainment opportunities.

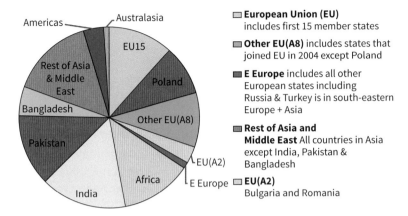

European Union (EU) includes first 15 member states

Other EU(A8) includes states that joined EU in 2004 except Poland

E Europe includes all other European states including Russia & Turkey is in south-eastern Europe + Asia

Rest of Asia and Middle East All countries in Asia except India, Pakistan & Bangladesh

EU(A2) Bulgaria and Romania

Figure 11.16 Pie chart showing the country of origin of immigrants in Birmingham in 2009.

Migration to Birmingham

Birmingham grew as a free-trade city and drew in people from the surrounding countryside, nearby cities and increasingly further afield. Migrants continue to come to the city to work and add to the economic and cultural growth of the city.

How has migration shaped the character of Birmingham?

Birmingham has a long history of welcoming migrants from overseas. In the late 19th century Jewish people fleeing persecution from Russia arrived in the city and this was repeated during the Nazi Holocaust in the 1930s. Many Polish refugees arrived in Birmingham during the Second World War and the post-war period saw an era of economic migration from British Commonwealth countries in the Caribbean and the Indian sub-continent. More recently the city has seen economic migrants from newer members of the European Union.

Migration has given the city a rich cultural heritage and has seen the development of successful multi-cultural communities. Evidence of this is seen throughout the city with ethnic restaurants (Figure 11.17) and cultural events as well as in specific locations, for example, skill-related business areas such as the Jewellery Quarter and the Polish Catholic Centre, built in Digbeth in 1947. Over the years migrants have added to Birmingham's growing prosperity and sense of tolerance and have helped to shape the city.

ACTIVITY 11.6

1 Examine the evidence to show that Birmingham is an important city, both nationally and internationally.
2 Why is Birmingham seen as a good place to start up a new business?
3 Why are national and international migrants often attracted to cities?
4 What is meant by 'socio-economic opportunities'?

Further research

For a short history of migration to Birmingham, visit the the **Birmingham Is Our Home website** (www.cambridge.org/links/gase40078).

Did you know?

In general, migrants are important to cities because:

- 78 per cent of immigrants locate in cities
- 77 per cent of all immigrants with a degree locate in cities
- migrants form a significant part of the highly skilled workforce in most cities
- cities with a higher percentage of migrants have stronger economies

Source: www.centreforcities.org

Figure 11.17 Birmingham's Chinatown contains a large number of restaurants, cafes and clubs.

How urban change has created opportunities in Birmingham

Recent developments in Birmingham have created a wide range of social and economic opportunities. The following article identifies some of the factors that have made Birmingham such a vibrant multi-cultural society.

With five universities and over 50 000 students Birmingham is a young, dynamic city which offers a wide range of social and economic opportunities. It has the largest number of businesses and new business start-ups outside of London, offering a huge range of job opportunities, including retail, leisure, finance, manufacturing and technology and research and development. Birmingham University BioHub, a biomedical research hub, and the Innovation Birmingham Campus are examples of the type of education and employment opportunities found in the city. The excellent transport infrastructure, with rail links to all major UK cities and an international airport puts Birmingham at the heart of the country.

Not only is Birmingham a great place to learn and work, it is also a great place to live. It has a strong historical culture, the city centre museum and art gallery house over 2000 exhibits and the Birmingham Hippodrome is the most visited theatre in the UK. In addition there are a wide range of sporting and music venues, hosting international events and world famous stars. A number of the old, city centre industrial areas have been regenerated, the canal basins now provide a new focal point for cafes, bars and clubs. Also, the city has the most Michelin starred restaurants outside of London offering cuisines from around the world. For a shopping experience Birmingham is the place to come – the Bullring shopping centre is one of the biggest in the UK and is home to all of the major department stores (Figure 11.18).

Birmingham is one of the greenest cities in Europe, with over 500 parks and open spaces, including the world famous Botanical Gardens. As a local city centre resident commented 'There is such a wide variety of things to do in the local area, partly because of the large communities of immigrants, all of which generate a lot of community events which everyone can enjoy. And the range of ethnic restaurants available is amazing.'

Figure 11.18 The Bullring shopping centre is the curving building on the right. It contains more than 160 shops and has an average of 750 000 visitors each week.

An integrated transport system for Birmingham

As the population of Birmingham has increased, there has been greater pressure on its transport systems. To cope with this, in 2014 Birmingham City Council announced a 20 year plan to create a fully integrated transport system. An integrated transport system involves coordinating different forms of transport, such as rail, bus and tram, to provide a single, easy to use system. The 'Birmingham Connected' transport plan will include:

- improved rail links across the city and beyond
- increased connectivity between road, rail and bus links
- a new train system
- development of cycle routes and walkways.

The improved transport system should benefit Birmingham residents and visitors.

Environmental opportunities: making Birmingham green

Urban greening is the process of creating and maintaining green space in urban areas. In order to increase the amount of green space in Birmingham, the Birmingham City Council have put forward a plan called 'Birmingham's Green Vision'. The plan aims to make Birmingham a leading green city which will:

- make the city a more attractive place in which to work and live
- encourage business and create prosperity
- use open space to encourage social interaction and good health.

What is being done to achieve this?

- tree planting and the development of urban greenways (Figure 11.19)
- development of green roofs and walls
- enhancing walkways and cycle tracks
- creating blue corridors alongside canals and rivers (Figure 11.20)
- developing parks and recreational spaces
- encouraging walking and cycling and other outdoor leisure activities.

Figure 11.19 Trees along the Eastside City Park. Eastside City Park is a six acre park that opened in 2013.

Figure 11.20 Blue corridors in the Longbridge redevelopment scheme.

ACTIVITY 11.7

1 Explain how Birmingham has created a wide range of employment opportunities.

How urban change has created challenges for Birmingham

Urban change can create a number of socio-economic and environmental challenges. Declining industry can lead to the development of **brownfield sites**, which fall into **dereliction**. Rising unemployment and poverty can create a 'spiral of decline' where living conditions fall, people move away because of a lack of opportunities and the reputation of an area declines. This discourages new industry from setting up in the area, resulting in the situation becoming worse. This leads to what is known as urban deprivation.

What is meant by urban deprivation?

Urban deprivation is often defined as having a quality of life below what is recognised as acceptable in a particular country. It is easy to recognise urban deprivation in cities in poor countries, where large slum areas with sub-standard housing and problems of disease are easily seen. However, there are also significant differences in living conditions within urban areas in wealthy countries. In urban areas in the UK it is possible to see modern office buildings and expensive apartments alongside run-down housing, boarded-up shops and areas of dereliction (Figure 11.21).

What determines a person's quality of life?

The term 'quality of life' is often used to describe the conditions in which people live. It includes information about economic conditions (e.g. income, unemployment rate, percentage of home ownership), social conditions (e.g. housing quality, healthcare access, quality of schools) and environmental conditions (pollution levels, amount of green space, amount of vandalism).

How is urban deprivation measured?

Every country measures urban deprivation in a slightly different way. In England, a 'multiple deprivation index' is used, where information is collected about a number of quality of life factors and then put together to create an index. This information is then shown on a deprivation map.

In 2015, Birmingham was ranked the third most deprived city in England, after Liverpool and Manchester. While all areas show signs of deprivation, it is most heavily clustered near the city centre.

Environmental challenges

Urban growth leads to a number of environmental challenges.

Building on brownfield and greenfield sites

As the population in Birmingham increases, so does the demand for housing. In 2015 it was estimated that Birmingham needed 89 000 new houses, but the city has only identified space for 51 100 new houses. Building new houses usually means building on either brownfield or greenfield sites.

Brownfield sites are sites that were previously used for industry but have fallen into disuse. They are valuable for building on because they are often large sites in urban areas and developing them usually improves an unsightly area. However, it can be expensive to build on brownfield sites because they often need clearing first.

Greenfield sites are sites that have not previously been built on, often farmland or countryside. They are usually cheaper to build on than brownfield sites, but their use leads to urban sprawl and the problems that come with it, such as traffic congestion.

Key terms

brownfield sites: land previously used for industry, which has fallen into decay

dereliction: land and buildings that have fallen into disuse

Figure 11.21 Urban deprivation in Birmingham.

Factfile

Quality of life factors in the English Multiple Deprivation Index

- income
- employment (rate and type)
- levels of health and disability
- levels of education/skills/ training
- quality of housing and services
- quality of the environment
- level of crime

Further research

Visit the **Birmingham County Council website** (www.cambridge. org/links/gase40080) and search for the Deprivation in Birmingham report from December 2015.

Waste disposal

As the population and number of businesses in Birmingham increase, the amount of waste generated also increases. Waste disposal is an environmental challenge. In 2015, only 30 per cent of Birmingham's waste was recycled, but Birmingham City Council are aiming to reach 40 per cent by 2026.

Birmingham City Council use a range of strategies to manage waste, including:

- **Household waste recycling centres.** There are collection points for recyclable materials across Birmingham and from these points recyclable waste is taken to one of five specially built recycling centres. In 2014, over 52 000 tonnes of rubbish was recycled, composted or diverted from landfill for further use.
- **Energy recovery.** Birmingham operates a state-of-the-art Energy Recovery Facility (ERF) which takes 350 000 tonnes of rubbish per year and converts it into electricity. However, the facility uses incineration to convert the waste, which produce huge amounts of carbon dioxide.

The impact of urban sprawl on the rural-urban fringe

The rural-urban fringe is the area on the edge of the urban landscape when urban and rural land uses are often mixed. It is an important area because it is under pressure from **urban sprawl** and is seen as a prime area for development.

There are a range of land use demands on the rural-urban fringe, including:

- the development of transport networks
- the growth of suburban residential developments (**suburbs**) and **commuter settlements**
- edge of town retail parks
- modern industrial estates the development of greenfield sites, which were previously farmland or open countryside
- leisure developments such as golf courses.

The growth of commuter settlements

Over the past 40 years most UK cities have seen a drift of population from the city centre towards suburbs and commuter settlements. In the last 10 years it is estimated that 42 000 people have moved from the centre of Birmingham to the outskirts of the city. The result of this has been a general increase in the number of people commuting to Birmingham each day for work, which puts increasing pressure on road networks and public transport and leads to traffic congestion and air pollution. The 2011 Census reported that the number of daily commuters was between 150 000 and 200 000.

ACTIVITY 11.8

1. Explain what is meant by the 'spiral of decline'.
2. Explain why quality of life is a mixture of social, economic and environmental factors.
3. Explain how quality of education, income and level of health might be linked.
4. Why might measuring urban deprivation be difficult?
5. Suggest how the Smethwick Action Plan might help to deal with the 'five key areas' of concern in Smethwick.

Did you know?

The World Health Organization identified these requirements to reduce urban deprivation:

- improve health and education
- improve housing quality
- reduce the inequalities in income.

Key terms

urban sprawl: the expansion of an urban area into surrounding, less populated areas

suburbs: largely residential (housing) areas in the outer part of a city

commuter settlements: towns where a significant proportion of residents work elsewhere

Usual residence	Number
Solihull	29 458
Sandwell	28 088
Walsall	16 037
Dudley	14 057
Bromsgrove	9 996
Lichfield	6 076
Wolverhampton	5 842
Tamworth	4 672
Coventry	4 472
North Warwickshire	4 238

Table 11.3 Top ten commuting areas into Birmingham (2011 Census).

Urban regeneration in Birmingham

Birmingham – the UK's second largest city

The West Midlands conurbation developed as a major industrial area, with the city of Birmingham known as 'the city of a thousand trades' because of the wide range of manufacturing industries found there. An example of the scale of manufacturing in the area can be seen in the former car assembly plant at Longbridge which produced cars for nearly a hundred years, reaching a peak of production in 1965 when over a third of a million cars were produced. During this time over 25 000 people were employed at the assembly plant and it had its own rail link which was used for moving goods in and out of the area. From the 1970s increasing competition from abroad led to the decline of a number of industrial and commercial areas in the West Midlands, resulting in significant job losses and some areas becoming increasingly derelict. An example of this was the Longbridge car assembly plant which closed in 2005 (Figure 11.22).

In the last 20 years, Birmingham has been transformed by a number of **regeneration** projects, with some being completed at the moment or planned for completion by 2030. Part of the transformation of Birmingham includes the Big City Plan, which will include the regeneration of five areas in or close to the city centre (Figure 11.23).

Figure 11.22 The former Longbridge car assembly plant.

Key terms

regeneration: improving the economic, social and environmental conditions of previously run-down areas

mixed land use: a mixture of uses such as business, leisure, residential

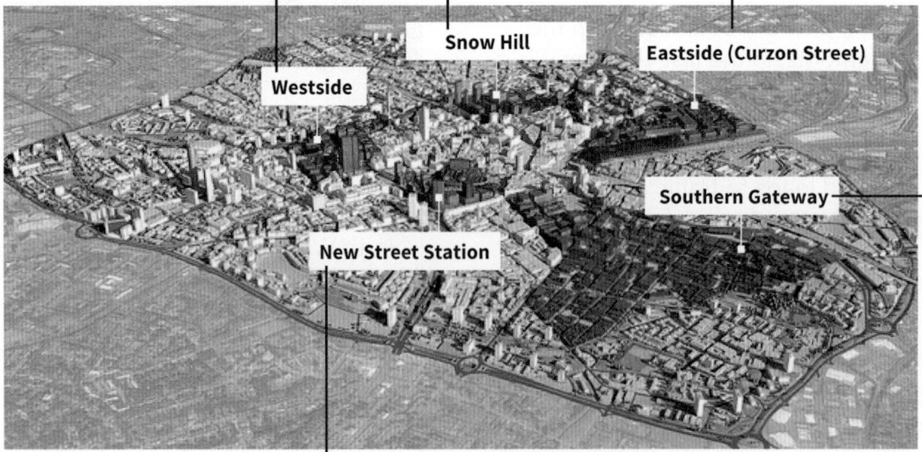

- mixed-use office/commercial space
- new 'library of Birmingham'
- 4* hotel with 250 bedrooms

- 200 000 m² office space
- improved public spaces and transport links
- up to 4000 new homes
- creation of 10 000 jobs

The area where the HS2 (High Speed 2) station will be built.
- 600 000 m² of new business space
- development of Birmingham City University
- a city park and recreational facilities
- 2000 new homes
- creation of over 30 000 jobs

Snow Hill

Eastside (Curzon Street)

Westside

Southern Gateway

A £1 billion **mixed land use** regeneration project
- development of retail markets
- 1000 new homes
- city centre park/walkways
- leisure/recreation facilities
- creation of 3000 jobs

New Street Station

'Turning New Street Station into a world-class station.'
- a large concourse with more commercial facilities
- improving links with the surrounding area
- improving the physical environment of the area
- a new John Lewis store creating 650 jobs

Figure 11.23 Birmingham regeneration schemes.

Further research

These are just some of the regeneration schemes in Birmingham. Use a web search to investigate others. An example is the Fort Dunlop building to the north of the city of Birmingham, which in the 1950s was the largest factory in the world. In the 1970s it closed down and then lay empty for 25 years. Describe how the area was regenerated and the types of new business established. Describe how socio-economic and environmental conditions were improved.

The Longbridge Regeneration Plan

The Longbridge Regeneration Plan aims to completely regenerate the former MG Rover car plant site on the southern edge of Birmingham (Figure 11.22), bringing life back to a former derelict brownfield site by building over 2000 new homes and creating 1000 jobs.

The plan will include:

- a technology park (Figure 11.24), including an innovation centre suited to small technology businesses (Figure 11.25)
- a £70 million town centre with a number of large national stores
- hotel and leisure developments, including restaurants and cafés
- Bourneville College (Figure 11.25) – a £66 million learning facility (Figure 11.24)
- a range of office accommodation to suit different size businesses
- residential developments to suit different age ranges
- large industrial and distribution centre buildings (warehouses), within easy reach of the local road networks.

ACTIVITY 11.9

1 Why are regeneration projects needed in some areas?
2 Why are regeneration projects called 'mixed-use projects'?
3 Suggest why planners are trying to encourage people to live in city centres.
4 Explain how the Longbridge Regeneration Plan will improve social, economic and environmental conditions for local people.

Investigate 'Urban Splash' (see www.cambridge.org/links/gase40081), a company that specialises in regeneration projects, turning disused buildings into business or residential spaces.

Factfile

The Birmingham Big City Plan aims to create a 'world class city centre'. It will include:

- 1.5 million m² of commercial floor space
- 50 000 new jobs
- 5000 new city centre homes
- improved leisure and recreational facilities
- walkways and cycle paths

Figure 11.24 Longbridge Technology Park and town centre.

Figure 11.25 An artist's impression of Bourneville College, Longbridge.

Further research

Find out more about the Longbridge Regeneration Plan on the **St Modwen Longbridge website** (www.cambridge.org/links/gase40082)

11.5 Sustainable urban living

The challenge of making urban areas more sustainable

Urban areas are often seen as the 'engines for growth', especially in LICs and NEEs. They offer the opportunity for employment, wealth and access to services. However, urban growth can also result in traffic congestion, increasing pollution, the growth of slums and the exploitation of resources such as energy and water. The aim of sustainable urban living is to manage all of these factors by taking advantage of the positive points about urban living while trying to reduce the negative points. In order to achieve this, urban areas will need to:

- have planned **green space**
- conserve water and energy resources to ensure that they are used efficiently
- create manageable amounts of waste and have effective waste recycling facilities
- have efficient public transport
- have high quality working and living environments.

The following examples describe a number of sustainable urban living initiatives.

Urban greening

Urban greening is about increasing the amount of green space in an urban area. It can be done in a number of ways, including:

- developing green gardens on the roof of buildings
- tree planting alongside roads, railway lines (Figure 11.26), canals and waterways
- developing open spaces and parks.

Why develop green spaces in urban areas?

Developing green space in urban areas is not just about making an area more attractive. It also has a number of other advantages, including:

- reducing heat in the summer
- reducing the risk of flooding
- increasing biodiversity
- improving health
- creating social areas
- providing more attractive areas for business
- creating recreation and tourism opportunities.

The following example describes the development of Tianjin Eco-city in China.

Tianjin Eco-city – China

In the last 30 years China has seen massive urban growth as millions of people have moved towards cities seeking employment opportunities. China faces the challenge of providing urban areas that are economically and environmentally sustainable. Tianjin Eco-city is one example of how they are doing this (Figure 11.27).

The chosen site for the Eco-city was a derelict industrial area with high levels of ground and water pollution. Cleaning up the area took three years and has been very successful. On a recent visit to the city a reporter said, 'As you enter the city you notice the green spaces and tree-lined streets. Solar panels can be seen on many buildings and there are a number of wind turbines. All of the public buildings have the most up-to-date technology to save energy and water, including motion sensitive lights and timed running water systems. All water is recycled and methane is collected from the sewage plant and used to produce energy.'

Key terms

green space: parks and vegetated areas and walkways

carbon neutral: does not add carbon dioxide to the atmosphere

sustainability: meeting the needs of today without harming future needs

Did you know?

Urban areas can be as much as 5 °C hotter than the surrounding countryside. This is called the 'urban heat island effect'.

Figure 11.26 New York High Line – a 2 km linear park built on an elevated section of disused railway line.

Further research

Investigate the sustainable cities index for UK cities and the green City Index for Global Cities.

Figure 11.27 Tianjin Eco-city.

Did you know?

The Tianjin Eco-city project, 150 km from China's capital Beijing, will be **carbon neutral** and energy efficient. Its aims are environmental preservation, conservation and **sustainability**. The project is expected to be completed by 2020 at a cost of nearly £101 billion. When completed it will be home to 350 000 people.

Find out more sustainable building projects such as BedZED on the Bioregional website (see www.cambridge.org/links/gase40083), the zed factory (see www.cambridge.org/links/gase/40084) and greenhouse in Leeds (see www.cambridge.org/links/gase40085)

ACTIVITY 11.10

1 What does an urban area need to make it increasingly sustainable?
2 Explain how urban greening is good for people and the environment.
3 a How does Tianjin Eco-city conserve water and energy?
   b How does Tianjin Eco-city promote waste recycling?
   c Do you think that Tianjin Eco-city is completely sustainable?
   d Would you like to live in Tianjin Eco-city? Explain your answer.
4 Suggest how the amount of green space might be increased in your local area.

Factfile

Key features of Tianjin Eco-city

- 20 per cent of all energy will be renewable, including solar and geothermal.
- All buildings will be energy and water efficient.
- A minimum of 50 per cent of water supply will come from desalination and recycled water.
- Over 60 per cent of waste will be recycled.
- 90 per cent of local journeys will be taken by walking, cycling or public transport.
- Electrical vehicle use will be encouraged.
- There will be planned parks and green walkways throughout the city, encouraging wildlife and recreation.
- A mix of ages and professions will be encouraged to create a positive social mix.

Assess to progress

1 Describe the 'push-pull' theory of migration. **4 MARKS**
2 Suggest why the growth of megacities is largely in LIC/NEE countries. **4 MARKS**
3 Explain how urban transport strategies are being used to reduce traffic congestion. **6 MARKS**
4 Using an example of an LIC/NEE city you have studied, describe the challenges of urban growth. **6 MARKS**
5 Using an example you have studied, explain how urban planning is improving the quality of life for the urban poor in an LIC or NEE. **9 MARKS**

Download Worksheet 11.2 from Cambridge Elevate to help you find out more about the Greenhouse eco-development in Leeds.

11.6 Managing traffic congestion in cities

The growth of traffic is a major challenge for many cities around the world. As cities grow, both in population and size and levels of car ownership increase, there is increasing pressure from traffic congestion (Table 11.4).

Traffic congestion causes delays which increase business costs and puts increasing stress on commuters. It also adds to the problem of air pollution and the pressure to build more roads.

Cities in LICs and NEEs		Cities in HICs	
Beijing (China)	100	London (UK)	36
Cairo (Egypt)	80	Los Angeles (USA)	28
Jakarta (Indonesia)	120	Madrid (Spain)	46
Mumbai (India)	85	Milan (Italy)	52
São Paulo (Brazil)	76	New York (USA)	20
Mexico City (Mexico)	100	Stockholm (Sweden)	16

Table 11.4 Commuter pain – the average time spent stuck in traffic each day, in minutes.

Traffic congestion adds to pollution problems in Los Angeles

Los Angeles is a sprawling city of 16 million people on the west coast of the USA. On most workdays the roads are choked with traffic as commuters drive to work towards the city centre or the large industrial areas across the city (Table 11.5). At certain times of the year smoke and fumes from vehicles and industry get trapped under the warm, calm air, causing photochemical smog (Figure 11.28). This appears as a grey-green haze over the city and can affect people's health and damage buildings.

Figure 11.28 Smog over Los Angeles, USA.

Date	Vehicle numbers in Los Angeles (approx.)
1980	3 800 000
1985	4 400 000
1990	5 100 000
1995	5 900 000
2000	6 500 000
2005	7 500 000
2010	7 500 000
2015	7 400 000

Table 11.5 Number of vehicles in Los Angeles, 1980–2015.

Key terms

public transport: shared methods of travelling, such as buses, trams and trains

integrated transport system: where all parts of the transport system link together to make journeys more efficient

Further research

Investigate the problems of congestion and air pollution in urban areas in LICs and NEEs.

Reducing traffic congestion in urban areas

The following examples illustrate different methods used in urban areas to reduce traffic congestion.

Park and ride

Park and ride is a system used in many cities for reducing urban traffic congestion. Drivers leave their cars at car parks on the outskirts of a city and travel to the city centre using a dedicated **public transport** system.

Advantages of park and ride:

- It reduces traffic in city centres.
- Public transport has less of a negative impact on the environment.
- It reduces the need to build more roads and car parks in the city centre.

Disadvantages of park and ride:

- It may increase traffic in the areas of the park and ride car parks.
- It requires the building of large car parks on the outskirts of towns and cities.

Manchester Metrolink

Manchester Metrolink connects a number of towns and suburbs to Manchester city centre and links up commercial areas such as Salford Quays and Old Trafford, the home of Manchester United football club. Trams run every 6 minutes (Figure 11.29) and extra units are coupled together to take extra passengers during the rush hour. All the stops have shelters and electronic signboards. In 2014 there were 80 000 daily journeys on the system, covering a total of nearly 200 million km. In 2015 the Manchester Airport line was opened and a second line through the city is planned.

Curitiba's (Brazil) Bus Rapid Transport (BRT) system

Curitiba, a city of 3 million people in the south of Brazil was the first Brazilian city to have dedicated bus lanes as part of its **integrated transport system**.

The Bus Rapid Transport (BRT) system has four elements:

- direct line buses operate from key points and run directly into the city without stopping
- speedy buses operate along five main routes into the city and have limited stops
- inter-district buses join up districts without going through the city centre
- feeder mini-buses pick up people from residential areas and take them to terminal points where they can connect to the main services (Figure 11.30).

The system has a smartcard payment system and bus terminals all have shelters with shops, a café and local bank. Increasingly the buses run on biofuel which creates less pollution. So, how effective is the system?

- Nearly 1.5 million passengers travel per day.
- 80 per cent of Curitiba's commuters use buses.
- There are an estimated 35 million fewer car journeys per year.
- Roads are less congested and safer, and the air is less polluted.

Figure 11.29 Manchester Metrolink tram.

Figure 11.30 Bus terminal in Curitiba.

ACTIVITY 11.11

1. a On a blank world map, construct a proportional bar graph to show the 'commuter pain' information in Table 11.4.
   b Explain why traffic congestion causes economic and environmental problems.
2. Suggest why some cities have a greater traffic congestion problem than others.
3. Manchester Metrolink was 'designed for both commuters and leisure users'. Explain this statement.
4. Why is the Bus Rapid Transport system in Curitiba called an integrated transport system?
5. How would you judge the success of a traffic management system?

 Further research

Investigate other methods of reducing traffic congestion such as cycle networks or congestion charging, for example the proposed Leeds Trolleybus system (see www.cambridge.org/links/gase40086).

 Download Worksheet 11.3 from Cambridge Elevate for help with Activity 11.11, question 1.

11.7 City centre regeneration: Birmingham, UK

In this issue evaluation you will consider whether or not the Jewellery Quarter regeneration in Birmingham city centre meets the needs of the local population.

The Big City Plan is a major regeneration plan for the Central Business District (CBD) of Birmingham, UK. The plan has five key areas and outlines how Birmingham will be developed over the next 20 years at a cost of £10 billion.

Five key areas of regeneration:

- **Westside** – to include the redevelopment of Paradise Circus, Baskerville Wharf and Arena Central. There will be a new Birmingham library and an enhanced setting for the Town Hall and Council House.
- **Snow Hill District** – there will be better pedestrian links across Great Charles Street so that people can get from the City Core to the Jewellery Quarter more easily.
- **Eastside** – the City Core will be expanded in the east. The focus will be the Eastside City Park and, the planned terminus of HS2 at Curzon Street will allow further regeneration in this area.
- **Southern Gateway** – here, the City Core will be expanded in the south and will mainly be a retail area specialising in markets and food.
- **New Street** – the redeveloped New Street station opened in 2015 and will act as a catalyst for wider regeneration.

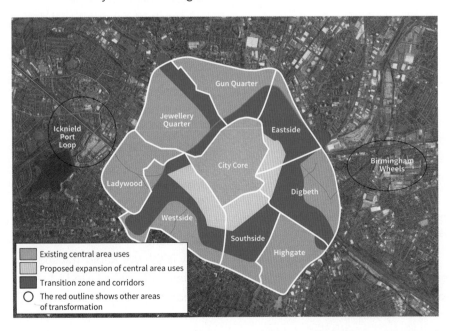

Figure 11.31 The location of the development districts in Birmingham's Big City Plan.

Factfile

The Big City Plan

- The plan aims to improve transport links to ensure that people can travel to Birmingham efficiently.
- New Street station has recently been redeveloped at a cost of £600 million.
- There is a proposal for a new HS2 terminal station to be built at the currently derelict Curzon Street station. The size of the CBD core will increase by 25 per cent. The inner ring road, which acted as a barrier to expansion, has been removed allowing the development of seven distinct districts (Figure 11.29): the City Centre Core, Eastside, Digbeth, Southside and Highgate, Westside and Ladywood, the Jewellery Quarter and the Gun Quarter.
- It is estimated that the proposed plan will involve the building of 5000 new homes and the creation of 50 000 new jobs, mainly in construction.

An introduction to the Jewellery Quarter

The Jewellery Quarter is located in the south of the Hockley area of the city centre (Figure 11.32) and makes approximately 40 per cent of the UK's jewellery. At its peak in the early 1900s over 30 000 people were employed in the area.

However, in the 1920s, foreign competition and the lack of demand for quality jewellery led to industrial decline. Before the removal of the inner ring road the area was isolated from the commercial centre of Birmingham.

Figure 11.32 An aerial image of Birmingham.

However, this artificial barrier meant that the Quarter was protected from previous waves of redevelopment and much of the Georgian and Victorian architecture remains and is listed. This means that it is legally protected in order to protect it for the future and so there may be restrictions on how the buildings can be developed. The area has a high percentage of derelict and vacant sites but due to the large number of grand buildings it offers great regeneration potential.

Jewellery Quarter redevelopment

The Jewellery Quarter has already seen the creation of the £1.5 million Golden Square which was designed to be a focal point for tourists and shoppers, and Newhall Square which is a mixed-used development fronting the canal. The latest proposed development in the Jewellery Quarter, which began in 2015 and will take at least a year to complete, has been spearheaded by the company Seven Capital who are financing an £80 million residential scheme, St George's urban village. This is likely to create more than 600 new homes, including 300 'loft-style' residential apartments in the Kettleworks, a disused warehouse, and 600 construction jobs over two years (Figure 11.33).

Locals oppose the redevelopment

While many are pleased that the once derelict brownfield site is being redeveloped, there are others, mainly locals, who are concerned that the exclusive apartments will not offer an affordable option for first-time home buyers.

'With a typical rent of £700 per month for a one bedroom flat in the area, there is no way that I'll be able to afford to live here. I expect that there will be an influx of affluent, young professionals which, in turn, will encourage the

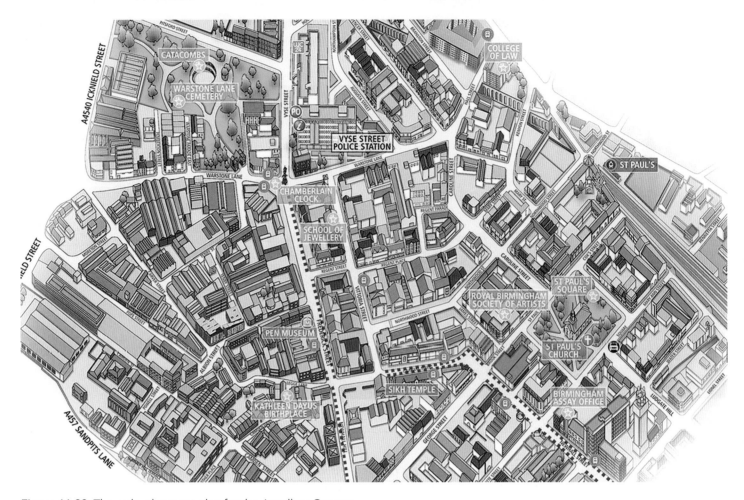

Figure 11.33 The redevelopment plan for the Jewellery Quarter.

opening of expensive bars and restaurants. I'd rather have a laundrette and a supermarket but I know that those services will not be popular with the people that move in.'

Local resident.

Already, several designer coffee shops, restaurants and gyms have opened on St Paul's Square. Despite assurances from architects developing the Jewellery Quarter that the visual character of the historic buildings will be maintained there is concern that the social character of the area will change irreversibly.

Plans for community services

Currently, approximately 3000 people live in the Jewellery Quarter. The needs of the local community include services such as schools, nurseries, healthcare and supermarkets. It is also important that the area offers jobs once construction is completed. An interesting policy of creating live/work units (where people live and work in the same place) is in place and concentrated in the Jewellery Quarter. There are 187 live/work units which are mainly one or two bedroom properties located on upper floors of buildings. A 2013 survey suggested that this policy had been fairly successful given that the primary aim was to limit the impact of residential development in the Jewellery Quarter.

There has been minimal residential development and much of the current development is industrial or commercial. However, there is now an oversupply of one or two bedroom properties and a shortage of high quality commercial and workshop space.

Despite the concerns of locals, there are some developments in the Jewellery Quarter that are likely to have a community focus. The Ruskin Mill Trust has started work on turning the previously derelict Standard Works building into a community hub for 16 to 25 year olds with Special Educational Needs. Named Argent College, it will allow students, many of whom have Asperger's Syndrome, to learn a trade at the same time as developing social and personal skills. The building will be mixed use and include a café, bakery, jewellery making facilities, a performance space and a therapy area.

Assess to progress

1 Outline two aims of the redevelopment of Birmingham's CBD. `2 MARKS`

2 Suggest one concern that locals may have about the Jewellery Quarter redevelopment. `1 MARK`

3 What is meant by the following terms:
   - brownfield site
   - Central Business District? `2 MARKS`

4 'Birmingham's Jewellery Quarter should be redeveloped to ensure that it serves the whole of the local community.'
   To what extent do you agree with this statement?
   Give reasons for your answer. `9 MARKS + 3 SPaG MARKS`

Tip

For question 4, you first need to decide the extent to which you agree with this statement. Do you completely agree, completely disagree or are you somewhere in between? Make sure you can use the available evidence to explain why you think this.

For more information see the **Jewellery Quarter website** (www.cambridge.org/links/gase40087)

Tip

It is always a good idea to make sure that you understand all of the terms that are used in the issue evaluation resource booklet. You could make an illustrated glossary of words you are not familiar with.

The changing economic world

12 The development gap
In this chapter you will cover:
- how we classify the world based on economic development and quality of life
- why development is uneven
- the causes and consequences of uneven development
- how the development gap can be reduced
- the impacts of rapid economic development in Malaysia

13 Economic futures in the UK
In this chapter you will cover:
- the causes of economic change in the UK
- how UK industry has changed
- how UK's transport infrastructure has changed
- how we can reduce the impact of industry on the physical environment
- strategies used to resolve regional differences in the UK
- how the UK fits into the global jigsaw

The changing economic world – an overview

There are tremendous global variations in the level of economic development. In some parts of the world people have high average incomes, access to facilities such as clean water, sanitation and energy and access to services such as education and healthcare. In other parts of the world people live on very low incomes and do not have access to these facilities and services. This is known as the development gap. The United Nations are committed to reducing this gap and ensuring that everybody has appropriate living conditions.

The last 30 years have seen increasing levels of globalisation as lower income countries (LICs) and newly emerging economies (NEEs) move towards becoming major industrial producers and exporters. This has created social and economic opportunities in these countries and allowed living conditions to improve. At the same time highly developed countries like the United Kingdom have gone through a period of industrial adjustment as traditional industries decline and they move towards a post-industrial period of regeneration with the development of science/technology based industries and expansion of service activities.

Figure S5.1 Technology has a massive impact on economic development. Automation increases productivity but can also lead to a loss of jobs. Advances in communications have allowed companies to reach more consumers and also access employees around the world. What other examples can you think of?

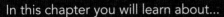

12 The development gap

In this chapter you will learn about...

- the different ways of measuring a country's level of development
- the different ways of classifying stages of development
- development indicators and their limitations
- the link between the Demographic Transition Model and development levels
- the causes and consequences of uneven development
- strategies for reducing the global development gap.

12.1 How do we measure levels of development?

What do we mean by development?

Development refers to the way that social and economic conditions improve in a country over time. All countries are developing – 70 years ago few people in the UK could afford to own a car, and yet today most families own at least one car. This is a measure of how wealth in the UK has increased and how improvements in technology have made manufactured goods relatively less expensive. In the UK average incomes are high and virtually everyone has access to basic services such as clean water and electricity. At the same time people in the UK also have free access to education and healthcare. This is not the case in all parts of the world. While some countries are wealthier than the UK, the majority are poorer, and in many of these countries people live in extreme poverty, with little access to many of the facilities that people in the UK take for granted.

The **development gap** therefore exists between countries, which means not only differences in economic wealth but also in quality of life – access to education (Figures 12.1 and 12.2) and health, the position of women in society, safety and welfare.

 Key term

development gap: differences in level of people's total well-being and happiness, physical standards of living and national wealth (Physical Quality of Life Index, GDP per head) between countries

Figure 12.1 Children going to school in Kibera, the biggest squatter settlement in Africa, Nairobi, Kenya.

Figure 12.2 Hi-tech classroom.

We can look at development indicators other than economic indicators, including:

- **gross national income (GNI) per head**
- birth and death rates – the number of deaths/births per 1000 people per year
- infant mortality – the number of babies (<12 months) dying per 1000 live births
- life expectancy – the number of years a person is expected to live, from birth
- people per doctor – population divided by number of doctors available (Figures 12.3 and 12.4)
- adult literacy rate – the percentage of adults able to read and write
- access to safe water
- the **Human Development Index (HDI)**

Key terms

gross national income (GNI) per head: in simple terms, the total value of a country's goods, services and overseas investments, divided by the number of people in that country

Human Development Index (HDI): development indicator combining life expectancy at birth, education and income

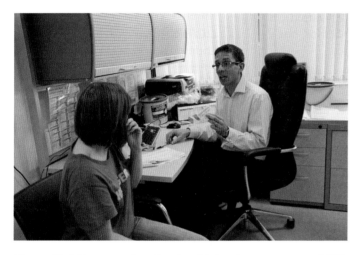

Figure 12.3 Doctor and patient in a higher income country (HIC).

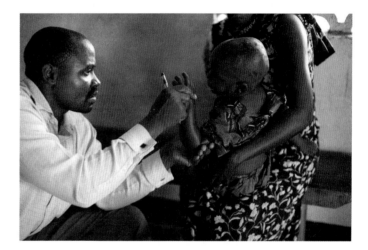

Figure 12.4 Doctor and patient in a lower income country (LIC).

Discussion point

Discuss the links between the different indicators: lack of access to safe water leads to disease, disease can lead to absences from school, a lack of education contributes to a lifetime of poverty. What other links can you see?

ACTIVITY 12.1

1 Look at Figures 12.1 and 12.2.
  a Discuss the positive and negative aspects of each situation in pairs or as a class.
  b Which figure shows the wealthier country? Give your reasons.
2 a Use Figure 12.5 on pages 364–365 to identify ten richer northern and ten poorer southern countries.
  b Compare their gross national income per head (GNI) values. The **Students of the World website** (www.cambridge.org/links/gase40088) is a good place to start.
  c Why might just using GNI to compare countries not always be useful?

Did you know?

The North/South divide is a simple line dividing richer northern countries and poorer southern ones, proposed by Willy Brandt, the ex-German Chancellor, in 1971 as a basis for future equal development.

Tip

Figure 12.5 can be found at the end of the book.

12.2 Classifying countries according to their level of development

Countries can be grouped according to their level of development. In the past this meant **economic development**. The first classification of countries was eurocentric: Europe was the 'first world' and its early rapidly developing colonies (USA and Australia) were labelled 'second world'. Poorer colonies (India and Pakistan) were grouped as 'third world', with the very slowest developing later classified as 'fourth world'. However, all countries are unique and develop differently. It is not realistic to put Brazil, with an average income 20 times that of Bangladesh, in the same third world category so this system was rejected.

In 1971 the world was split into two parts, the North/South divide, with richer countries in the north and poorer countries in the south (see Figure 12.5 on pages 364–365 at the back of the book), but this was too simple. Then the **LIC/HIC** division was used to separate poorer and wealthier countries. A fivefold economic division, including oil-exporting and heavily indebted countries, has been proposed, but is difficult to know where some countries fit.

12.3 What are development indicators?

GNI/head (Gross National Income per person) GNI measures the economic growth of countries. It is a bit different from GNP (Gross National Product), which you may read about in some books and websites. GNI for the UK means all the income earned by UK companies and people wherever they are in the world. It includes the incomes of UK people abroad. For example, a UK citizen living and working in France but paying UK tax counts in the GNI calculation, but would not do so in the GNP sum. Ultimately, what matters is that we have a system of comparing the progress of the world's countries.

Death rate is low across the world. The highest in 2014 was South Africa with 17.49 per 1000 per year. Healthcare is improving in poorer countries, so people are less likely to die of everyday illnesses. However, death rate is not always a good indicator of development as it is linked to the age of the population – a youthful population of a poorer country could have a lower death rate than a wealthier country with an ageing population.

Birth rate is a better indicator of development – as a rule, the more developed a country, the lower its birth rate. People have access to family planning and believe small families have more opportunities. If **infant mortality** is high due to poor health and nutrition, birth rate stays high. Poorer governments cannot provide the education essential for development.

ACTIVITY 12.2

1 Draw a flow diagram showing the consequences of not having clean water. Compare with other members of your class. Combine your ideas in a more complex diagram.
   a Explain why the doctor : patient ratio matters.
   b How does access to healthcare affect quality of life?

Key terms

economic development: a change in the balance between primary, secondary, tertiary and quaternary economic production. Poorer countries rely on primary (raw materials) production. As countries develop economically there is a move towards secondary activities (manufacturing) and then to tertiary (services) and quaternary (research and development activities). As a country develops economically, people's standards of living increase

LIC/HIC: lower/higher income countries

Did you know?

The number of people per doctor varies greatly. In the UK we visit a doctor when we need to – the doctor : patient ratio is 1 : 357 (2013). Malawi and Tanzania are much worse off with a ratio of 1 : 50 000 (2013).

Source: World Health Organization.

Key term

infant mortality: the average number of deaths of babies under 12 months old per 1000 live births

Download Worksheets 12.1 and 12.2 from Cambridge Elevate for additional questions.

The Human Development Index (HDI)

This complex indicator includes **life expectancy** (health), **adult literacy** (education) and GNI per head (living standard). Vietnam and Pakistan have similar GNI per head, but Vietnam's literacy and life expectancy are better. Vietnam's HDI is therefore higher (Table 12.1).

Country	Life expectancy at birth (years) (2013)	Mean number of years in school (2012)	HDI (2013)	HD1 ranking
Norway	81.4	12.6	0.944	1
Canada	81.5	12.3	0.902	8
UK	80.5	12.3	0.892	14
Czech Republic	77.7	12.3	0.861	28
Saudi Arabia	75.5	8.7	0.836	34
Romania	73.8	10.7	0.785	54
Venezuela	74.6	8.6	0.764	67
Kazakhstan	66.5	10.4	0.757	70
Brazil	73.9	7.2	0.744	79
Jamaica	73.5	9.6	0.715	96
Indonesia	70.8	7.5	0.684	108
South Africa	56.9	9.9	0.658	118
Vietnam	75.9	5.5	0.638	121
Pakistan	66.6	4.7	0.537	146
Nigeria	52.5	5.2	0.504	152
Haiti	63.1	4.9	0.471	168
Afghanistan	60.9	3.2	0.468	169
Mali	55.0	2.0	0.407	176
Niger	58.4	1.4	0.337	187
World	**70.8**	**7.7**	**0.702**	**n/a**

Table 12.1 Development indicators for selected countries
Source: UNDP (United Nations Development Programme)

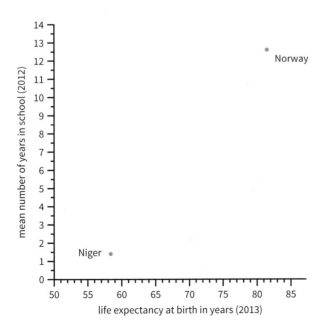

Figure 12.6 A correlation graph showing development indicators.

Key terms

life expectancy: the average number of years a person is expected to live from birth in a particular society at a certain time

adult literacy: the percentage of adults who are able to read and write to a basic functioning level

Further research

1 Look up data for more countries on the Human Development Reports page of the **United Nations Development Programme website** (www.cambridge.org/links/gase40089).
2 Put 'Millennium Development Goals' into a search engine.
   a Make a chart with the eight Millennium Development Goals (MDGs) showing their successes and failures.
   b Now search for 'Sustainable Development Goals (SDGs)' which will build on the MDGs when the programme ends in 2015.

ACTIVITY 12.3

1 Using the data in Table 12.1, copy and complete the graph in Figure 12.6. Two points are done already; choose at least eight countries. Draw a best-fit line to test the relationship between life expectancy and education.
2 Are these indicators correlated? Think about a positive/negative, strong/weak relationship.
3 Explain the pattern on your graph.

12.4 What are the limitations of development indicators?

We use indicators of development to compare the quality of life, economic status and progress of countries. Not all, however, are equally useful.

Life expectancy and associated indicators of development

People commonly assume that LICsmust have higher death rates, but this is often untrue. Although healthcare may be lower quality and more difficult to access, much progress has been made. LICs have young populations who, despite poverty, are less likely to die than the elderly who make up a large proportion of HIC populations. Death rate is therefore a poor indicator of development.

Birth rate, on the other hand, can be useful. Many HICs in Central and Eastern Europe have very low birth rates (see 12.5 What is the Demographic Transition Model?). In contrast, LICs retain traditions of larger families, especially in Sub-Saharan Africa. Birth rate data can be affected by government policies such as China's one-child policy (Figure 12.7).

Figure 12.7 In 1979 China implemented its one-child policy. The Chinese government estimates that the policy has prevented about 400 million births since it began, although this figure is contested.

Infant mortality statistics can be useful. There is an inverse relationship between the wealth of a country's people and their infant mortality rate. This indicator reflects poor nutrition, access to health services and level of education. The main problem is the accuracy of data. Many LIC births are not registered and level of infant mortality may be considerably above published statistics.

Life expectancy links all these indicators together. In general, the higher the life expectancy, the more developed the country. However, data quality can be variable and circumstances change. Wars, natural disasters and disease affect figures in the short term. In HICs, life expectancy has increased steadily but soon may reverse and decrease due to problems associated with obesity. This would imply a country had become less developed, but that would be untrue.

Literacy rates

Literacy rates show the amount of education a government is able to offer its people, which is usually a matter of available funding. Education equals opportunity, improved economic status and quality of life. In the context of the UK, literacy means that a person can read and understand a tabloid newspaper as well as being able to write reasonably. In an LIC, adult literacy links to the percentage of children attending school and the length of time that they attend for. In the UK and other HICs, adult literacy is 99 per cent, but in Somalia adult literacy is as low as 24 per cent.

As an indicator, literacy rate ignores other skills such as farming and practical knowledge, which may be equally valuable to the economy.

GNI/head

GNI/head is a simple measure used to compare the amount that countries earn per year, but it does not tell us much about individual people's standard of living. The main value of this indicator is that it allows us to compare countries of vastly different sizes, such as China with 1.356 billion people and Tuvalu with only 10 782, because the country's income is divided by its population to produce a value per person.

However, GNI/head is a blunt tool for assessing development and standard of living. Many agricultural and craft product values are not included in GNI and similar statistics, and there is no built-in measure of how much a unit of currency will buy. £1 or $1 varies hugely in what it will buy in different countries. Food bought in large markets in West African or Indian cities costs remarkably little.

Which indicators of development are the most useful?

This is an impossible question! Ideally by using several sets of data together we achieve real understanding of people's lives and experiences. Indicators involving several ideas are now popular with geographers.

Increasingly used to assess development, the Human Development Index (HDI) combines life expectancy, education and income adjusted for purchasing power (in other words, it takes into account how much money will buy). It includes both economic and social elements so is more useful for discussing quality of life.

The Physical Quality of Life Index (PQLI) uses only social measures of well-being. People may be economically poor, yet they are educated and live to a good age with reasonable health.

12.5 What is the Demographic Transition Model?

The **Demographic Transition Model (DTM)** explains population change in a country. It has five stages:

Stage 1: Total population is low. High birth rates are balanced by high death rates.

Stage 2: Birth rates remain high but death rates start to fall because of improved healthcare and sanitation. Population increases.

Stage 3: Total population continues to increase but more slowly as birth rates start to fall due to availability of contraception, health education and a desire for smaller families.

Stage 4: Total population is high but balanced due to both birth and death rates being low.

Stage 5: Total population starts to decline due to an ageing population and very low birth rates. Only a few countries have reached this stage and many may never do so.

Key terms

Demographic Transition Model (DTM): the five stages through which a country passes in terms of birth, death rate and natural change; as a country passes through the DTM its economy becomes more sophisticated

newly emerging economy (NEE): a country in the process of change from an LIC (lower income country) to a highly developed, more complex economy

A clear relationship exists between the DTM and economic development (Figure 12.8) – the higher the stage, the greater the development. Stage 1 only exists within the most basic rainforest tribes. Afghanistan represents Stage 2, an extremely poor country with one of the world's highest birth rates (38.84 per 1000 in 2014 – *CIA World Factbook 2015*). With few resources, people need to produce children to support their future. Afghanistan's progress is also held back by political instability.

Malaysia, a **newly emerging economy (NEE)**, is developing rapidly economically (see Table 12.2). Improving living standards mean people can see the advantages of having fewer children. In the most HICs (the USA, the UK and Germany) it is the norm that both parents work, keeping family size small. People take advantage of the opportunities to invest the maximum possible into one or two children. Lithuania and Germany are Stage 5 countries with declining populations.

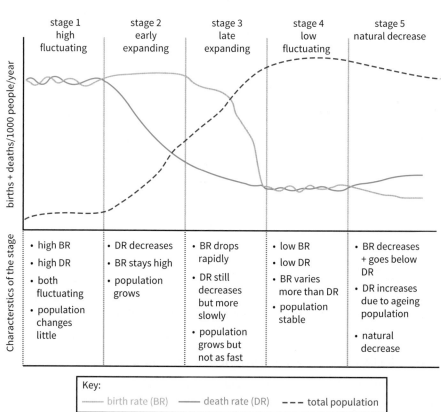

Figure 12.8 The Demographic Transition Model.

	2005	2008	2015
Total population (millions)	23.0	25.3	29.3
Growth rate in previous 5 years	1.93	1.69	1.60
% population aged <15	33.7	31.4	26.1
% population aged >60	6.5	6.7	8.5
Life expectancy M (yrs)	70.8	72.0	72.7
Life expectancy F (yrs)	75.7	76.7	77.3
Fertility rate (number of children/woman)	2.9	2.6	2.0
% population urban	58.1	63.7	75.4
BR (number of births/1000 people/year)	22.6	20.6	17.7
GDP/person ($)	4130	5150	10430
% of workforce in agriculture	18.0	15.0	12.6
% of workforce in industry/ manufacturing	32.0	30.0	28.4
% of workforce in service industries	50.0	55.0	59.0
Energy consumption/person in kg of oil equivalent	2168	2279	2639
% adult literacy	88.7	88.7	93.1
% in primary school	99	100	100
% in secondary school	70	76	No data
Number of hospital beds/1000 people	No data	1.8	1.9
Number of doctors per 1000 people	No data	0.7	1.2

Table 12.2 Malaysia's economic and social change 2005–2015
(Source: The Economist Pocket World in Figures (2005, 2008 and 2015)

Birth rate can indicate the stage a country has reached in the DTM:

Stage 1: birth rate = 40–50/1000/year – death rate = c.50/1000/year

Stage 2: Côte d'Ivoire (Ivory Coast): birth rate = 36.9/1000/year (2015) – death rate = 14.3/1000/year

Stage 3: Malaysia: birth rate = 17.7/1000/year (2015) – death rate = 4.7/1000/year

Stage 4: UK: birth rate = 12.2/1000/year (2015) – death rate = 9.4/1000/year

Stage 5: Germany: birth rate = 8.5/1000/year (2015) – death rate = 10.9/1000/year

ACTIVITY 12.4

1 Use Figure 12.8, the Demographic Transition Model, plus the birth rate data above, to describe:

a the change in birth and death rates as a country passes through the DTM

b the likely pattern for Malaysia's population growth in the future.

Skills link

- Looking at a large table of data like Table 12.2 can be confusing. The best way to identify changes in economic and social indicators is to draw graphs. Here are some suggestions of suitable methods:
- Use three pie charts, one for each year represented, to show percentages of workers in agriculture, industry and services. You could make the size of each pie proportional to the national population in that year. Changes over time will be easy to identify.
- Compound bars, one for each year, would reveal changes in percentages of population in each age category over time.

12.6 Why is development uneven?

Resources are not evenly distributed over the Earth. Physical resources are naturally available: oil, wind energy, water, soil, climate, vegetation, minerals. Human resources mean people's labour and their improving technology to enable fuller exploitation of physical resources.

Physical and historical causes of uneven development

Location

Physical geography in some countries does not favour development. Of the 54 countries in Africa, 16 are landlocked, including Mali, Central African Republic, Zambia and the continent's newest country, South Sudan (Figure 12.9). Not having a coastline makes trade difficult. **Landlocked countries** depend on neighbours to import/export goods. South Sudan can only sell its oil if North Sudan cooperates over the pipeline to the coast (Figure 12.10). From January 2012 to March 2013 South Sudan halted its oil production in a dispute over the amount of money it had to pay North Sudan to move its crude oil through North Sudan's pipeline. This act damaged the economies of both countries and nearly brought them to war.

Historical events

The development of a country is often affected by events in its past, such as conflict or colonialism. Wars can disrupt development in a country, both by destroying infrastructure and services and also causing potentially catastrophic loss of population. In 2013, an estimated 28.5 million primary school age children living in conflict-affected areas were unable to attend school. The disruption to education limits the opportunities of people long after the conflict has ended.

Colonialism has had both positive and negative effects on the development of some countries. Between the 1870s and 1900, many European countries began to take control of areas in Africa and Asia in order to gain access to abundant natural resources. This necessitated the building of infrastructure in the colonies, including railways, roads and port facilities. The local population also benefited from these, as well as hospitals and schools which were built for them.

Wealth derived from raw materials went mainly to the colonial powers, so the colonies remained poor. The colonies, now mostly LICs, still have economies based around raw materials rather than manufacturing. Niger is a former French colony and in 2014 it was the least developed country in the world, with a Human Development Index of 0.348. After its independence in 1960, Niger has experienced severe political instability.

Climate

Crops cannot grow well without water at the right time of year. Applying technology like pumps and irrigation systems increases yields. In Ethiopia, after the famines in the 1970s and 1980s, water conservation techniques such as **drip-feeding** made food supply more reliable. Indian farmers only get high yields if enough **monsoon** rain falls

Key terms

landlocked country: a country without a coast, so without access to the sea, affecting trade

drip-feeding: pipes laid across a cropped area have regular small holes to distribute water. Slow flow limits wastage through evaporation

monsoon: seasonal wind system in the Indian Ocean which controls the rainfall pattern

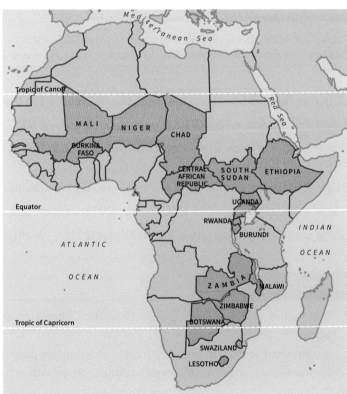

Figure 12.9 Landlocked African countries.

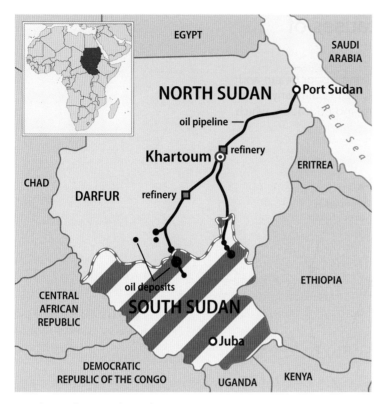

Figure 12.10 The Sudanese oil pipeline.

between June and September. Crop yields reduce if the monsoon rains begin late, finish early, or bring too little rainfall.

Climate change is a threat to food production in many countries, particularly those in drier and warmer parts of the world. Wheat is a staple crop grown around the world but it is sensitive to heat. It is estimated that because of global warming, wheat yields could drop by 2 per cent every decade. Declining crop yields mean that food prices rise, which can lead to poverty and political instability.

Tropical regions suffer more climate-related diseases than temperate zones. Ebola is a recent disease which can kill; malaria is debilitating, meaning people have little energy to earn a living.

Extreme events

Globally, **extreme events** like hurricanes and drought are becoming more severe and frequent. Tropical zones are worst hit. Hurricanes destroy crops and livestock as well as infrastructure necessary for growth and trade, like airports and ports.

Haiti was hit by a magnitude 7 earthquake (Richter scale) on 12 January 2010. The epicentre was 25 km west of Port-au-Prince, the capital, so did massive damage to existing infrastructure. With 3 million people affected (220 000 deaths; 300 000 injuries; 1.3 million homeless; several collapsed hospitals reducing aid capacity; businesses destroyed; airport, ports and roads damaged), the country's development was set back decades. Other countries and global institutions provide emergency funding, but investment to redevelop is harder to find. Countries like Haiti are already burdened by an enormous amount of debt to global financial institutions.

 Key term

extreme events: hurricanes, drought, earthquakes, volcanic eruptions – physical events causing serious impacts on life and economy

12.7 What are the economic causes of uneven development?

Poverty causes poverty! A person with a low standard of living – poor health and nutrition – is unlikely to improve their situation (Figure 12.11). Similarly, LICs often do not have the money to invest in infrastructure and education, which are vital for development.

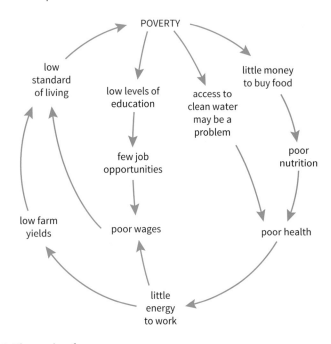

Figure 12.11 The cycle of poverty.

Health and water quality

Clean water is a basic requirement for health and development but is not always available (Figure 12.12). Charities like WaterAid prioritise clean water projects (Figure 12.13).

Bilharzia is a parasitic infection caught from freshwater snails in some tropical areas. Children playing in water, and adults working in wet rice fields, are susceptible (Figure 12.14). Symptoms include swollen belly, anaemia, stunted growth and poor intellectual development. Adults, if untreated, can die through severe liver damage and bladder cancer. Bilharzia is treatable if caught in time.

Figure 12.12 Woman in Ghana collecting unclean water.

Figure 12.13 WaterAid project – a child in village with clean water source.

Figure 12.14 Swimming in bilharzia infested water and the bilharzia parasite.

Education

Development needs a workforce sufficiently educated to be literate to the level of a young teenager in the UK. In Uganda primary education is free. Almost every young child goes to school, but secondary education means paying fees. Students arrive at school at any time in the year when they have enough money. Many never get there.

Trade

Efficient trade requires **infrastructure** and poorer countries lack this because it is expensive. Global **trading policies** do not favour the poor. Countries buying goods may add **tariffs** to imports, making them more expensive. Poorer countries usually export **primary goods** which command low prices. Richer countries process these, adding value and making profit.

Political factors

Not all countries are governed efficiently and not all politicians are completely honest. Sometimes corrupt government workers divert development funds into their own pockets (i.e. they take public money for themselves). Transparency International is an organisation that publishes an annual Corruption Perceptions index, which ranks countries in terms of perceived public sector corruption. In 2015, Denmark was ranked number 1 (least corrupt) and Somalia number 167 (most corrupt).

China is the biggest investor in Africa, its biggest trading partner, with an input of $163.9 billion in 2012. This is mainly on infrastructure, such as the Sudanese oil pipeline (Figure 12.10), mining, telecoms, transport, construction and power projects. Most profits go outside the country.

Key terms

infrastructure: the framework of transport and energy networks, including roads, railways, ports and airports, plus energy distribution

trading policies: rules decided by a country or group of countries to control imports and exports

tariffs: extra taxes put on goods imported, making them more expensive to buy

primary goods: raw materials – mining, oil/gas extraction, renewable energy, farming, fishing, forestry. In poorer countries these employ most people

ACTIVITY 12.5

1. Research the Haiti earthquake. How does the destruction of basic infrastructure (water, sanitation, education) hold back development?
2. Figures 12.12 and 12.13 show unclean and clean water supplies.
   a. Discuss the consequences of each situation.
   b. Create a table to compare the two situations. Compare your ideas with others in your class.
3. List the symptoms of bilharzia and explain their impacts on an infected child's development.
4. Find out about Oxfam's clean water projects. What is the 'Oxfam bucket'?
5. Look at the **WaterAid website** (www.cambridge.org/links/gase40090) and make a list of how people's lives can be improved with clean water.

Discussion point

In what ways do improvements in technology aid development?

Find out more:

Look at the **BBC website** (www.cambridge.org/links/gase40091) for facts on the Haiti earthquake 2010.

Visit the **WaterAid website** (www.cambridge.org/links/gase40092) for information on clean water projects in Ghana.

Read the in-depth article on tropical diseases limiting development on the **American Society of Tropical Medicine and Hygiene (ASTMH) website** (www.cambridge.org/links/gase40093).

12.8 What are the consequences of uneven development?

Disparities in wealth

There are immense variations in wealth globally, from nation to nation, family to family, and person to person. Wealth inequality across the world and within countries is consistently increasing.

Extreme inequality is not simply to do with income, it also involves opportunities and access to resources, which are largely controlled by where you are born – the country, region, and district. Attempts to even up these inequalities are being made nationally and internationally, but are still struggling to succeed.

In 2000, the richest 1 per cent of global citizens owned 40 per cent of global assets and the top 10 per cent controlled 85 per cent of the world's economic production. At the same time the poorer half of the world's population owned only 1 per cent of its assets (Oxfam data). Despite much global economic and social progress, which has brought millions out of poverty and increased opportunity, wealth disparity continues to grow, not just globally but also within countries. By 2014 the 85 richest individuals in the world controlled the same wealth as the poorer 50 per cent of the world's people (over 3.5 billion individuals). Oxfam predicts that figures will show that the richest 1 per cent will control 50 per cent of the global economy.

However, this is not just an issue of wealth versus poverty globally. Within both LICs and HICs the rich and poor are moving further apart economically. One in four Spaniards of working age is unemployed and many migrate to try to find work in other EU countries. There has been a significant flow of young Spanish workers into South West France, just across the border. In the USA, income distribution has consistently become more uneven. In 2015, the top 1 per cent of earners received 35 times the average national salary, and the trend is becoming increasingly unequal.

Average LIC income increased by 11 per cent between 1990 and 2010, yet family incomes are more unevenly distributed in 2016 than in the 1990s. Some people's economic situation is genuinely improving due to increased government-generated education and opportunity, plus NGO (charity) projects, both leading to national growth.

Nigeria is Sub-Saharan Africa's largest economy and the 26th biggest economy globally, mostly from oil production. Its number of millionaires grew by 44 per cent from 2007 to 2013, yet a majority of its people remain grindingly poor. Opportunity is extremely poorly distributed. Between 2015 and 2035, over 100 million young adults are due to enter the workforce because of continuing rapid natural increase in Nigeria, yet unemployment in 2014 in this age group was already 75 per cent! It looks as if inequality will remain for quite some time.

Disparities in health

Life expectancy is often lower in LICs than in HICs. For example, in 2015 the average life expectancy in Somalia was 52, while in Japan it was 85. Life expectancy reflects a number of factors, such as the incidence of war, but usually indicates the level of disease and malnutrition in a country.

Access to healthcare varies with development. HICs can offer better medical facilities, better access to treatments and preventative care (vaccines and screening).

HICs are often better able to prevent diseases occurring; for example, many diseases are caught by contact with dirty water and HICs often have better access to clean water.

Find out more:
Look at an article on the rapid population growth in Niger on the **IRIN news website** (www.cambridge.org/links/gase40094).

Read the article '**New immigrant crisis on the way: 185,000 migrants flood into Europe in just three months**' in the *Express* newspaper (www.cambridge.org/links/gase40095).

Migration

Uneven development can lead to migration as people in LICs migrate to NEEs or HICs. Migration can also occur within a country as people move to regions with greater economic opportunities.

People **emigrate** out of a country and **immigrate** into another country. Someone leaving one part of the country to move to another part is called an **out-migrant**, whereas they are considered an **in-migrant** by the area that they are moving into.

Differences between economic migrants and refugees

A refugee is a person who is forced to flee (escape) from their home country due to war, political unrest or natural disasters. These are classed as forced, or involuntary moves, since people's lives are at risk. All sections of the population are affected – men, women, children and the elderly.

Recent examples of countries from which a large number of people have been forced to flee include: Afghanistan, Syria, Ethiopia, Eritrea and Somalia.

An economic migrant is a person who makes a permanent move to another country or region to improve their standard of living. They search for a better job, more money and a higher standard of living. This is classed as voluntary migration – people choose to move, they do not have to do so. Certain types of people are more likely to move for economic reasons than others, for example:

- Individual men might move to earn money to send home to their families – young men try to migrate from Senegal to Italy, not always legally, for this reason.
- Single unmarried women may move for jobs – this is common in China, where young women from rural areas go to work in urban factories to send money home.
- Whole families move to take advantage of job opportunities and better wage rates, as well as good services. Families from Poland and other Eastern European countries move to the UK and other European Union (EU) countries.

International migration

International migration includes refugees and economic migrants. European Union countries, the chosen destination of many economic migrants from Asia and Africa, try to limit numbers of new entrants for economic reasons. Some people are concerned that there are not enough jobs for all those who want to move to the EU countries for a better quality of life.

The Syrian civil war has led to thousands of people paying to cross the Mediterranean sea to seek safety in the EU, but this also gives people who are economic migrants, not refugees, an opportunity to enter the EU.

The decision to migrate

There are many reasons why people migrate. It is a balance between PUSH and PULL factors. PUSH factors are things that make a person want or need to leave their home; PULL factors are things attracting them towards another location.

The differences between international and internal migration

International migrants cross borders: internal migrants do not. In both cases, people can be refugees or economic migrants. In 2015 thousands of potential migrants tried to cross the Mediterranean from North Africa to Europe for different reasons.

People move home for many different reasons. Activity 12.6 lets you work through several situations for yourself.

Key terms

emigrate: leave a country to live in another, with the intention of remaining at least a year

immigrate: enter a new country with the intention of living there at least a year

out-migrant: someone leaving one part of a country to move to another

in-migrant: someone moving into another region of their country

ACTIVITY 12.6

1 Make lists of push and pull factors resulting in migration.
2 Consider these situations and categorise the migrations as:
   - international/internal
   - refugee/economic/other (in this case, try to define the situation)
   a A retired UK couple, whose children are grown up and financially independent, moving to France
   b A Syrian family bombed out of their home and jobs in the civil war (2011 onwards)
   c A young, qualified UK doctor going to work in Australia
   d UK university graduates moving to London and South East England
   e The Central/Eastern European person who serves your coffee/food in a coffee shop/restaurant in the UK
   f Residents of Montserrat (Caribbean) moving to the UK after the 1995–1999 eruptions of the Souffrierè Hills volcano

12.9 What strategies exist to reduce the global development gap? (1)

There are a number of ways in which the development gap can be reduced and quality of life improved. These include:

- investment and industrial development
- aid
- debt relief
- fairtrade
- intermediate technology
- village savings and loan associations (VSLAs) and microfinance loans
- tourism.

There are two approaches to narrowing the development gap:

'Top-down'	'Bottom-up'
Large-scale **capital intensive** projects with high inputs from governments, global institutions like the **International Monetary Fund (IMF)** and **transnational corporations (TNCs)**. Benefits should trickle down throughout the economy so everyone gains.	Small-scale investments to individuals or small businesses. People earn more which they will spend in the local economy, hence others benefit too.

Large-scale investment and industrial development

Large-scale investment involves governments, organisations or companies investing in large projects such as infrastructure (e.g. power stations, roads, airports) or industry (e.g. factories, hotels).

Hydropower is the fastest-growing energy source in the world, encouraged by the World Bank, which sees it as the basis for economic growth in Africa and South Asia. A 40 000 megawatt (MW) plant is planned to be built in the Democratic Republic of Congo (2016–2020). However, this is a fragile, war-torn state, which may prevent the project from going ahead. Rapidly developing LICs – China, India and Brazil – all have huge potential to build hydropower projects.

China

The Salween River valley in eastern China is called 'the Grand Canyon of the East' due to its deep valley and incredible views. Thirteen new dams on the Salween were part of China's five-year plan in 2013.

Some people argue that water power is preferable to China's main energy resource, coal, which is much more polluting. But hydropower projects can have huge human and environmental costs. Around 40 000 people will be uprooted and resettled as a result of the Salween projects, as was the case with the Three Gorges Dam project.

Water pressure can increase earthquake risk. The weight of the Zipingpu reservoir may have contributed to the Sichuan earthquake (2008), with 80 000 deaths.

India

India's hydropower plans concentrate on Himalayan rivers, where steep gradients mean huge electricity generation potential. India's rapidly growing economy and many millions of jobs depend on industries using hydropower, but there are costs. Sikkim is India's most species-rich state and the large-scale dam-building planned will threaten its rich biodiversity, especially the endangered leopard population.

Key terms

capital intensive: an economic system with high inputs of money, expensive equipment and technology; people's physical labour is less important

International Monetary Fund (IMF): an organisation of 188 countries that aims to promote global economic stability

transnational corporations (TNCs): TNCs link several countries together in the production and marketing of goods

hydropower: electricity generation from moving water: damming rivers, wave and tidal power

Brazil

The Itaipu Dam (Figure 12.15) on the Paraná River (opened 1984) has the world's second largest capacity after China's Three Gorges Project. This plus other dams provide 90 per cent of Brazil's energy, fuelling its economic growth. Tucuruí Dam (also 1984), the largest in a rainforest zone, supplies huge industrial projects with the necessary electricity.

Figure 12.15 The Itaipu Dam, Brazil.

ACTIVITY 12.7

1 Define what is meant by 'top-down' and 'bottom-up' approaches to development.
2 What type of person would be in favour of each approach?
3 Research either of the Itaipu or Tucuruí dams. Who benefited and who was affected negatively by this development?

Large-scale aid

Aid means funding coming into a poorer country from various sources for development purposes. On the large scale the idea is that new infrastructure boosts an economy and the benefits trickle down from top to bottom in the economy. Everyone's quality of life improves. For large-scale aid the funding comes from other countries or global financial institutions like the World Bank or IMF. In some cases the funding is a gift, in others a loan which must be repaid and with interest.

Debt relief

Many poorer countries are in debt through borrowing from richer countries and **global financial institutions** to fund development. Their economic growth barely repays the interest and their debt often increases. Debt relief sometimes means writing off debts, but more often making the repayments lower and the term of the debt longer – in other words, making debt repayments more affordable. The IMF/World Bank HIPC (highly indebted poorer countries) initiative helped 41 countries, mostly in Africa.

Fairtrade

Figure 12.16 shows the Fairtrade logo. You may recognise this logo from grocery products – chocolate, coffee, tea, sugar, bananas and flowers are most common. It represents greater independence for farmers in poorer countries. It prevents exploitation by large companies by giving realistic prices and better working conditions. By trading more independently and in cooperatives, farmers have more control over their businesses when dealing with large buyers. Fairtrade improves standard of living, gives dignity and helps future planning. Workers' children have better education and healthcare. Customers are sure of high environmental standards.

 Read the article '**Hydro dams could jeopardise "Grand Canyon of the east"**, say green groups' on the **Guardian website** (www.cambridge.org/links/gase40096) to learn about the impact of Chinese dams on the environment.

Read the article '**Nature "dammed" in the Himalayas as large-scale dam building threatens diversity**' in the *Daily Mail* about the impacts of Himalayan dams (www.cambridge.org/links/gase40097).

Visit the **WWF website** (www.cambridge.org/links/gase40098) to find more information about the environmental impacts of dams in the Amazon Basin.

 Key term

global financial institutions: financial institutions which cover the world in terms of borrowing and lending – the two main ones are the World Bank and the IMF – wealthier countries contribute and any country can borrow or be funded by these institutions

 Did you know?

1.5 million farmers and workers are certified by Fairtrade.

1210 producer organisations are certified by Fairtrade.

25 per cent of all Fairtrade farmers and other workers are women.

Figure 12.16 Fairtrade logo.

12.9 What strategies exist to reduce the global development gap? (2)

Small-scale aid projects

If someone cannot see, they cannot earn a living. Sightsavers is a UK-based charity restoring eyesight and therefore opportunity. Before Sightsavers worked in Nampula Province, Mozambique (Figure 12.17), 29 000 people were blind or partially sighted due to cataract. Limited healthcare and poor sanitation had to be overcome. Sightsavers opened their eye clinic at the local hospital in 2014, serving northern Mozambique's 7 million people (28 per cent of national population). Around 40 000 patients were seen in 2015 (Figure 12.18).

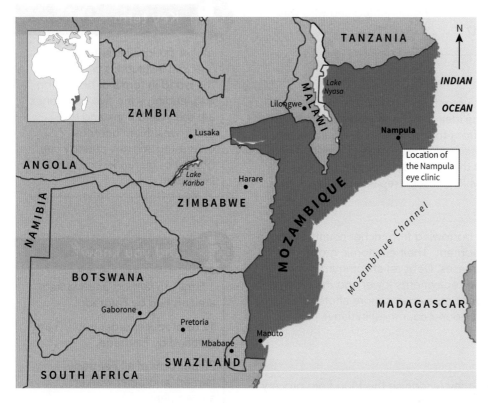

Figure 12.17 Location of Nampula eye clinic, Northern Mozambique.

Figure 12.18 Nampula eye clinic, Mozambique.

Intermediate technology

Intermediate technology combines quite sophisticated ideas with cheap, readily available materials, appropriate for use in poorer countries. Local knowledge and tools are put to good use. This is usually **labour intensive**, creating employment.

Solar cookers are used all over the world today. They focus the Sun's rays on the cooking pot. Cheap to make and easy to transport, they have great potential, especially in remote areas like the Himalayas (Figure 12.19).

Figure 12.19 A solar cooker in Ladakh, India.

Small-scale investment

Care International gives cash grants to get people back on their feet after a disaster. Funding comes from charity donations, mostly in the UK. Here are four examples from across the world:

1 Over 50 000 people in South Sudan were given fishing equipment, vegetable seed kits and tools to make them economically secure.

Key term

labour intensive: an economic system dependent on people's physical labour – there is little technology, but that does not mean this economy cannot thrive

2 Gaza is a tiny strip of land on the eastern Mediterranean with serious water supply problems. Care International helped small-scale farmers improve their land with essential irrigation systems after the 2014 conflict with Israel. Food production was able to continue (Figure 12.20).

3 Helping women be economically productive can improve quality of life for the whole family. Several Gazan women have been used these grants to set up small manufacturing and retailing businesses.

4 After a disaster, many households are headed by women. Aileen Militante, who lost her home in Typhoon Haiyan (Philippines, 2013), was given cash to restock her small livestock farm and support her family.

In 2014 Care International helped over 437 000 people affected by conflict or natural disaster to restore or expand their means of earning a livelihood to support their families.

Figure 12.20 A farmer attends to his irrigation system in the Gaza Strip.

VSLAs and microfinance

In a village savings and loan association, members pool their savings to create a bank of money and, whenever one person needs finance, they borrow from this. People invest in each other's businesses. Care International and other charities offer business training to give women management skills. Microfinance loans are also provided by private investors in richer countries to entrepreneurs in poorer regions.

ACTIVITY 12.8

1 What difference does having good sight make in terms of making a living and supporting a family?

2 Are large or small-scale aid projects more useful? Give reasons for your answer.

 Discussion point

'Large-scale energy projects can provide the electricity needed for development but can also have high costs for people and the environment.' Explain this statement.

Visit the **Fairtrade website** (www.cambridge.org/links/gase40099) to see how the Fairtrade system operates.

Visit the **Sightsavers website** (www.cambridge.org/links/gase40100) to see how saving sight can help people support themselves and their families.

Look on the **Care International website** (www.cambridge.org/links/gase40101) for a charity advertisement to attract donations.

Visit the **Care International website** (www.cambridge.org/links/gase40102) to read case studies of people helped by Care International.

Visit the **Action Aid website** (www.cambridge.org/links/gase40103) for information on child sponsoring projects.

 Watch a video on Cambridge Elevate about a self-help housing project in Bangalore, India.

Download Worksheet 12.3 from Cambridge Elevate for additional questions.

 Assess to progress

1 Describe how loans are different from charity aid. 3 MARKS

2 What are the problems for recipient countries of becoming dependent on aid? 4 MARKS

 Tip

Aim to refer to an example whenever it's applicable; quoting dates of events helps too.

How tourism in Tanzania is reducing its development gap

Tourism brings many benefits to LICs. Foreign currency enables a country to trade freely. Globally, tourism earnings represent 5 per cent of GNI and 1 in 12 jobs; it is the first or second most important source of foreign earnings in 20 of the 40 poorest countries, reducing poverty.

Tourism requires infrastructure – roads, railways, airports, a range of hotels and other accommodation. The more educated the workforce, the better, especially in terms of global languages like English. Many tourists want to experience the local culture, including food, but others demand global cuisine. This leads to the formation of a supply chain to source the ingredients and skilled personnel from abroad. Sports and other facilities are developed, so new products and services become available to local people, if they can afford them. This is the **multiplier effect**.

Tanzania offers high cost holidays to HIC tourists. For example, a seven day safari can cost around £2000 per person. Figure 12.21 shows the growth in tourist numbers which in 2010 represented 12.7 per cent of the national economy.

11 per cent of Tanzanians worked in tourism in 2013. The sector is growing rapidly, worth $4.48 billion in 2013.

There are two holiday experiences:

1 **The Northern Safari Circuit** 300 000 foreigners visit Tanzania's northern national parks like Ngorongoro annually. Local people provide tourist services such as accommodation, food and transport (Figure 12.22). Wildlife is an important sustainable resource, if protected.

2 **Climbing Mount Kilimanjaro**
Kilimanjaro National Park earns more than the rest of Tanzania's national parks. Climbing Kilimanjaro is increasingly popular with Westerners (Figure 12.23). A total of 47 per cent of the cost of any climbing trip goes to the national park; most of it goes towards paying wages (Figure 12.24). Most visitors give decent tips directly to guides and porters. Day wage rates are between $3.50 and $10.60, well above average earnings.

Key term

multiplier effect: the 'snowballing' of economic activity, for example, if new jobs are created this gives people more money to spend which means that more workers are needed to supply the goods and work in the shops

Skills

Using different kinds of graphs: Remember the key differences between Figure 12.21 (a continuous line graph) and Figure 12.22 and Figure 12.24 (pie charts). The line graph indicates change over time and the pie diagrams are a snapshot – the data is from one specific year.

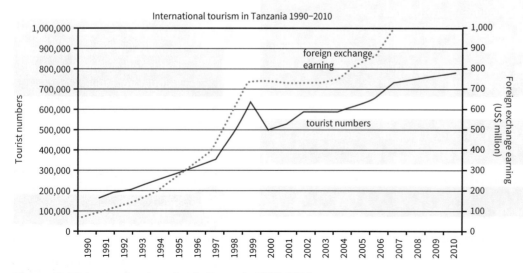

Figure 12.21 International tourism in Tanzania 1990–2010.

The Tanzanian government recognises the potential of tourism to reduce poverty. Local people do benefit from tourism; some are lifted out of poverty, but a significant proportion of profit still goes outside Tanzania via international companies.

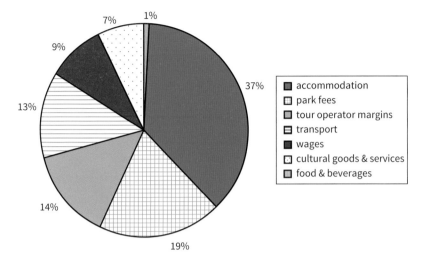

Figure 12.22 The cost components of a typical safari holiday.

Figure 12.23 Tourists climbing on Kilimanjaro.

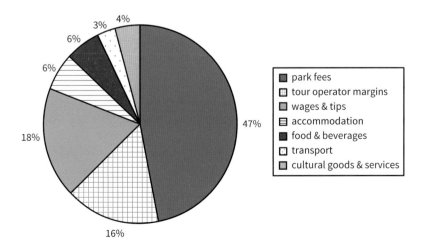

Figure 12.24 The cost components of a typical mountain-climbing holiday.

Further research

For more details on tourism in northern Tanzania try reading 'Making success work for the poor: Package tourism in Northern Tanzania' (Overseas Development Institute).

Did you know?

85 people, about as many as would fit on a double-decker bus, own as much wealth as the bottom half of the world's population. (*Source: Oxfam*)

ACTIVITY 12.9

1 Discuss the benefits and difficulties of relying on tourism as a key part of a country's economy.

To find out more:

Look at the **United Nations Population Fund article** about population and poverty (www.cambridge. org/links/gase40104).

Read the report on the **Overseas Development Institute website** (www. cambridge.org/links/ gase40105), 'Making success work for the poor: Package tourism in Northern Tanzania'.

Rapid economic development in Malaysia

The location and importance of Malaysia

Malaysia is a newly industrialised market economy (NEE) located in Southeast Asia, bordering Thailand, Indonesia and Brunei (Figure 12.25). Malaysia is actually a federation of 13 states and three federal territories. These are split across two regions that are separated by 640 miles of the South China sea.

Malaysia's regional and global positions have grown in recent decades. Despite its relatively small population of 31 million, Malaysia's economy is the third largest in Southeast Asia after those of Singapore and Brunei, and the 35th in the world (2015). It is one of the largest global producers of palm oil, rubber, timber and electronic goods.

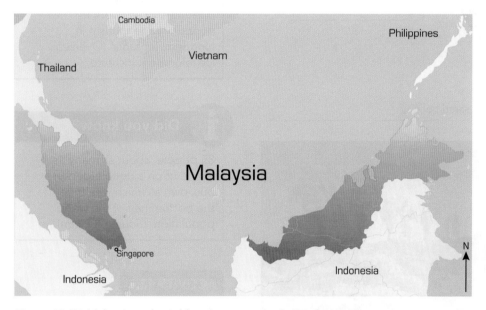

Figure 12.25 Malaysia and neighbouring countries in Southeast Asia.

Politics, society, culture and the environment in Malaysia

Of the 31 million people in Malaysia, more than 1.7 million live in the capital city of Kuala Lumpur. Malaysia is one of the most urbanised countries in East Asia, with an urban population of 15 million (nearly half of its total population). Malaysia's sprawling territory includes part of the island of Borneo, which is the location of one of the oldest rainforests in the world.

The Malaysian population is made up of several ethnic groups:

- Malay: 60 per cent
- Chinese: 26 per cent
- Indian and indigenous groups: 14 per cent

Since 1971, the ethnic Malay population have benefited from positive discrimination in business, education and civil service, and are dominant in politics. However, the ethnic Chinese population has more economic power. The Indian and indigenous groups are the poorest communities.

The dominant religion is Islam, although there are a number of other religions.

Malaysia has been criticised about its record on human rights. Its security laws allow suspects to be detained without charge or trial. Homosexuality is condemned and can be punishable by imprisonment.

Changing industrial structure in Malaysia

Since the 1970s Malaysia has been transformed from a low income country relying mainly on the export of **raw materials (primary products)**, such as rubber and palm oil, to an economy more heavily dependent on manufacturing and services (secondary and tertiary activities); for example, manufacturing computer disk drives. Its markets for products and services are largely global – Malaysia has increasing international trade links. However, as people become better off and have a larger disposable income, the home market also expands. The tourism industry has also grown and is estimated to have made up 14.9 per cent of Malaysia's GDP in 2014.

In 2005, agriculture employed 18 per cent of Malaysia's working population, dropping to 12.6 per cent in 2015; at the same time, services grew from 50 to 59 per cent. This reflects a more educated workforce and increased global trade. Manufacturing forms most of the rest of Malaysia's industry. Unemployment, already low compared with many countries in 2005, has shown a small but regular decline in recent years. In 2014, unemployment stood at 2.9 per cent. As the country is becoming more industrialised, increasing numbers of people are moving from rural areas to urban-industrial centres in order to find employment and have an opportunity to improve their quality of life and social conditions – i.e. health and education services.

Malaysia's aim is to reach high-income status by 2020. High income status is defined by the World Bank as a country with a gross national income per capita of US$12 735. The government is encouraging growth in technology, biotechnology and service industries. The Economic Transformation Programme is encouraging the growing middle class to buy manufactured goods, in order to increase the home market.

The Central Business District (CBD) of Kuala Lumpur, Malaysia's capital, with its architect-designed high-rise skyscrapers reflects the national attitude towards economic development (Figure 12.26).

Key term

raw materials (primary products): unprocessed material collected from the Earth: examples include fossil fuels, metal ores, agricultural produce

Figure 12.26 The high-rise Central Business District in Kuala Lumpur.

How the manufacturing industry can stimulate economic development in Malaysia

Malaysia's rapid rural-urban migration flows have provided a huge workforce for its growing manufacturing and service sectors. Manufacturing industries create export products which earn money for the Malaysian economy. During the latter 20th century rural-urban migrants worked mainly in manufacturing, resulting in Malaysia starting to compete on a par with the economies of Europe and North America.

Electrical and electronic businesses are a key part of Malaysia's manufacturing sector. Its products create 32.9 per cent of the country's total exports, provide millions of jobs at all skill levels, and encourage new research and development. Oil-based businesses are also very important, representing four of the top five manufactured exports (Table 12.3).

	Percentage of manufactured exports
Refined petroleum products	9.1
Liquefied natural gas	8.2
Oil-based chemicals and associated products	6.6
Palm oil (vegetable oil)	6.4
Crude petroleum	4.4

Table 12.3 The top five manufactured exports in Malaysia.

The government has identified key economic regions (economic corridors) to concentrate assembly and research and development work within the manufacturing sector, as well as reducing economic differences across the nation. The two main growth areas are the Northern Corridor Economic region and the East Coast Economic Region. Agri-biotechnology and food processing are both important. Key centres such as these pull in new businesses, create new jobs and thereby boost national economic growth.

Pay in the manufacturing sector is relatively high within Malaysia, though compared with neighbouring Singapore, pay is low. Some Malaysian companies have opened manufacturing sites in neighbouring Thailand and Indonesia to cut manufacturing costs, just as wealthier countries used to do in today's NEEs like Malaysia.

High value manufacturing industries, especially IT, electronics and multimedia, remain within Malaysia using its increasingly educated workforce. These are its exports for the future.

The role of transnational corporations (TNCs) in Malaysia

Malaysia's government policies since the 1960s have encouraged foreign investment into the country. This has often been by TNCs (transnational companies), which have played an important role in the economic development of Malaysia by exploiting its resources for the global market, bringing wealth into the country as well as to the TNC itself.

ConocoPhillips (USA) currently has four deep-water oil and gas projects underway in Malaysia, jointly with Shell (UK/Dutch) and Petronas (Malaysian). Note the mix of nationalities involved here. Profits go to all countries involved.

Global banks such as Citibank and HSBC (Hong Kong and Shanghai Banking Corporation – a bank also found in UK high streets) provide loans for regional and local business investment.

TNCs are attracted to Malaysia because of the increasingly available well-educated workforce for relatively low labour costs, as well as for the opportunities given by the primary resources available.

Advantages to Malaysia of TNCs

TNCs bring investment into the host country, which creates economic growth. A positive multiplier effect brings new roads, airports and services (Figure 12.27), built to serve the TNC business, but at the same time benefiting the whole economy and population. There is also a transfer of technology and managerial skills from HIC to LIC. Living standards in the LIC are, on average, raised. TNCs create jobs in the host country (usually an LIC). Wages paid to local workers are often low, but this must be balanced against what money will buy in that local economy. Graduates may work in call centres, a role which might be seen by inhabitants of HICs as relatively unskilled, but the salary is reasonable in an LIC economy in terms of what it can buy.

TNCs increase globalisation, the connections and contacts between countries, in every possible way. They can be seen as building economic and social bridges.

Disadvantages to Malaysia of TNCs

Most TNC profits do return to their country of origin (usually an HIC) and the shareholders there.

The presence of TNCs may stifle the growth of local businesses which might not be able to compete. Due to their size, TNCs often have lots of power and can access discounts that smaller businesses might not be able to, meaning that smaller businesses cannot match the prices of goods and services.

As living standards rise, labour costs in Malaysia might also rise, which could prompt the TNCs to relocate to a cheaper country.

Figure 12.27 Global service links – a Malaysian Airlines airbus on the tarmac at Kuala Lumpur airport.

Changing political and trading relationships with the rest of the world

Malaysia has one of Southeast Asia's most booming economies, after decades of industrial growth and political stability. Its manufacturing exports create important links with the rest of the world. Electronics, electrical machinery and transport equipment together made up over 85% of Malaysia's exports in 2015. Chemicals are becoming increasingly important. Singapore (taking 13.6 per cent of Malaysia's total exports), China (12.6 per cent), Japan (11.8 per cent) and the USA (8.7 per cent) are Malaysia's main markets. These are also the four main countries from which Malaysia buys good and services, which is not at an uncommon situation – about half of the UK's exports go to the EU and the same percentage of imports come in from these countries too.

Rapidly growing tourism attracts people from other countries to Malaysia. Rainforest environments such as Northern Borneo are a popular destination and to date, tourism is less environmentally damaging than palm oil plantations and illegal logging.

Malaysia's trade has helped its relations with many countries, but Malaysia is also active in international organisations that promote regional co-operation. It was a founding member of the Association of Southeast Asian Nations (ASEAN) and is an active member of many other organisations, including the Asia Pacific Economic Cooperation (APEC) and the United Nations. However, Malaysia has experienced some difficult political relations:

- China, Indonesia and the Philippines have all expressed concern about the treatment of members of different ethnic groups within Malaysia.
- Relations with Singapore and Brunei have been strained by disputed territorial claims over areas.

International aid and Malaysia

Aid strategies are categorised in different ways.

- Short-term aid and long-term aid: short-term aid aims to save lives in emergency situations such as floods, earthquakes and famine. Long-term aid is intended to finance larger-scale structural projects which will help the economy to grow; dams for water and electricity supplies as well as big agricultural projects fit into this category.
- Top-down aid and bottom-up aid: top-down aid involves large-scale projects with the benefits 'trickling down' through the economy and society to benefit all. Hydroelectric power gives ordinary people electricity in their homes, improving their standard of living. It fuels small businesses and increases their output, creating jobs and boosting local economies. Bottom-up aid means starting with the basics, such as small charities providing poor people with seeds and basic farming tools to help them feed themselves. As they earn some money, they can spend it in other local businesses, leading to overall growth.

Aid brings benefits and difficulties. Genuine benefits occur – small and large businesses benefit, bringing a higher standard of living and new job opportunities. Aid that leads to new developments can be more effective than aid that simply solves a short-term problem. The Band-Aid and Live-Aid projects helping Ethiopia out of famine in the 1980s and beyond initially prevented people from starving but then also set up agricultural projects to save water and increase crop yields, improving food security and quality of life.

In contrast, if a region or country becomes dependent on aid which does not lead to self-sufficiency, real progress can never be made. Much of Africa relies on foreign aid. Ethiopia has made great progress, for example in healthcare and university provision, but it remains one of the poorest countries on the planet.

Aid projects in Malaysia

Malaysia has a relatively strong economy but still receives and donates aid. For example, in 2012 the US government gave $10 091 376 to Malaysia. This was divided into:

- economic support
- refugee assistance
- anti-terrorism strategies
- grants for smaller economic projects
- military assistance

In 1991 the UK government gave Malaysia £234 million to fund the building of the Pergau dam. This was highly controversial because the aid was in exchange for a major arms deal. The dam was criticised for not being a cost-efficient way to increase the production of electricity. This deal prompted a review and change in the way that the UK gave aid to countries.

Japan, the wealthiest Asian country, has a strategy to help poorer Asian countries develop. Malaysia and Japan have close ties – economically and culturally. The 'Look East' policy transferred Japanese work ethics and management styles to Malaysia and elsewhere. Thousands of Malaysian students have graduated from Japanese universities. Trade links between the two countries remain strong.

The effects of economic development on quality of life

Economic development generally improves the quality of life for most of the population. The government has more money to invest in infrastructure and services, such as roads, access to clean water, sewage facilities, education and healthcare.

Education

An educated workforce supports a growing economy. Adult literacy, already quite high in Malaysia in 2005, rose from 88.7 per cent to 93.1 per cent by 2015. To compare, literacy rate in the UK is 99 per cent. All Malaysian children haveprimary education and numbers completing secondary stage continue to increase. Over one third then continue to some form of tertiary education.

Healthcare

The number of hospital beds and doctors per 1000 people is a good indicator of the quality of the healthcare service. Malaysia made some progress between 2008 and 2015:

- the number of beds increased from 1.8 per 1000 people in 2008 to 1.9 per 1000 people in 2015
- the number of doctors increased from 0.7 per 1000 people in 2008 to 1.2 per 1000 people in 2015.

However, compare this with UK data of 2.9 beds per 1000 and 2.8 doctors per 1000 people.

The environmental impacts of economic development

Malaysia has a history of environmental damage as a result of waste from industry. Its traditional industries of tin mining, natural rubber and palm oil produced waste that polluted rivers and seas. Industrialisation from the 1960s worsened the problem by generating factory waste. More recently, air pollution has increased in urban areas as a result of increasing traffic.The Malaysian government has been implementing air pollution controls and investing in the construction of environmentally beneficial schemes such as sewerage systems to handle household wastewater. However, as the middle class continues to grow and consumption increases, the environmental problems could worsen.

Data from the UN found that the deforestation rate in Malaysia is accelerating faster than any other tropical country in the world. The rainforests of northern Borneo are threatened by palm oil plantations and illegal logging, and in February 2007 the three governments of Borneo – Malaysia, Brunei and Indonesia – signed a declaration to protect and conserve an area called the 'Heart of Borneo'.

Assess to progress

1 Define these terms:
   a economic development   `2 MARKS`
   b social change.   `2 MARKS`
2 How might TNCs help a poorer country become more involved in world trade?   `6 MARKS`

ACTIVITY 12.10

1 Explain how Malaysia's exports illustrate its changing level of economic development.
2 Explain why the following happen as a country develops economically:
   a percentage of people >60 increases
   b adult literacy improves and more children go to school
   c health services expand.

Did you know?

Malaysia's main exports are: semi-conductors and electronic equipment; petroleum and gas, solar panels; chemicals; textiles; wood and wood products; palm oil and rubber.

Tip

Two excellent sources of statistical data on most countries, which are updated annually, are:

- *The Economist Pocket World in Figures*
- *The CIA World Factbook*

To find out more:

Visit the *CIA* **World Factbook** (www.cambridge.org/links/gase40106).

Look at the **World Health Organization Global Health Observatory (GHO)** data (www.cambridge.org/links/gase40107).

Look at the statistics on the **UNICEF website** (www.cambridge.org/links/gase40108).

13 Economic futures in the UK

13.1 What are the causes of economic change in the UK? (1)

Globalisation

Globalisation is the increasing interconnection between countries. It influences a number of aspects of life (Figure 13.1) but is often thought of in terms of international trade. In the 1970s there were 7000 transnational corporations (TNCs) and this had risen to over 60 000 by 2008.

Globalisation began in the 19th century as colonialism increased trade. **Interdependence** between combinations of nations grew. European and North American economies expanded, followed in the 20th century by newly emerging economies (NEEs) such as India, China and Malaysia. The UK has moved from a labour-intensive, manufacturing base towards an increasingly service-based economy, with a growth in research and technology. Labour costs in manufacturing industries in higher income countries (HICs) have increased, and NEEs have taken their place in the global market in part because wage rates are lower for skilled workers in these countries. For example, as manufacturing has grown outside the UK it is cheaper to build ships in South Korea than in the UK.

Key terms

globalisation: the integration of economic, financial, social and cultural ideas and contacts between countries; increased trade and labour migrations are a big part of this

interdependence: two or more things that depend on each other, in this case countries in terms of trade

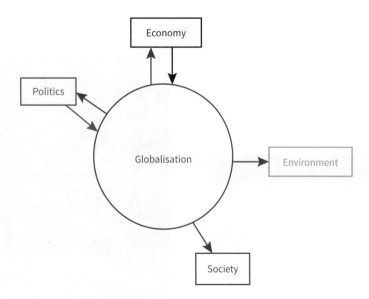

Figure 13.1 Influences of globalisation.

In the 1960s there was rapid economic growth in the 'Asian Tiger' countries: Hong Kong, Singapore, Japan and Taiwan. The 1980s saw free market ideas grow, allowing the UK and USA to trade with many more countries. The fastest growing economies in 2015 were Brazil, Russia, India and China (BRIC countries), followed by Mexico, Indonesia, Nigeria and Turkey (MINT countries).

Change in demand for goods

Richer countries have long demanded goods from across the world. The key recent change is growing demand for luxury goods by NEEs. For example, Scotch whisky is one of the UK's fastest-growing exports. Mexico imports £110 million of Scotch whisky per year and Brazil spends £99 million on it.

Globalisation and the UK government's policies

In 2015, the UK economy was the fifth largest in the world and the second largest in the EU. Its three biggest imports are: machinery, oil products and vehicles. The three main exports are: crude oil, cars and medicines. The financial services sector makes London's economy larger than that of any other European city, as well as making the UK one of the top financial hubs in the world.

The UK has always been open to trade. Europe is one of the UK's most important trading markets and in 1973 the UK joined what was then the **European Economic Community (EEC)**. There were no trade boundaries between member countries. Goods became cheaper, job opportunities wider and travel more possible. The EEC became the EU in 1990 and citizens of all 28 EU nations in 2015 can move and work freely across countries in the union.

New global markets opened giving access to a wider selection of goods at lower prices. For example, the electronics industries in China and South Korea have grown in response to demand globally and prices of electronic goods have fallen. A small flat-screen TV that cost £800 in 1996 cost around £200 in 2015.

Key term

European Economic Community (EEC): the name for an organisation that links European countries through trade and, more recently, political agreements. In 1993 it was renamed the European Union (EU)

ACTIVITY 13.1

1 Research online websites of electrical retailers.
   a Choose a particular product such as TVs or laptops. For the product you choose note where they are manufactured. Which countries dominate?
   b Draw a bar graph or pie chart to show numbers/percentages of products on offer in the UK and where they come from.
   c What does this tell you about the globalisation of industry?

Skills link

See Section 19.1 Graphical skills – graphs and charts for help with drawing graphs.

13.1 What are the causes of economic change in the UK? (2)

Decline of the traditional industrial base and de-industrialisation

By the 1980s, some traditional UK resources, such as coal, were running out or were beginning to become too expensive to obtain. The 1978 postage stamp (Figure 13.3) represents deep coal mines from the 19th century until that date, but these are now in significant decline in the UK. Deep coal reserves are being replaced with a mix of imports and coal from new UK opencast mines.

Employment in traditional industries like coal, iron and steel, and shipbuilding dropped dramatically as a result of **de-industrialisation**. Between 1978 and 1987 431 000 jobs were lost in these industries combined. The north of England, Scotland and Wales were worst hit; coal-mining had 107 000 workers in these regions in 1978 (Figure 13.2) but less than 10 000 by 1996. Across the UK the steel industry decreased by 129 000 workers (1979–1993) across the UK.

On a more positive note, although some existing oil reserves are declining, new resources are also being located. Overall it is a case of moving from labour-intensive primary and manufacturing industries to capital-intensive service industries.

Other countries can often produce similar goods more cheaply. Lower income countries (LICs) such as Cambodia and Thailand have lower labour costs. Labour-intensive industries such as textiles and clothing manufacture took advantage of this to lower production costs. Even with extra transport, manufacturers' costs remained lower overall. Here are two examples:

Marks & Spencer

Marks & Spencer used to manufacture all their clothes in the UK, but today most are made in countries such as Indonesia and Morocco.

> **Key term**
>
> **de-industrialisation:** the decline of traditional forms of industry, often accompanied by increased unemployment

Figure 13.2 A miner working on the coal seam at Granville Colliery, Shropshire, in the 1970s.

Figure 13.3 A UK postage stamp representing the UK coal industry, circa 1978.

M&S switch to foreign suppliers threatens 16 000 UK textile jobs

Marks & Spencer plans to cut the proportion of its clothes made in the UK to less than one-third. Unions claimed the move could mean the loss of 16 000 jobs to countries such as Morocco and Indonesia, where wages and other costs are low … A spokeswoman for Marks & Spencer said that the store would still make more clothes in the UK than any other firm. 'In order to stay in the high street we have to be competitive. To do that, we have to source more overseas, just as our rivals do. We haven't made any secret out of that,' she said.

The Independent, 2 November 1999

Dyson

Dyson vacuum cleaners, invented in the UK and manufactured in Malmesbury, Wiltshire, for many years, have been made in Malaysia since 2003 due to cheaper labour. With 865 UK jobs being lost, the company was accused of betraying UK workers. The research and development side was retained in the UK and savings in manufacturing costs enabled the company to employ 100 more people in this section.

Further research

Look at the article 'UK could be Europe's "largest" economy by 2030', on the **BBC News website** (see www.cambridge.org/links/gase40109), which predicts considerable growth for the UK economy and compares it with the rapidly growing economies (NEEs). Read this and assess its contents.

ACTIVITY 13.2

1 What is the importance of UK coal mining being represented on a stamp?
2 Explain the relationship between de-industrialisation and globalisation.
3 Write a paragraph to explain the term 'interdependence'.
4 Study Figure 13.1 showing the factors surrounding globalisation. Notice that some factors have single arrows to the 'globalisation' circle, but others have two.
   a Why might this be?
   b Discuss your ideas with others in your class.

- Article on Marks & Spencer moving jobs from the UK abroad (see www.cambridge.org/links/gase40110)
- Dyson moving from Wiltshire to Malaysia (see www.cambridge.org/links/gase40111)
- The impact of globalisation on UK politics (see www.cambridge.org/links/gase40112)

13.2 How has industry in the UK changed?

Classification of industrial activities

Figure 13.4 shows economic activities subdivided into sectors in the early 21st century, with examples of activities in each. The balance of numbers employed in each sector has changed significantly over time. Originally we worked with only primary, secondary and tertiary sectors, but as the economy has matured tertiary has been subdivided to create first quaternary and then quinary sectors. Not everyone thinks the quinary sector is valid.

Sectors of industry

Although there are hundreds of different jobs or occupations, they can all be classified into four categories:

 Primary industry: producing natural resources from the Earth i.e. raw materials – farming, fishing, forestry, mining, oil and gas extraction.

 Secondary jobs involve making things from those raw materials or assembly from already-manufactured components. For example, rubber and metals are used to make a car wheel; that wheel and other parts are then put together to make the finished car. Also food processing, plastics and chemicals.

 Tertiary e.g. theatre staff, restaurants tertiary jobs involve providing a service: retailing – stacking supermarket shelves, shop manager; entertainment services including hairdressing; professionals like surgeon or solicitor.

 Quaternary sector – a recent category based on knowledge and skill: research and development, like creating new medicines; information technology, e.g. web designer; consultancy, government and education.

 Some people identify a **quinary,** or fifth, sector for the top decision-makers in government, society and economy.

Figure 13.4 Sectors of industry.

The move towards service industries

As a country develops economically, the proportions of people working in each sector change.

- **Primary sector:** the percentage working in primary jobs steadily decreases. This does not mean the primary sector becomes less important, indeed it generates essential raw materials to support secondary and tertiary industries, and food, oil and gas products are exported. Increasing timber production (1.9 million tonnes in Wales alone in 2015) reduces the UK's need for imports.
- **Secondary:** the secondary sector rises due to growth in manufacturing and traditional industries, such as the UK in the 19th century, and later falls due to increased mechanisation.
- **Tertiary:** as the economy matures and people have more money to spend, they want to buy **services**. Also, the government spends more on education and healthcare.
- **Quaternary:** improved education and technology lead to greater research potential.

Figure 13.5 compares employment sectors in the UK with Brazil and Cameroon. Brazil is an NEE and Cameroon is just above the income threshold for an LIC ($3000). LICs offer people fewer economic opportunities and lack of access to

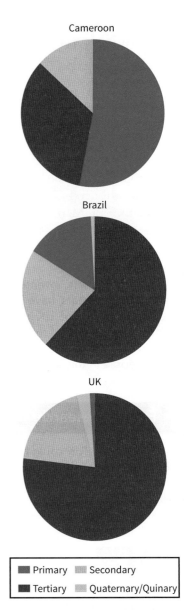

Figure 13.5 Changes in employment structure as countries develop.

 Key term

services: another term for tertiary industries, the sector of the economy providing services for individuals, other industries and the community

education leave many under-qualified. Many people must grow their own food. In an expanding economy, commercialised farming frees people to do other jobs and provides more opportunities in manufacturing and services.

Moving towards a post-industrial economy

During the later 20th century people had increasing **disposable income** in the UK. This fuelled the growth in the tertiary and quaternary sectors and the term **post-industrial economy** was coined.

Information technology

The IT sector generated £58 billion for the UK economy in 2014, and this is increasing annually. Around 2 million people work in IT across all businesses. *Note: counting employees is difficult because they are classified in other sectors. Someone working on Tesco's IT systems would be classed as working in retailing rather than IT.*

Finance industries and UK exports

Around 3.4 per cent of the UK's workforce is engaged in the financial sector – banking, insurance and the stock market. Larger banks are **TNCs**, meaning UK banks do business overseas and foreign banks have bases in the UK (Figure 13.6).

The financial services sector provides 10 per cent of UK **GNI** and 29 per cent of national exports. Taxes paid to the UK Government from these businesses amounted to £21.4 billion in 2013/14.

Over half of all business in the finance sector in the UK is based in London (50.5 per cent), attracting foreign workers; for example, 350 000 French workers live in London, many of whom work in finance.

Research and development

Research is essential to future economic growth and improving our quality of life. Pharmaceutical companies, such as Pfizer and GlaxoSmithKline, fund new drug research in a competitive market. Research centres are often located close to universities from which they recruit graduates as their labour force. Cambridge Science Park has grown steadily as new investors like Amgen have moved in. The Cambridge Science Park began on a small scale in 1968 but by 1982 it was one of only two, the other being Heriot-Watt near Edinburgh. In 2015 there were over 100 research parks employing over 42 000 people.

Business parks are also hubs for development. Manor Royal Business District in Crawley is the largest business park in South East England, covering an area of 240 hectares. The park hosts over 500 businesses and 30 000 people. The businesses are involved in areas such as IT security, healthcare technology, and energy technology.

Figure 13.6 Canary Wharf, part of London's financial district, which holds two international banks' headquarters, HSBC and CitiBank.

 Key terms

disposable income: the amount of money a person has left over to spend freely after paying for the essentials like food, housing and taxes

post-industrial economy: a period of growth in an industrialised economy in which the relative importance of manufacturing decreases and the relative importance of services, information and research increases

transnational corporations (TNCs): TNCs link several countries together in the production and marketing of goods

Gross National Income (GNI): the total value of goods and services produced within a country, usually measured per year; it is a measure of wealth and standard of living

ACTIVITY 13.3

1 Decide which of the five economic sectors the following jobs belong to:
   a the prime minister
   b a hospital receptionist
   c an electrician
   d a school librarian
   e a horticulturalist
   f a geologist working in oil exploration
   g a chef
   h a film producer.
   Think of other examples to test your class.

2 a Describe the differences between the employment structures of Cameroon, Brazil and the UK.
   b Explain the reasons for these differences.

3 Study the photograph in Figure 13.6. In pairs or small groups decide what it tells you about the quaternary sector.

13.3 How has UK infrastructure changed? (1)

Infrastructure, including transport networks, allows economy and society to function. The UK has a small area with high population density, especially in South East England. With an increasing population and a growing economy, there are more people with greater wealth. Car ownership grows year on year, placing pressure on the roads. Increasing commuting and leisure activity place pressure on public transport too.

UK transport networks are already responding to growing demands. The UK is an 'island nation', meaning we also need to work harder at connecting our networks with those abroad.

Road network development

The Preston bypass in Lancashire was the UK's first motorway and opened on 5 December 1958. Since then, motorways have transformed the way we travel. Car ownership has grown from 4.5 million cars in 1958 to 31.4 million in 2014. Motorways are relatively safe – road deaths halved in that time period. Moreover, motorways opened up commuter and leisure opportunities.

However, improved roads attract more traffic. The M25, London's orbital motorway, completed in 1986, has expanded to four or five lanes in places and the Heathrow section remains prone to congestion. The Birmingham stretch of the M6 was bypassed by the M6 Toll motorway in 2003 (Figure 13.7) which drivers pay to use.

Key terms

infrastructure: the framework of transport and energy networks, including roads, railways, ports and airports, plus energy distribution

networks: interconnecting patterns of roads or railways, etc.

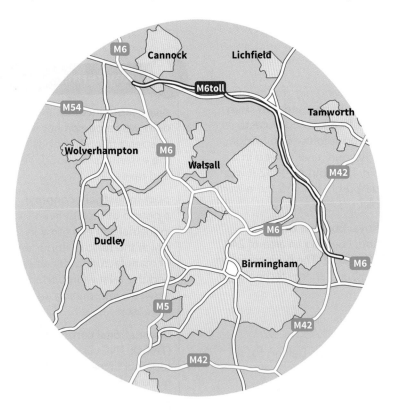

Figure 13.7 The M6 Toll route.

The Dartford Tunnel under the River Thames between Dartford (Kent) and Thurrock (Essex) opened in 1963, later becoming part of the M25. Since 1991 the tunnel has taken northbound traffic with the newer Queen Elizabeth II Bridge running southbound (Figure 13.8). The toll booths regularly caused traffic jams so in November 2014 Dart Charge, a pay-online system, was introduced to make the network more efficient.

Figure 13.8 Queen Elizabeth II Bridge spanning the Thames at Dartford.

Rail connections across London and the UK

Railways supported industrial growth in the 19th century and today carry 10 per cent of UK **freight** plus millions of commuters. The cross-Channel rail link connects our network to Europe, supporting trade and tourism.

Regional railway networks

As roads become increasingly congested, railways become more attractive. Many areas saw improved services during the 2010s. The line from Tunbridge Wells in Kent to London, an important commuter link, saw train frequency double from two to four trains per hour. The South East region has a greater concentration of railways than elsewhere in the UK.

Crossrail

By December 2019 Crossrail's 200 m long trains should allow east–west travel for 1500 people at a time across London without having to change stations and with multiple links to London Underground stations; 26 miles of tunnel construction make this the biggest such project in Europe.

HS1 and HS2

The intention of the High Speed Rail 2 (HS2) line is to improve the quality of rail networks across the country. Debate continues about its effectiveness in making outer regions more accessible. HS1, the fast line from London Euston to Glasgow, has already improved journey times.

London–Birmingham HS2 is planned for 2026, and by 2033 sections to Manchester and Leeds will be in place. Routes will be faster and more direct (Figure 13.9). Cities on the network are likely to benefit, but more distant regions may do worse.

The Channel Tunnel Rail Link (CTRL)

St Pancras International station (Figure 13.10) is the UK hub of the Eurostar network. Linking into the national rail system, it provides direct access to Paris, Lille and Brussels. Eurotunnel, a separate company, runs 'Le Shuttle', a train carrying all road vehicles, from Folkestone to Calais. Between 1994 and 2014 the Channel Tunnel carried 346 million passengers (Eurostar, 153 m; Le Shuttle, 193 m). In 2014 that meant 1.4 million trucks, 2.5 million cars and 58 500 passengers daily! Both services run through the Channel Tunnel beneath the English Channel.

Key term

freight: goods being transported by road, rail, container ship or plane

Download Worksheet 13.1 from Cambridge Elevate for questions on the Channel Tunnel Rail Link.

Key to journey times (Hours:Minutes)

	HS2	Now
London-Birmingham	0:49	1:24
Birmingham-Manchester	0:41	1:28
London-Nottingham	1:08	1:44
London-Manchester	1:08	2:08
London-Leeds	1:22	2:12
Reading-Manchester	1:46	3:10
Southampton-Leeds	3:35	4:17
London-Edinburgh	3:38	4:23

Figure 13.9 HS2 information.

Figure 13.10 St Pancras International station.

13.3 How has UK infrastructure changed? (2)

UK ports – key global connections

As an island nation, the UK has many ports – 95 per cent of our international trade is transported by sea (Figure 13.11). Dover (Kent) is the biggest for roll-on/roll-off ferries, carrying trucks, plus cars and passengers. Felixstowe (Essex) is the largest container port in the UK. Southampton (Hampshire) is the most efficient European container port and main centre for cruise ships.

The major ports of Southampton, Liverpool, London Gateway and Felixstowe have all increased their capacity since 2010. Felixstowe acquired two new berths for megavessels in 2011. Around 37 000 people are employed in ports with thousands more jobs being created by the multiplier effect.

Figure 13.12 Planes queuing to take off at Heathrow.

Figure 13.11 The main sea ports in the UK.

Airport traffic and capacity

London's airports carry 133 million passengers (2011), using four of the UK's five busiest airports – Heathrow, Gatwick, Stansted and Luton. Manchester Airport is the third busiest in the UK. Heathrow has five terminals and Gatwick has two. Demand grows annually (Figure 13.12). Stansted increased **capacity** to 17.5 million passengers/year after its £50 million extension was completed in 2008. Regional airports regularly increase their range of destinations.

Heathrow is the largest global **hub** airport, supporting 220 000 jobs, making it a key economic generator. It wants to expand to maintain its status and remain the largest European airport. It has opposed the concept of a brand new London airport in the Thames Estuary.

ACTIVITY 13.4

1 Look up details of the busiest airports in the UK by total passenger traffic (see www.cambridge.org/links/gase40113).
   a Identify five regional airports and their destinations.
   b How can the development of regional airports expand local economies?

- The history of motorways (see www.cambridge.org/links/gase40114)
- London's new railway: Crossrail (see www.cambridge.org/links/gase40115)
- The debate around the HS2 development (see www.cambridge.org/links/gase40116)
- Reinventing Heathrow (see www.cambridge.org/links/gase40117)

Key terms

capacity: in terms of transport, the number of people or amount of goods that can be carried by a network or hub, for instance, the number of people passing through a particular airport in a certain time

hub: in terms of airports it means one of the most important globally; millions of passengers change planes at global hubs to reach their destinations

Did you know?

One plane takes off from Heathrow every 45 seconds, or 80 per hour.

Discussion point

Should future airport development be concentrated on existing sites round London or be spread across the country?

Example

Teesside: impacts of industry on the physical environment

How UK industrial development is becoming more environmentally sustainable

Teesside is a well-known heavy industrial region. Located in North East England and centred on the towns of Middlesbrough and Stockton-on-Tees, it has 58 per cent of the UK's chemical industry and steel manufacturing remains important. The North East's industries contribute £26 billion annually to the UK economy, but this comes at a significant cost. These industries, plus electricity generation from fossil fuels, produce significant air pollution (Figure 13.13). Teesside is home to five of the UK's worst offending 25 companies in terms of **carbon dioxide emissions**; 5.6 per cent of UK emissions come from this region's industries, resulting in a huge carbon footprint.

Making Teesside industry cleaner

Permits to output high volumes of pollution will soon become so expensive (four times in 2030 what they were in 2015) that these businesses will lose their competitive edge. Millions of tonnes of carbon dioxide can be saved by retro-fitting carbon capture technology to chemical, steel and processing industries. This has been begun elsewhere in the UK, including Drax power station in North Yorkshire and Scottish Power's site at Peterhead, Aberdeenshire. The extracted carbon dioxide must be stored. So far this has been placed in containers and stored under the bed of the North Sea, so Teesside's proximity to the North Sea is a benefit.

Such a **carbon capture and storage (CCS)** process could help the region by:

- saving 5 million tonnes of carbon dioxide annually by the 2020s – this represents 90 per cent of emissions
- contributing to the UK meeting its target of reducing carbon dioxide output by 80 per cent by 2050
- maintaining the region's industrial competitiveness while securing 12 000 jobs and creating 2600 new ones between 2015 and 2025; this should generate greater spending power in the region.

Teesside remains a less well-off region. However, the growing University of Teesside with over 20 000 students has increased employment and retail potential. Guaranteeing the future of Teesside's heavy industry should enable new retailing and leisure projects to be developed.

Key terms

carbon dioxide emissions: carbon dioxide being emitted into the atmosphere as a by-product of industrial processes or other economic activities

carbon capture and storage (CCS): the process of capturing carbon dioxide emissions and storing them in a way that they are unable to affect the atmosphere.

Figure 13.13 Teesside heavy industry.

13.4 Social and economic changes in rural areas

Many rural areas are undergoing change – some are growing and doing well economically but others are in decline.

Kidlington – a growth area

Kidlington, Oxfordshire, is located on the edge of the city of Oxford, close to the M40 motorway and its mainline railway station to London and Birmingham. Kidlington's population has grown as shown in the census data in Table 13.1.

The village has become increasingly urbanised over this time. It grew by 30 per cent between 1981 and 2001. Small housing estates have been built to house people who work in the settlement itself, in the city of Oxford and in nearby urban areas. Today more people are working from home via broadband which gives them a greater choice of where to live – they no longer need to be in major urban areas. Typical new-build houses are medium to large, higher priced and are aimed at families. Their occupants are relatively well-off and have lived in the village less than ten years. Many have moved outwards from Oxford itself.

Village services have generally prospered. There are about 50 shops, banks and building societies, a public library, village hall and a weekly market. London Oxford Airport, established in 2009, originally Oxford Airport, has recently provided extra jobs.

There are issues for local long-term residents – the ability of local people, especially younger adults, to buy homes in the village is being eroded. The lack of affordable housing is acute.

Wrotham – a declining village

The economy of Wrotham near Tonbridge, Kent used to be based on rural industries, primarily agriculture, jobs which no longer exist. It tends to be the smaller, more remote villages which suffer decline; Wrotham is certainly small, but located in the densely populated South East of England – it is not remote (Figure 13.14). However, 2006 saw the closure of the post office, a crucial factor in decline, with social as well as economic consequences.

Regional change – the Isle of Purbeck

The Isle of Purbeck in eastern Dorset shows evidence of both growth and decline, which is true in many rural areas. Purbeck's many pretty villages attract relatively wealthy retirees and second home owners. This has led to a dramatically ageing population, with characteristics of Stage 5 of the Demographic Transition Model (birth rate = 10.1/1000/year; death rate = 11.9/1000/year). Only new in-migration saves the region from natural decrease (population 2001 = 44416; 2011 = 45000).

Local job opportunities are few and wages low, yet house prices have risen steadily, well beyond the reach of local people, leaving them little choice but to move to the cheaper towns nearby.

Public transport is limited in Purbeck. The only station is Wareham and there are four bus routes serving an area of 200 km². Three out of four villages do not have a general store and many have lost their post office. By 2014, 35 petrol stations had closed and eight pubs. Langton Matravers (population 853 in 2011) has virtually no services yet Corfe Castle (population 1398 in 2011), a tourist honeypot site, is much better served (Figure 13.15).

Date	Population
1901	1300
1961	4400
1981	10553
2001	13719
2011	13723

Table 13.1 Kidlington's population 1901–2011.

Figure 13.14 Despite its location in the prosperous South East England, the village of Wrotham is in decline.

Figure 13.15 Corfe Castle village, Isle of Purbeck, Dorset.

ACTIVITY 13.5

1 Describe the benefits of carbon capture and storage technology in Teesside to:
   a the industries concerned
   b the UK Government
   c local people.
2 Explain why village post office closure is so controversial.
3 Design a flow diagram to show the consequences of lack of shops and public transport for poorer families in Purbeck.

- Information on CCS projects in Teesside: Teesside collective (see www.cambridge.org/links/gase40118); business Green (see www.cambridge.org/links/gase40119); article in the *Northern Echo* (see www.cambridge.org/links/gase40120)
- Health impacts of heavy industries in Teesside (See www.cambridge.org/links/gase40121)
- Village decline in Wrotham, Kent (See www.cambridge.org/links/gase40122)
- Rural change in the Isle of Purbeck, Dorset (www.cambridge.org/links/gase40123)

Discussion point

Should people have complete freedom to move into an area even though their presence may cause economic and social difficulties for residents already living there?

13.5 What strategies have been used to resolve regional differences in the UK?

What is the North–South Divide?

Figure 13.16 shows the North–South Divide in the UK. The North and Midlands are generally less well off than the South and South East. However, poor areas exist within the South like Thanet (Kent) and wealthy ones such as Wilmslow (Cheshire) in the North. Cardiff and Swansea are just below national employment levels, but Rochdale lost 9300 (12.2 per cent) jobs between 2004 and 2013. Blackpool had fewer businesses in 2013 than in 2004. The North–South Divide reflects the differences between the **periphery** and **core** of the UK. London, as the capital and a global city, is an important driver of the economy.

Key differences between the regions:

- In the 19th century the North was the economic hub of the UK.
- Northern landscapes are hillier with poorer farmland.
- Health tends to be worse in deprived areas and these are more common in the North as Figure 13.17 illustrates, even though healthcare spending is higher.
- The South has higher population densities.
- Government spending on infrastructure is higher in the South.
- Earnings are higher in the South, as are house prices.

There are indications that overall growth in the UK's economy has been driven by the South of England. Even in the decade 2005–2015 the gap between the economies of northern and southern cities in England has widened significantly. For every 12 jobs created in the South, only one new job has been created in the North. In Scotland the economic divide is clearly urban/rural, with Glasgow and Edinburgh growing economically, but smaller towns and rural areas lagging behind.

Solutions to North–South differences: UK Government policies

Local devolution

In November 2014, Manchester became the first city region outside London to sign a deal with the government to have its own directly elected mayor, who will have increased powers to plan and run the city.

City Deals and new investment

An additional 27 northern towns and cities have been given limited extra powers in planning and transport through the City Deals system. £7 billion has been promised in future investment for the North.

Infrastructure investment – Salford Quays

By the early 1990s, £280 million of government and private funding had been invested in the regeneration of Salford Quays. High quality materials and designs were used throughout, even for street furniture. It is not the norm for businesses to move from south to north but in 2011 the BBC began a compulsory move of its departments from London to MediaCityUK, Salford. The Lowry arts venue and Imperial War Museum North attract local people and visitors.

EU regional funding

The UK Government has access to EU funds to promote regional development. The European Regional Development Fund (ERDF) is aimed at any region, core or periphery, which needs an economic boost. Its main aim is to support projects and activities that reduce regional disparity by stimulating economic development and increasing employment. Companies or local authorities can apply for funding to move their business or developments forward.

Key terms

periphery: the weakest regions of a country economically; they tend to be physically remote, making it more difficult to attract industry, for example, the Scottish Highlands

core: the strongest regions of a country economically; these are usually main cities and their surroundings, such as London and the South East

Figure 13.16 The UK North–South Divide.

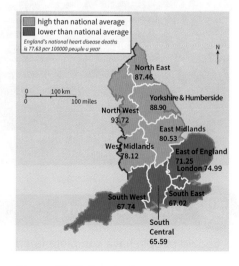

Figure 13.17 Deaths from coronary heart disease in England.

A renewal of northern cities

Many northern cities like Leeds, Liverpool and Manchester have grown in importance. Leeds is home to company headquarters of businesses such as British Telecom and the Royal Mail, as well as being the only Bank of England location outside London. Google's main offices are in Manchester.

Improving the North's image – the Angel of the North

Since 1998 Gateshead in North East England has been home to the Angel of the North sculpture (Figure 13.18). At 20 m tall and costing £800 000 of National Lottery money, one of its aims is to be a symbol of northern pride.

ACTIVITY 13.6

1 What does Figure 13.17 tell you about people's wealth and standard of living?
2 Put 'North–South Divide UK' into a search engine and click on 'images'.
   a Choose four images, fairly representing the North and the South.
   b On a copy of each, write annotations to show evidence of economic growth and problems.
3 How can the ERDF and other policies reduce the North–South Divide?
4 Explain the image of northern pride in Figure 13.18.

 Download Worksheet 13.2 from Cambridge Elevate for additional questions.

Figure 13.18 The Angel of the North sculpture, Gateshead.

 Further progress

Read the articles 'The North–South Divide – Where is the line?' (see www.cambridge.org/links/gase40130) and and 'Economic growth "dramatically" diverges between North and South' (see www.cambridge.org/links/gase40131).

 Did you know?

- While the City of London tends to grab more headlines, Edinburgh is the UK's second largest financial centre, employing thousands of people in financial services.
- The Centre for Cities ranked Glasgow second to London for people's skills.

 Discussion point

Should the UK Government:

- Invest in all regions equally?
- Invest more in less well-off regions?
- Concentrate on the wealthier regions which generate more GNI for the country as a whole?

- Facts and figures on the North–South Divide on the **BBC website** (see www.cambridge.org/links/gase40124) and in the *Guardian* (see www.cambridge.org/links/gase40125)
- Detail on the Salford Quays developments (see www.cambridge.org/links/gase40126)
- The iCon centre, Daventry and the ERDF (see www.cambridge.org/links/gase40127 and see www.cambridge.org/links/gase40128)
- Edinburgh as a financial services centre (see www.cambridge.org/links/gase40129)

13.6 The place of the UK in the wider world

Links through trade

The UK trades globally; for instance, February to April 2015 was typical: 49.6 per cent of exports went to EU countries and 50.4 per cent to the rest of the world. Exports to China were increasing rapidly and were 26.3 per cent higher than in the same three months in 2011.

The wider transport network

The UK's situation as an island country influences our transport patterns. Rail and ferry transport links across the English Channel are crucial (Figure 13.19). For example, imports such as out-of-season salad crops travel by road through France and then across the Channel by ferry or Eurotunnel.

Heathrow places the UK centrally in global air travel, resulting in a boost to the UK economy of £6.58 billion in 2013. Over the next 40 years world economic growth will be driven by Asia, North and South America. Without world-class travel connections the UK will lose out on attracting this business. Already 25 per cent of our exports to Brazil and China fly from Heathrow freight depots.

Electronic communications

Electronic communications overcome international and physical barriers in sharing information. The internet allows us to access information from most places. The number of businesses trading through the internet is growing year on year. The *Telegraph* newspaper reported on 14 December 2015 that 8 million people, almost one in six, run some kind of internet business from home. People employed outside the home are highly likely to utilise electronic communications. The day-to-day functioning of our economy has become reliant on the internet.

Much of the world has internet access. People in sub-Saharan Africa can have mobile phones. This helps health and education services as well as the economy. However, there are a few limitations – China, for example, censors its internet access.

Culture

UK culture is extremely mixed. Television represents this: *Coronation Street* is seen as representing northern culture and *EastEnders* as the South, but these scenarios are not the norm. Distinctive accents are characteristic, for example the North East, Liverpool and London's East End, but the situation is more complex. We are a multicultural society, due to migration between regions and centuries of immigration.

UK culture has spread globally through the Commonwealth. Many countries' government systems, health and education services are modelled on the UK's. India is an example. Students abroad may sit UK public examinations. UK media productions often have a global audience. The BBC exports many programmes abroad, just as we import American productions.

Key economic and political links between the UK and the rest of the world

The EU

The European Union began as the European Coal and Steel Community in 1951 with only six member countries – Germany, France, Italy, Netherlands, Belgium and Luxembourg. The most recent member is Croatia which joined on 1 July 2013, making 28 in total in 2015.

Factfile

The UK in the EU

- The UK was the fastest growing economy in Europe in 2015, because UK manufacturing and services were strong.
- Economically weaker countries like Spain and Italy were in economic difficulties.
- UK exports to stronger EU countries were reliable.
- UK exports to rapidly growing economies in Asia and Latin America were increasing.

ACTIVITY 13.7

1 Look at the **Your Heathrow website** (www.cambridge.org/links/gase40132) for a map of UK exports from Heathrow.

   a Make a list of products exported from London Heathrow and their destination countries outside the EU. Make a note of the importance of these products in terms of percentages.

   b Decide whether these are high or low value products.

   c Write a paragraph to explain why Heathrow is the UK's busiest cargo port by value.

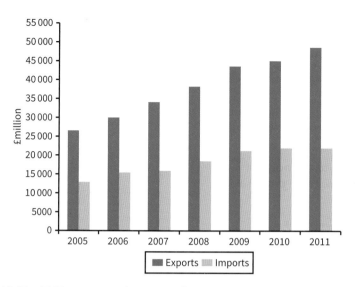

Figure 13.19 Total UK imports and exports of services to Europe, 2005–2011.

The EU accounts for around 50 per cent of our exports of services and 51 per cent of imports of services. Figure 13.19 shows the trend in value of both exports and imports has been upwards. The joint EU economy is larger than any individual national economy in the world – bigger than the USA or China. Nevertheless, as the NEE economies continue to grow, this situation is likely to change.

EU support for farmers through the Common Agricultural Policy (CAP) and funding for urban regeneration like the Urban II Fund have served the UK well.

We all have the right to live and work in any EU country, to free healthcare on holiday and consular protection all over the world from any EU consulate or embassy. Some UK students take advantage of lower university fees in EU countries.

The Commonwealth

The Commonwealth of Nations is an association of former UK colonies founded in 1931. In 2015 it comprised 53 states spread across the world. The Commonwealth's total population is 1.8 billion people which is between a quarter and one-third of the world's population. The Queen is Head of State of 16 Commonwealth countries today as well as overall Head of the Commonwealth. She attends the summit meetings and broadcasts to all the countries on 2 March, Commonwealth Day, every year. Commonwealth Headquarters remains in London.

Nevertheless, the UK is no longer as important within the Commonwealth. The secretary-general of the Commonwealth is elected by all the heads of government in the group.

The main benefit of membership is cooperation between countries. It promotes democracy, good government and human rights; Pakistan, for example, was suspended twice when under military rule and Zimbabwe over many years for violent elections.

The Commonwealth also promotes economic development. The UK trades with its previous colonies and the wealthier countries give preferential aid to the poorer ones.

Assess to progress

1 What are the benefits and challenges of globalisation for the UK?
2 What is meant by the term 'post-industrial economy'?
3 In what ways does the UK's infrastructure support and encourage its economy?

ACTIVITY 13.8

1 What are the advantages for the UK of being part of the EU?
2 Describe the trends shown in UK trade with the EU shown in Figure 13.19.
3 The UK has the most frequent online shoppers in Europe, spending £52.25 billion in 2015, twice as much as in 2005.
   a Use a search engine to find out 'How does the internet help the UK economy?'
   b Then write an answer to this question making sure you quote some evidence.

Download Worksheet 13.3 from Cambridge Elevate for additional questions.

13.7 Expansion of an industrial estate in Bournemouth

In this issue evaluation you will consider the importance of both tourism and the digital economy to Bournemouth's economic growth and consequently whether or not the Aviation Business Park should be expanded.

The Aviation Business Park is located approximately 3 km to the north of Bournemouth on the Dorset/Hampshire border (Figure 13.20). A proposal has been put forward to develop 35 of its 200 acres.

The park currently houses over 140 businesses including Siemens and Honeywell, and provides 2500 jobs. It is one of the most sought after business destinations in the region and also provides accessible starter units with the advantage of being located in the vicinity of an airport. A figure of £45 million has already been spent upgrading the adjoining airport with new facilities such as a departure terminal and arrivals hall.

Figure 13.20 An aerial view of the Aviation Park, Bournemouth.

The proposed plan by MAG Development will cost £60 million and will be phased over 15 years. It is hoped that 1200 jobs will be created. Many of the existing poor quality buildings will be demolished and the extension to the Business Park will include provision for a small supermarket, bank and a crèche to ensure more onsite services to help workers and reduce car trips. £1.1 million will be paid to Dorset County Council to improve transport links to the site. Opinions from local residents about the plan include:

- 'I'm pleased that this is going to provide an opportunity for more decent, permanent and non-seasonal jobs.'
- 'My biggest issue with the expansion is that it is going to result in the development of greenfield land. Bournemouth is built on tourism and this isn't going to help its image.'

The importance of tourism to Bournemouth's economy

Tourism is an important part of Bournemouth's economy and has remained relatively stable in recent years (Table 13.2 and Figure 13.21).

Type of visitor	How much do they spend?
Day visitor	£208.9 million (average spend per day £35.87)
UK tourist*	£190.9 million (average spend per day £63.28)
Overseas tourist*	£85.8 million (average spend per day £72.73)

*Of those that stayed overnight, 69% were on holiday which contributed a total of £189.9 million to the economy but there were other reasons for visiting too (Figure 13.21).

Table 13.2 Contribution of tourists to Bournemouth's economy.

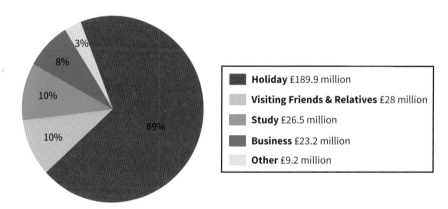

Figure 13.21 A pie chart showing the visiting reasons of overnight-staying visitors to Bournemouth in 2013.

In 2013 there were 6.88 million visitors to Bournemouth compared with a total for Dorset of 30 million. Most of these were day visitors (5.82 million) although 1.06 million stayed overnight. Table 13.2 shows the average spend of different types of visitor. Of the 1.06 million that stayed overnight, 890 000 were UK visitors who stayed an average of three nights each and 171 000 were overseas visitors who stayed an average of eight nights each. Figure 13.21 shows the visiting reasons for the overnight-staying visitors.

In the same year, tourism supported 12 345 jobs in Bournemouth, 9190 of which were directly related to tourism and a further 3155 of which were part of the tourism supply chain.

Tourism was also valued at £501 million which represents 15 per cent of Bournemouth's total economy. Most of this (30.2 per cent) came from the £151 million that tourists spent on food and drink, but other things that tourists spent their money on include shopping (£108 million or 21.6 per cent of the tourist economy), accommodation (£96 million or 19.2 per cent of the tourist economy) and travel (£77 million or 15.3 per cent of the tourist economy). Those that stayed overnight contributed most to the economy

The need to accommodate the increase in the digital economy

While tourism is a central part of Bournemouth's economy, in recent years the importance of the digital economy in the conurbation has increased. The digital economy is one that involves digital computing technologies such as telecoms and computer networks. Despite the 2008 global recession, 50 per cent of digital companies in the UK were formed after 2008 and 15 per cent of all UK companies formed in 2013–2014 were digital.

'74 per cent of digital businesses are actually based outside Inner London. What does that mean for the UK? The digital sector is growing and creating jobs, economic growth and national prestige. The UK has the fastest growing economy in the G7 and the digital industry is at the heart of our growth story.'

Baroness Joanna Shields, Chair of Tech City, UK

Location	Growth in new digital companies between 2010 and 2013	Digital companies formed in 2013–2014 as a % of total companies
National average	53%	15%
Bournemouth	212%	24%
Liverpool	119%	21%
Inner London	92%	20%
Brighton and Hove	91%	20%
South Wales	87%	18%
Belfast	73%	17%

Factfile

Total population of Bournemouth: 183 500

Number of people in Bournemouth of working age: 125 000 (66.2%).

Number of people employed in Bournemouth's digital economy: 7272 people (5.8% of Bournemouth's working age population)

% of people working in the digital economy nationally: 7.5% of the national working age population (2015); 12.9% (estimated) by 2020.

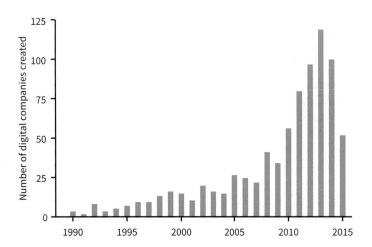

Figure 13.22 The number of digital companies created each year in Bournemouth, from 1975 to 2015.

Bournemouth has the fastest growing digital economy in the UK with the number of digital companies increasing by 212 per cent between 2010 and 2013 (Figure 13.22).

The top three industries in Bournemouth's digital economy are advertising and marketing, e-commerce, and games development and production. These, in turn, promote the capabilities of mobile and tablet development, visual and audio design, content and media production and digital marketing.

The seaside location with its associated lifestyle is an attraction to those wanting to set up business in the town. Similarities have been drawn to California's coastline and temperate climate. Bournemouth, and Dorset more widely, has already started to market itself as the 'Silicon South' in an attempt to mirror California's success as a 'Silicon Valley'. Development needs to take place in order to accommodate this rapid growth in the digital economy.

The advantages and disadvantages of developing a greenfield site

Advantages of developing greenfield sites	Disadvantages of developing greenfield sites
The development is new and so buildings and facilities will not require maintenance in the short term.	It can take longer for councils to approve plans compared with brownfield development.
The buildings can be designed to meet current and future needs.	There can be a lot of local opposition to greenfield development.
There is space for the design of the new buildings and facilities to be flexible.	The greenfield sites that are available may have physical issues such as sloping land or unstable ground conditions.
It provides an opportunity to improve the image of the area.	Transport links need to be built to connect the greenfield development to the nearby town.
If it is an extension to an existing project, many of the services such as sewers and water will already be in place.	If it is a completely new development sewers and water will need to be put in and this will lead to additional costs.

Tip

It could be useful to use the information that you have already put forward in previous questions to help you answer question 4. You might want to consider whether you think tourism or the digital economy is more important to Bournemouth's economic future. If you think that expansion is needed to accommodate the digital economy is the Aviation Business Park the best place to do this?

Assess to progress

1 Using Figure 13.20, describe the location of the Aviation Business Park. `3 MARKS`
2 Explain the importance of tourism to Bournemouth's economy. `4 MARKS`
3 Explain the importance of the digital economy to Bournemouth's economy. `4 MARKS`
4 'The Aviation Business Park should be developed in order to accommodate growth in Bournemouth's economy.' To what extent do you agree with this statement? Give reasons for your answer. `9 MARKS + 3SPaG MARKS`

Tip

Make sure that you include facts and figures to support the points that you make.

You will study:
- resource management
- where the world's resources are found

14 Demand on resources in the UK

In this chapter you will cover:
- how the demand for food resources is changing
- how is the UK's water supply managed?
- how the UK's energy demands and supplies are changing

You will also study one of the following chapters.

15 Food resources

In this chapter you will cover:
- the global patterns of food supply
- the global patterns of calorie intake
- how and why food consumption is increasing over time
- why food supply and consumption are not spread out evenly
- the problems caused by food insecurity
- ways that food supply can be increased
- how Kilombero Plantations Limited (KPL) benefits Tanzania
- what is meant by sustainable food supplies
- how the DRC is attempting to tackle food insecurity

16 Water resources

In this chapter you will cover:
- where our water resources come from
- where the world's water resources are consumed
- the human and physical impacts on water availability
- the factors that affect water consumption
- the different impacts of water insecurity
- the advantages and disadvantages of a water transfer project in China
- how we can move towards a sustainable water resource future
- the social and economic advantages of a local water conservation scheme in Kenya

17 Energy resources

In this chapter you will cover:

- where the UK get its energy from
- how global energy consumption is changing over time
- how a country's location affects its energy supplies
- why energy supplies change over time
- how energy insecurity affects different countries
- what the UK could do to improve its energy supply
- the advantages and disadvantages of oil exploitation in the North Sea
- measures that can be taken to reduce energy use
- how a Solar Mini Grid has improved life for people in rural Tanzania

Resource management – an overview

Population growth and economic development have increased the demand and competition for global resources. Food, water and energy are recognised by the World Bank as fundamental resources because they are vital for human development and without access to them countries will remain trapped in a cycle of poverty. Reliable and affordable food and water supplies are needed in order to lead a healthy life while energy supply is an important part of the process of industrialisation, which creates opportunities and can move people out of poverty.

The need to increase the supply of food, water and energy puts pressure on environments and has the potential to create political conflict. Moving towards a more sustainable resource future, where everyone has access to these fundamental resources is one of the major challenges of the 21st century.

Figure S6.1 Nuclear power is one of the most controversial power sources. The production process emits much less carbon dioxide than coal and gas, but it does produce radioactive waste. Do you think we should build more nuclear power stations in the UK?

Earth's resources

The world relies on a vast range of resources to fuel national economies and people's living standards. So far the Earth has delivered these, though at a cost. However the same Earth must produce enough food calories, water and energy for 9.6 billion people by the year 2050, based on the 2013–2014 World Resources Report. In addition, the aim is to increase standards of living of the world's poorest people. At the same time the United Nations and global governments are aiming to advance rural development, reduce greenhouse gas emissions, protect valuable ecosystems, create more water resources and explore energy possibilities. Taken together this is immensely demanding. The production of food, water and energy resources are tightly intertwined, both practically and economically.

Distribution of food supply and demand

Food production potential is controlled by many factors, including:

Climate

Temperature, precipitation and the annual patterns of these in different regions affect food production.

Soil potential

Large-scale river valleys, with their wide, flat valleys and silt loam soils have huge potential, but in 2015 are often already used to their maximum. Examples are the Ganges basin, the Indus (Indian subcontinent), the Nile (North East Africa), the Chinese Hwang Ho and Yangtze, and the Mekong which runs right through Southeast Asia (China, Myanmar, Laos, Thailand, Cambodia and Vietnam).

Level of technology

The USA has used water from western rivers like the Colorado to produce huge amounts of food crops, but that usage has been so heavy that the river's capacity to provide water has been overstretched and sections of the channel regularly run dry; in many poorer regions such as sub-Saharan Africa, simpler sustainable technology has much more chance of producing more food for the future.

Food demand is driven by:

Changing diet

As poorer countries improve the quality of life for their populations people have more disposable income and demand a higher quality diet including more meat and dairy products which take more land to produce. Beef is the least efficient source of calories and protein, as well as generating six times more greenhouse gases than pork, chicken or egg production. Fish-based diets would save millions of hectares of forest and savanna being sacrificed to new food production. Farmed fish production has great potential.

Increasing population

The greatest population increase is likely to occur in parts of the world where many people are undernourished and food supply is least secure (Figure S6.2). The world will need to produce 69 per cent more calories in 2050 than in 2006. The most productive but sustainable technology for each region will need to be employed.

Figure S6.3 shows the average number of kilocalories available per person per day across the world in 2013. Which areas had the lowest available calories? Which areas had the highest available calories?

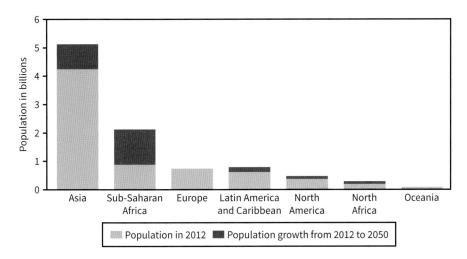

Figure S6.2 Projected population growth (in billions), 2012–2050.

Figure S6.3 Available food supply in kilocalories per person per day in 2013.

Water supply and demand

In the mid 20th century water was seen almost as an infinite resource. That was when global population was 4 billion – half of what it will be by 2025 (8.1 billion, according to UN data). There is increasing competition for water from food production (which is becoming increasingly intensive and therefore thirsty for water), biofuel agriculture, industry and urbanisation.

The processes of urbanisation concentrate an increasing number of people into relatively small areas, creating a huge demand for water from a limited area. Whilst early settlements were located close to water supplies, the scale of modern urban development, especially in newly industrialised countries (NICs) and newly emerging economies (NEEs), does not always think ahead in terms of water. Much development is not planned, yet people require water supplies.

More than one-fifth of the world's population (7.3 billion in 2015), over 1.2 billion people, live in areas of water scarcity, where there is simply not enough water to meet all their demands. The land on which Bangkok is built has physically sunk due to the dramatic drop in the water table due to over-extraction of groundwater for urban water supply. If there is an annual drop of 10 mm significant parts of the city could be underwater by 2030.

247

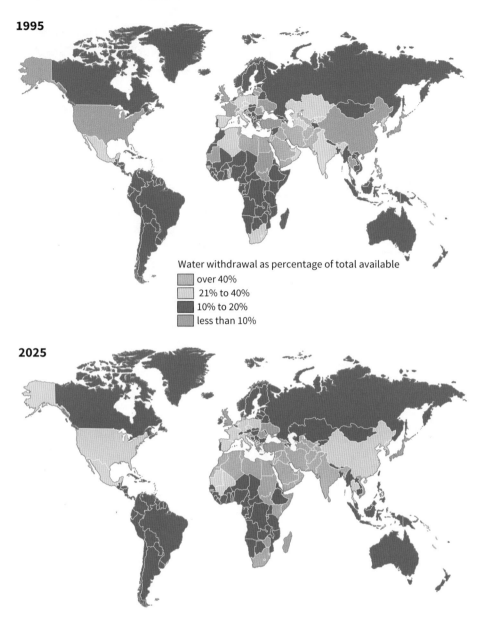

1995

Water withdrawal as percentage of total available
- over 40%
- 21% to 40%
- 10% to 20%
- less than 10%

2025

Figure S6.4 Global water stress.

One-third of the world's population does not have access to clean drinking water – over 2.4 billion people. You may have seen charity advertisements aiming to make people aware of people in this situation, especially in sub-Saharan Africa.

Figure S6.4 shows the distribution of water stress in 1995 and what is predicted for 2025. Note the differences – the shading indicates the significant changes. The USA, Central Asia, China, India and parts of North Africa have all increased water use threatening their supplies.

Energy – future challenges

Energy is needed for transport, economic development, technology and providing the power for everyday lives. Around 10 per cent of all transport fuels are predicted to come from biofuels by 2050. This is a way of reducing a country's carbon dioxide impact globally so governments are keen to use this resource to help them achieve their nationally agreed goals. However, such massive production of biofuel crops like sugar cane and maize competes for land with food crops as well as using up large quantities of water.

Figure S6.5 shows the UK and part of mainland Western Europe at night. It can be argued that a great deal of unnecessary energy is used at night when it is not really needed. Nevertheless, shops and businesses choose to be lit up to promote their businesses. In the future higher energy costs may preclude this.

Figure S6.6 shows the global distribution of energy production in 2014, while Figure S6.7 shows the global distribution of energy consumption in 2014. Can you see any patterns in the distributions? Which countries must be importing energy? Which countries must be exporting energy?

Figure S6.5 The UK and part of mainland Western Europe at night.

Below 50
50 to 100
100 to 200
200 to 1000
Above 1000

Figure S6.6 The global distribution of energy production in 2014.

Below 50
50 to 100
100 to 200
200 to 2000
Above 2000

Figure S6.7 The global distribution of energy consumption in 2014.

ACTIVITY

1 You can use a search engine to find many maps similar to Figure S6.5 of the Earth at night.

2 What causes the clusters of lights and what are the consequences of this in terms of energy demands?

249

In this chapter you will learn about...
- the changing demand and provision of food resources in the UK
- the changing demand and supply of water in the UK
- how to safeguard water resources and supply in the UK
- how the UK's energy demands and supplies are changing
- the problems caused by the exploitation of energy sources.

14.1 How is the demand and provision of food resources in the UK changing? (1)

What food resources does the UK produce?

The UK's climate is cool temperate western margin. July average temperatures vary from 18 °C (South East England) to 15 °C (North Eastern Scotland). Precipitation varies from west to east as westerly winds bring rain and drier air masses affect the east (Figure 14.1). Farm produce therefore varies by location.

Dairying requires good pasture so dominates in the west (Figure 14.3). Wheat cannot tolerate too much rain so is especially productive in East Anglia. Relief matters too: high land in the west has relief rainfall which promotes sheep farming (Figure 14.4). Easterly regions are in a rain shadow.

During the 19th century the UK's global empire provided a huge resource-base to feed its growing population and industries, providing a comparative advantage. Food resources not available in our temperate climate became accessible, including the UK's favourite drink, tea.

Having been almost self-sufficient in food during the Second World War, the UK has since produced a lower percentage of its needs. We rely partly on imports and some traditional crops have declined due to foreign competition, for example cheap apples from France led to reduced production in Kent.

Demand for high-value food in the UK

Demand patterns for food in the UK are changing in the 21st century. The growing economy has given people a greater disposable income and some choose to buy high-value, specialist food products; for example, It is now easy to find specialist Japanese mushrooms. Some people are also choosing to buy organic produce, which is more expensive because production costs are higher.

Globalisation of food

UK supermarkets supply foodstuffs from around the world. Many are **primary products** from lower-income countries. Importing provides tropical foods we cannot grow ourselves and out-of-season fruits and vegetables. Without these imports, food choices would be more limited. Some people think the UK is

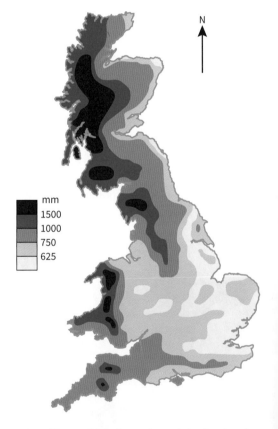

Figure 14.1 Annual precipitation levels in the UK.

 Key term

primary products: raw materials from the Earth, including fossil fuels, uranium (fuel for nuclear power), metal ores, crops, fish and timber

jeopardising its **food security** by relying on vast quantities of imported fruit and vegetables, many of which could be grown in the UK (Figure 14.2).

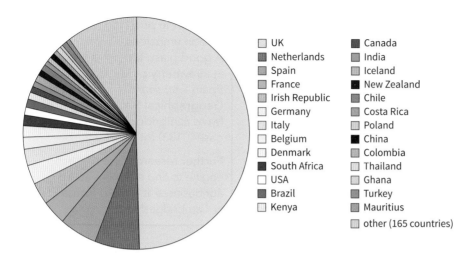

Legend:
- UK
- Netherlands
- Spain
- France
- Irish Republic
- Germany
- Italy
- Belgium
- Denmark
- South Africa
- USA
- Brazil
- Kenya
- Canada
- India
- Iceland
- New Zealand
- Chile
- Costa Rica
- Poland
- China
- Colombia
- Thailand
- Ghana
- Turkey
- Mauritius
- other (165 countries)

Based on the farm-gate value of unprocessed food

Figure 14.2 Origins of UK food.

Key terms

food security: having a reliable, affordable food supply

carbon footprint: the amount of carbon dioxide released into the atmosphere by an economic activity (e.g. person, business or event)

food miles: the distance food is transported between producer and consumer

agribusiness: highly-intensive, large-scale commercial farming

- >30 000 tonnes each of apples, onions and tomatoes were imported each month in 2012
- 20 000 tonnes of broccoli and cauliflower were imported in May 2014
- 50 000 tonnes of potatoes were grown outside the UK in April 2015

The UK's food industry is part of global trade – we import, we export. Much animal feed comes from abroad, while we export many food and drink products, especially salmon, chocolate, cheese, beef and whisky.

Food miles and the carbon footprint

The **carbon footprint** of the UK's food is increasing because food is being transported for greater distances. **Food miles** also add to the cost of the UK's food. Food imports travel 18.8 billion miles (30 000 billion kilometres) annually each year. Around 68 per cent of food imports come from the EU, the rest from further away (Figure 14.2). This means that 19 million tonnes of carbon dioxide are added to the atmosphere by transporting food in the UK.

In the UK we choose from a wide range of foods all year, yet in France seasonal crops are the norm. Buying local food in season cuts food miles and carbon dioxide emissions dramatically. Today, in the UK, more people want to buy locally produced fruit, vegetables and meat. Many small businesses are growing based on this demand.

At the same time, more food is produced in huge, intensive operations. Commercial, large-scale farming is called **agribusiness** and it describes the increasing inputs into farming in the developed world, for example specialist seeds, agricultural chemicals and large-scale machinery. This trend has grown since the 1970s – farming has become big business with profit as the main motive. However, the cost of food produced by agribusiness is lower because overall production costs are lower.

In the UK, agribusiness has changed the landscape of East Anglia, especially Cambridgeshire. Farm size has increased dramatically due to farms merging to achieve economies of scale. Fewer farm-workers are needed as mechanisation has taken over.

14.1 How is the demand and provision of food resources in the UK changing? (2)

Hedgerows have been removed to create huge fields to make the big equipment efficient, but this has led to lack of protection from the wind and hence to soil erosion. Hedgerows were important habitats for wildlife, resulting in a loss of biodiversity. There is evidence of agricultural chemicals getting into the water table and into the food produced, and important pollinating insects like bees being killed by chemical spraying. Some people object to the large-scale rearing of animals indoors.

Figure 14.3 Dairy cattle grazing in lowland SW England.

Figure 14.4 Sheep grazing in the Lake District mountains in NW England.

Agribusiness has led to considerable social changes. Many villages used to be dependent on farming for jobs, but people have had to leave to find work in towns. On the other hand, these villages have become dormitory settlements and have lost many services.

ACTIVITY 14.1

1. Explain the links between climatic conditions and agricultural production.
2. Identify a selection of fruits and vegetables in your local supermarket.
   a. Note their origins.
   b. Map these locations on a world map.
   c. You could construct a desire line map or a choropleth, with darker shading to represent more products from a particular country.
3. List the advantages and disadvantages of importing food into the UK.
4. Compare and contrast the advantages to the UK and to lower income countries of food trade.

 Visit the **Climate Choices web page** (www.cambridge.org/links/gase40134) relating to food miles and carbon dioxide which contains some good diagrams.

Read an article from *The Independent*, **'Food miles: The true cost of putting imported food on your plate'** (www.cambridge.org/links/gase40135).

 Further research

Investigate the costs of food transport in price per mile and carbon emissions per mile. A good place to start is looking at strawberry production in Spain – try searching the **Royal Geographical Society website** (www.cambridge.org/links/gase40133) for information.

Further research: Use a search engine to find out more about agribusiness at Lynford House Farm in Cambridgeshire.

 Discussion point

Many imported food products are grown in poorer countries. What impact would richer countries reducing imports have on them?

Why do food miles matter?

14.2 How has the demand for water changed?

Only 2.7 per cent of the world's water is fresh and 77 per cent of that is unavailable, stored in ice sheets and glaciers. Most fresh water is stored as **groundwater** in the **aquifer** and only some is accessible for people to use. Globally 70 per cent of accessible fresh water is used in farming and 23 per cent in industry.

The hydrological cycle shows our water supply sources (Chapter 16: Water Resources, Figure 16.1). Depending on where you live in the UK, your water comes from a river (surface flow), a **reservoir** (surface storage) or the groundwater store.

Five litres is needed per day for a human being to survive, but many people in poorer countries have less. The UK average daily domestic water consumption is 175 litres per person compared with 45 litres per person in Bangladesh. Water demand grows due to population increase and higher standards of living and individual use has increased by 1 per cent annually since 1930.

Only 4 per cent of drinkable water in the UK is actually drunk. The rest is used for toilet flushing (30 per cent), baths and taps (21 per cent), showers (12 per cent), washing clothes (13 per cent), washing up (8 per cent), outdoor (7 per cent), and other (5 per cent) (*Source: Waterwise*). Gardens and leisure activities in summer are a huge drain on supplies when water stores are lowest.

Water quality and pollution management

Pollution and water quality management is a constant challenge in the UK. The UK water supply is amongst the best quality in the world. Nevertheless, sources of environmental pollution include:

- nitrates and phosphates from fertilisers
- pesticides and weedkillers
- heavy metals from industrial waste
- acid rain.

UK and EU guidelines mean water quality is regularly tested and 99.7 per cent of our water complies with standards, unlike poorer countries. The UK needs to improve the quality of its water bodies – rivers, streams, lakes and estuaries – and in some cases its groundwater. Only 27 per cent of open water in England is up to EU standards in this respect. Making improvements will create safer water for leisure, and preserve jobs and habitats. Do not forget that all water for drinking and domestic use is cleaned by the water authorities before being fit for use.

The Environment Agency responds to pollution incidents. Along with Defra (the Department for Environment, Food and Rural Affairs) and the Coal Authority it monitors abandoned mine sites which can leak dangerous metals like cadmium, lead and zinc. Urban water pollution is a permanent problem because the sources of pollution are our everyday activities such as transport, car washing and construction. This is a particularly difficult challenge.

More success has been achieved with phosphates. Phosphates enter rivers through sewage from detergents used in washing machines and dishwashers. They cause algae to grow in open water, reducing oxygen and choking the plant and animal life. The amount of phosphate allowed in laundry detergents was reduced by law in 2013, and will be for dishwasher products in 2017.

Phosphates from agricultural fertilisers are more difficult to manage because their pollution can spread a long way. It is the farmer's responsibility to control pollution but mistakes happen. The government gives advice and financial help to limit farm pollution.

Key terms

groundwater: water found underground in soil, sand or rock

aquifer: a body of permeable rock that can store water and through which water can easily move

reservoir: a man-made lake, or natural one which has been adapted, used for collecting and storing water supply

14.3 Matching supply and demand – areas of deficit and surplus

The UK has a mismatch between water **supply and demand**. The higher, wetter North West is less populated than the lower-lying, drier South East, with its high average population density. This creates a North/South divide in **water security** (Figure 14.5): secure areas have a surplus while insecure areas have a deficit. Figure 14.6 shows evidence of water insecurity in South East England.

Water transfer?

There are several **water transfer schemes** in England and Wales (Figure 14.7). Built between 1897 and 1925, the 95-mile-long Thirlmere Aqueduct carries 250 000 m³ of water per day from Thirlmere reservoir in the Lake District to Manchester, using pipelines and concrete-covered channels.

Further large-scale transfers are planned to fulfil future demand. There are cost and environmental issues. One huge scheme proposes to take water from Kielder Reservoir (Northumberland) to restock London reservoirs, a distance of around 350 miles.

Did you know?

A running tap wastes 6 litres per minute. If everyone turned the tap off while cleaning their teeth, this would save enough water to supply 500 000 homes.

Key terms

supply and demand: supply means how much of something is available; demand is how much of it people would like

water security: having a safe and regular supply of a key resource; in this case, water

water transfer schemes: systems of pipelines and aqueducts to move water long distances from areas with plentiful supply to those of high demand

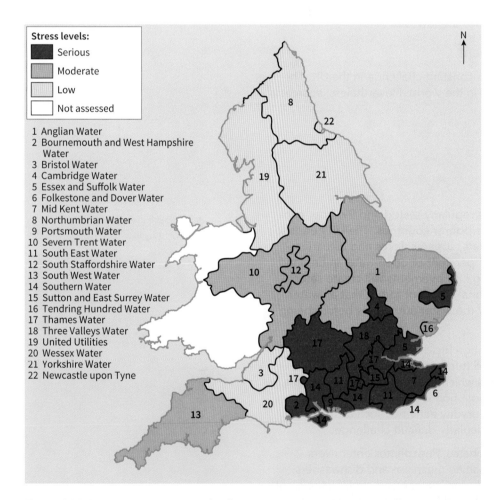

Stress levels:
- Serious
- Moderate
- Low
- Not assessed

1 Anglian Water
2 Bournemouth and West Hampshire Water
3 Bristol Water
4 Cambridge Water
5 Essex and Suffolk Water
6 Folkestone and Dover Water
7 Mid Kent Water
8 Northumbrian Water
9 Portsmouth Water
10 Severn Trent Water
11 South East Water
12 South Staffordshire Water
13 South West Water
14 Southern Water
15 Sutton and East Surrey Water
16 Tendring Hundred Water
17 Thames Water
18 Three Valleys Water
19 United Utilities
20 Wessex Water
21 Yorkshire Water
22 Newcastle upon Tyne

Figure 14.5 Water insecurity in England.

Figure 14.6 Bewl Water reservoir, Kent, in a dry summer.

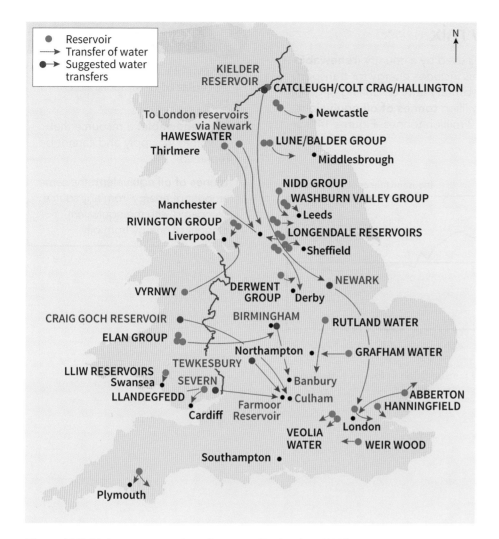

Figure 14.7 Major water transfer schemes in England and Wales.

Further research

- Investigate the Thirlmere Viaduct system and other existing/future water transfer networks such as Kielder or the London Circular system.
- Find out where your local water supply comes from. Is it reservoir, river or groundwater? Are you in a region of surplus or deficit?
- Use the Waterwise website to see how you can conserve water. What is 'grey water'? How can it be useful in conservation?

Discussion point

Calculate how much water you and your household use in 24 hours. Compare with other members of your class. How might you cut down?

One solution to water deficit is to use less water per person. How might people be persuaded to do this?

Skills link

Activity 14.2 provides practice in atlas use, diagram design, field sketching and annotation skills.

ACTIVITY 14.2

1 Use the Waterwise data to design a diagram of proportional water use in the UK.
2 Use your atlas to research population density and precipitation maps for the UK. Figure 14.1 will also help.
   a Describe the pattern of each and identify areas of possible water stress.
3 Explain why water supplies must be treated and monitored.
4 What are the problems caused by large-scale water movement systems? Are they worth it? You decide!

To find out more:

Read the Water fact sheet on the **Waterwise website** (www.cambridge. org/links/gase40136) which explains water use in the UK and tips on water efficiency.

Read the article in the *Guardian* **'New water future flows from engineering's past'** (www.cambridge.org/links/gase40137) which discusses the need to move water from the north and west to the south and east.

Look at details on **Wikipedia** of the system taking water from Thirlmere in the Lake District to Manchester (www.cambridge.org/links/gase40138).

14.4 The changing energy mix

Energy mix means the range of energy sources used by a country (**renewable** and **non-renewable**) and their proportions. This includes energy for transport. The UK was dependent on coal from the 19th century until oil-drilling began in the 1950s. In 1970 coal consumption was 57 million **tonnes of oil equivalent**, but by 2012 that had plummeted to less than 3 million tonnes. Figures 14.8 and 14.9 show patterns of UK energy use.

Key terms

renewable: a resource that can be replaced or replenished over time

non-renewable: a resource that is limited in supply and cannot be replaced

tonnes of oil equivalent: the same amount of energy from any source compared with the equivalent amount of energy from oil

Total energy consumption (1970–2030)

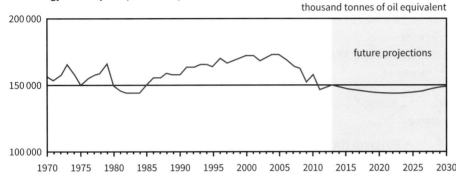

Figure 14.8 A line graph showing the total energy consumption in the UK between 1970 and 2030.

UK energy mix (2013)

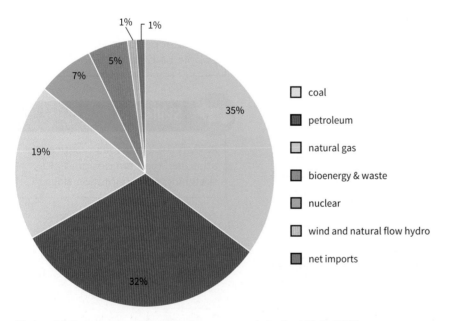

Figure 14.9 A pie chart showing the energy mix in the UK, in 2013.

Electricity is the source of lighting, some heating and household equipment. It can be generated from coal, oil, gas, wind, nuclear fission, and flowing water. Most UK coal now goes to make electricity. In 2014 just 6 per cent of UK electricity was generated from renewable sources. The EU had told the UK Government that this figure needs to rise to 15 per cent by 2020 and keep on rising thereafter. The UK Renewable Energy Strategy aims to reach 30 per cent by that date.

It is projected that the total energy use in 2030 will be similar to that in 1970, but will be from a very different balance of sources. Use of petrol and diesel for vehicles stays steady because, although there are more vehicles, they are more efficient.

Fossil fuels v. renewable energy

Most countries rely heavily on non-renewable **fossil fuels** – coal, oil and gas. In 2005, 85 per cent of global energy used was from these sources. Demand for energy increases as population grows and quality of life improves, especially in rapidly developing NEEs like China, India and Brazil. Fossil fuel reserves will not last forever.

There are three solutions:

- find new fossil fuel resources, including shale gas
- use energy more efficiently
- develop renewable energy.

The UK's fossil fuel reserves

The 1980s and 1990s saw many coal mines run out of profitable coal; there were strikes and closures. Cheaply produced open cast coal was imported from Poland, Russia and even Australia to feed UK power stations. UK coal producers are responding where they can. With only ten deep mines in the UK today, surface mining began again in 2000; by 2014 there were 22 sites including Shotton in Northumberland with 6 million tonnes reserve. New mines are proposed in areas such as Highthorn near Druridge Bay, Northumberland, but they are meeting opposition on environmental grounds.

Oil and gas come from UK North Sea offshore fields, tanker deliveries from the Middle East and the European gas pipeline. In 2008 the UK was the 14th largest global producer of oil and gas, but yields have decreased. In 2008, gas production in the UK was 68 202 million cubic metres. In 2014 this had fallen to 34 849 million cubic metres. New finds have been located near Shetland, but sea conditions are very difficult. The UK needs to develop alternatives such as nuclear power and green energy.

Issues of energy exploitation

In any consideration of the energy mix, there are issues that need to be understood:

- UK energy bills are high and people increasingly struggle to pay.
- Renewable energy technology like wind turbines is expensive to set up, but cheaper to operate.
- Emissions from burning coal caused serious air pollution, but today this is reduced by **carbon capture** and **sulphur scrubbing**, which are processes that remove pollutants when coal is burned to make electricity.
- Transporting oil can cause massive pollution. The *Torrey Canyon* tanker spilt 120 000 tonnes of crude oil off Cornwall in 1967, affecting beaches and wildlife from the UK to Spain.
- Some people fear radiation leaks from a nuclear power accident like Chernobyl (Ukraine 1986) or Fukushima (Japan 2011). However, nuclear power is reliable. By the late 2020s, it will contribute more to the UK's energy mix as the new generation of nuclear power stations come online.

Key terms

fossil fuels: energy from plant and animal remains, such as coal, oil and natural gas, including shale gas

carbon capture: where carbon dioxide produced by power stations is stored underground

sulphur scrubbing: removing sulphur from flue gases before they escape from the power station chimney

Watch a video on Cambridge Elevate about the European gas crisis in 2009.

14.5 What is the future of energy in the UK?

The UK has entered a period of declining outputs of North Sea oil and gas, although these energy sources will still dominate the UK's energy mix in 2020. Coal-fired power stations will be closing from 2015 onwards, partly because some are at the end of their useful lives and partly to fit in with emissions standards set by the Large Combustion Plant Directive. The UK is moving towards being a low-carbon economy. Nuclear energy generation produces virtually no carbon dioxide (only in the initial construction) but several nuclear power stations are approaching the end of their productive lives.

The aim is to be as self-sufficient as possible or rely on 'friendly' countries as some governments use access to energy supplies as a political tool, as in the case of Russia putting pressure on Ukraine in 2008–2009. The UK's supply margin (the excess of supply over demand) is only 2 per cent in a cold winter such as that predicted for 2015–2016. Fortunately, despite a growing population, peak demand has fallen significantly from 60 gigawatts (2005–2006) to 54 GW (2013–2014), which is an important factor in keeping the lights on.

Figure 14.10 shows further reduction in demand for electricity by 2020 and a significant increase in the contribution made by renewable sources, projected to be least 30 per cent.

Download Worksheet 14.1 from Cambridge Elevate for an activity on North Sea oil.

Did you know?

Oil is one of the most political resources in the world. Wars have been fought over it in the Middle East. North Sea oil started flowing in 1975, making the UK much more energy secure.

Tip

It is essential to offer a balanced set of arguments. You can state your own opinion, as long as you can support it with reasons/ evidence.

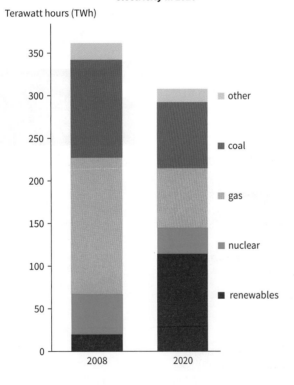

Renewables would need to be the largest single source of UK electricity in 2020

Terawatt hours (TWh)

Key:
- other
- coal
- gas
- nuclear
- renewables

Figure 14.10 Changes in the mix for electricity generation in the UK 2008–2020.

Source: 'Key issues for the New Parliament 2010', www.parliament.uk

Economic issues of energy exploitation

The main economic issue is the cost of building so much new infrastructure, including:

- new nuclear power stations
- new pipelines to import gas from Norway
- new terminals for imported liquid gas
- new gas storage
- a new 'smart' (more responsive) National Grid
- new carbon capture and storage systems for coal-fired power stations

Is shale gas the way forward?

Shale gas, natural gas stored in permeable ground rock, is found in Bowland and Fylde (north Lancashire), the Weald (Sussex) and Northern Ireland. Exploration shows that Fylde might yield £6 billion of gas per year for 30 years.

Fracking, the extraction process, means deep wells are drilled (900 m in Sussex), and explosions are set off to break rock and release underground gas. There is evidence of earthquakes and water supply pollution from the USA drill sites. Demonstrations by residents and environmentalists against fracking took place in Balcombe, Sussex, in 2013 (Figure 14.11). The government remains determined to exploit this new form of fossil fuel to increase **energy security**.

Figure 14.11 Caroline Lucas (a Green Party MP) demonstrates at Balcombe, Sussex, in 2013.

Key term

energy security: availability of affordable energy; not being reliant on countries who may cut supply for political reasons (e.g. Russian gas supply to Ukraine in the 2000s)

Download Worksheet 14.2 from Cambridge Elevate for a revision quiz.

ACTIVITY 14.3

1 Discuss the reasons for and against the UK increasing opencast coal mining. Research on the internet if you need to.
2 Explain why people in the UK might be concerned about the expansion of nuclear power.

To learn more

Read about the redevelopment of UK opencast mining in **The Journal** (www.cambridge.org/links/gase40139).

Read about the beginning of North Sea oil exploitation on the **BBC website** (www.cambridge.org/links/gase40140).

Look at the impact of oil spills on wildlife on the **National Wildlife Federation website** (www.cambridge.org/links/gase40141).

Read about fracking demonstrations at Balcombe on the **BBC website** (www.cambridge.org/links/gase40142).

Assess to progress

1 Describe the UK's energy mix, commenting on the country's dependence on fossil fuels. **4 MARKS**

2 Suggest how the UK could reduce its food miles and carbon footprint. **4 MARKS**

3 Explain how the UK could increase its water security. **6 MARKS**

Fieldwork

- Design a questionnaire to discover people's attitudes to the different energy sources in the UK. Ask both open and closed questions. Try your questionnaire on people in school and your community.
- Draw suitable diagrams to present your results clearly. Draw conclusions from your diagrams and compare with the findings of your classmates.

14.6 Opencast coal mining in Northumberland

In this issue evaluation you will consider whether or not a new opencast mine should be developed at Druridge Bay, Northumberland.

Opencast mining is a surface mining technique where coal is extracted from an open pit (Figure 14.12). This technique is used where the coal deposits are found near the surface and the overburden – the surface material which covers the coal – is relatively thin and easily removed.

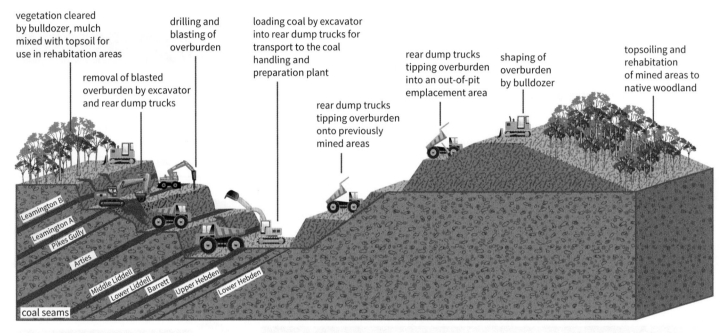

Figure 14.12 Opencast mining.

Coal is not a major part of the UK's energy mix, and accounts for about 20 per cent of electricity generation behind gas (30 per cent), renewables (25 per cent) and nuclear power (21.5 per cent). However, in 2012 there were 19 coal fired power stations providing 103.7 TWh of electricity per year. This required 41.9 million tonnes of coal, four times more than is being domestically mined. Most of this demand is met by cheap coal imports from the United States.

Factfile

- The UK mines approximately 10 million tonnes of coal from its 26 opencast mines each year.
- Most of this comes from Scotland.
- In 2014 there were four operational opencast sites in Northumberland which produced 1.7 million tonnes of coal (Figure 14.13).

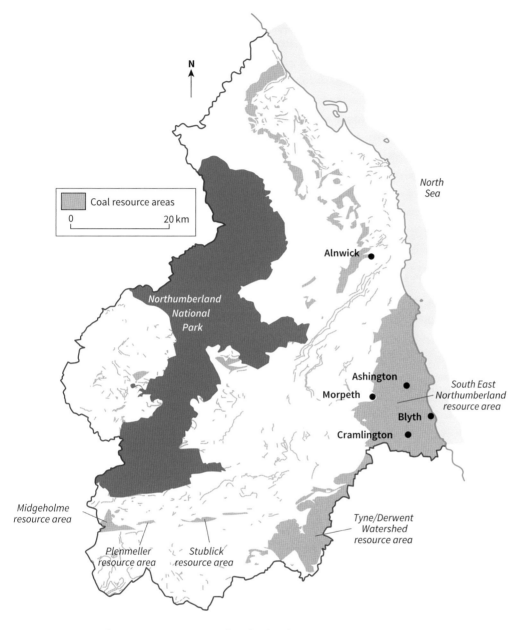

Figure 14.13 Coal resource areas in Northumberland.

The proposed site for a new opencast coal mine

Banks Mining – a company that develops and operates surface coal mines in the UK – has plans for a large opencast coal mine near Druridge Bay, Northumberland. Druridge Bay is part of the Northumberland Heritage Coast (Figure 14.14). It is popular in summer with tourists and is a popular location with nature lovers, dog walkers and photographers throughout the year. The proposed 1700-acre site, known as Highthorn, lies between the villages of Ellington in the south and Widdrington in the north (Figure 14.15). The company propose to mine 7 million tonnes of coal at the site over ten years and estimate that this will create more than 150 jobs.

Figure 14.14 Druridge Bay in Northumberland.

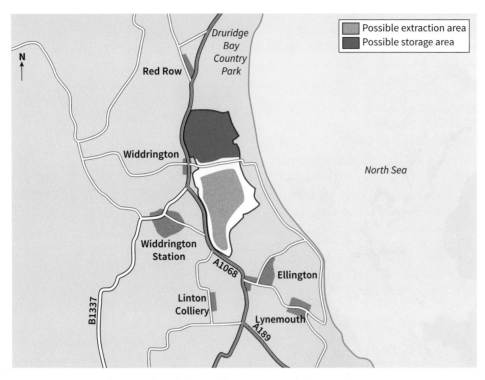

Figure 14.15 Northumberland Coast showing possible extraction area.

The benefits of developing Highthorn

Banks Mining have identified a number of social, economic and environmental benefits for the area if the proposal is accepted:

- £450000 community fund which will be split to fund local good causes and to help people overcome barriers to work. The proposal also includes the creation of 7.9 km of footpaths, bridleways and cycle paths to encourage healthy lifestyles.
- It will create and maintain jobs which will generate £13.2 million in wages over the life of the site. The proposal is estimated to contribute £6 million in taxes to the UK economy.
- Banks Mining has also negotiated the halt in extraction of 422000 tonnes of sand used for construction which will save an estimated £844000 for Northumberland County Council. Planning permission for sand extraction has been in place for 45 years. However, this is under review and Northumberland County Council would have been liable for compensation payouts had Banks Mining not intervened.
- When the area was first mined in the 1950s the landscape was left flat, featureless and designed for agriculture rather than natural wildlife. The company has held talks with Northumberland Tourism and Northumberland Wildlife Trust as well as local residents and parish councils to try to create a high quality land restoration when the project is completed in 2027.
- A total of 151 hectares of new wetlands and wet grassland habitats will be created around Druridge Bay and wildlife hides and a dark sky lookout will be constructed.

For more information see
Banks Group website
(www.cambridge.org/links/gase40143)

Local and global concerns

However, despite these perceived benefits, concerns have been raised by the local community who believe that increased drilling, blasting and transportation will create high levels of noise. They also believe that there will be increased fine particulate air pollution which can cause irritation to the eyes, nose and throat, contribute to respiratory diseases and is also linked to cardiovascular disease. The extraction and storage areas also may not be set back far enough from the beach to be out of sight for tourists which may deter them from visiting the area (Figure 14.16).

As well as local impacts, the 'Save Druridge' website also cites the global impact of climate change. In order to meet carbon emissions targets the UK needs to reduce the burning of fossil fuels by 3 per cent per year. In November 2015 the UK government announced plans to close the remaining coal-fired power stations by 2025. The International Energy Agency estimates that the global demand for coal will need to fall by 45 per cent by 2050, in comparison with 2009 levels. Residents argue that money should therefore be invested in sustainable, renewable energy technology such as solar and wind technology, rather than investing in unsustainable and old-fashioned solutions such as opencast coal mining.

Figure 14.16 The impact of open-cast mining.

Assess to progress

1. Study Figure 14.15 which shows the location of the proposed Highthorn site. Describe the location of the site. `2 MARKS`

2. Outline why Highthorn may be a suitable site for the development of an opencast mine. `2 MARKS`

3. Suggest two groups of people who may be against the development of an opencast mine at Druridge Bay and briefly explain why they are against the development. `4 MARKS`

4. Describe the social advantages of the proposed scheme. `4 MARKS`

5. Should Banks Mining go ahead with their proposal for opencast mining at Druridge Bay? Justify your answer. `9 MARKS + 3 SPaG MARKS`

Tip

Consider the characteristics that would make a site suitable for the development of an opencast mine. How many of these characteristics does Highthorn have?

Tip

Make sure you only focus on the social advantages here and not disadvantages or any other category, e.g. economic or environmental.

15 Food resources

In this chapter you will learn about...
- the global patterns of food supply
- the global patterns of calorie intake and identify which countries consume the most calories
- how and why food consumption is increasing over time
- why food supply and consumption are not spread out evenly
- the problems caused by food insecurity
- the ways that food supply can be increased
- what is meant by sustainable food supplies.

15.1 Where does our food supply come from?

Almost half of the usable land surface in the world is now being farmed. Different countries and continents specialise in producing different types of food, which they export to one another using aeroplanes, container ships and temperature controlled storage. A wide range of foods can be accessed throughout the whole year.

Table 15.1 shows where some foods originate.

	Cereal (millions of tonnes) 2015	Wheat (millions of tonnes) 2015	Rice (millions of tonnes) 2015	Oil crops (millions of tonnes) 2015	Meat (thousand tonnes) 2015	Milk and milk products (thousand tonnes milk equivalent) 2015	Fish and fish products (million tonnes live weight) 2013
Asia	1 123.2	321.1	452.8	134.5	135 727	313 370	62.5
Africa	165	26.7	18.5	17.7	17 286	46 612	1.6
Central America	41.4	4.1	2	1.8	9 030	17 367	0.4
South America	172.4	24.1	16.9	175.2	42 765	71 549	2.1
North America	481.9	85.5	7	141	47 481	104 738	0.6
Europe	488.5	233	2.6	67.5	60 324	220 100	2.8
Oceania	36.9	24.7	0.5	4.7	6 089	30 780	0.2
Total	2 509.3	719.2	500.3	542.4	318 702	804 516	70.2
% LICs	57.75	48.40	96.32	59.57	61.33	51.66	94.02
% HICs	42.25	51.59	3.64	40.41	38.67	48.34	5.98

Table 15.1 Global crop, oil, meat, dairy and fish production.

This data shows several patterns:
1. Asian countries are the biggest producers of wheat and rice.
2. South American countries grow the most oil crops.
3. Asian countries are the biggest producers of all types of meat, milk and milk products, and fish and fish products.
   - China produces the majority of each type of meat product and also fish and fish products.
   - India produces the majority of milk and milk products.
4. European countries are the second biggest producers of milk and milk products, and fish and fish products.

15.2 Food security

Food supply is uneven across the globe; there are areas of surplus and deficit. **Food security** occurs where people have access to an affordable and nutritious food supply. Many countries experience **food insecurity**, which is the opposite. The Food and Agriculture Organization of the United Nations (FAO), an agency working internationally to end hunger, produce a map of world **hunger** to monitor food insecurity. Figure 15.1 shows the world hunger map for 2014–2016.

> ### Key terms
>
> **food security:** having a reliable, affordable food supply
>
> **food insecurity:** having no reliable access to affordable and nutritious food
>
> **hunger:** prolonged or frequent undernourishment, where a person has insufficient calories to carry out light activities

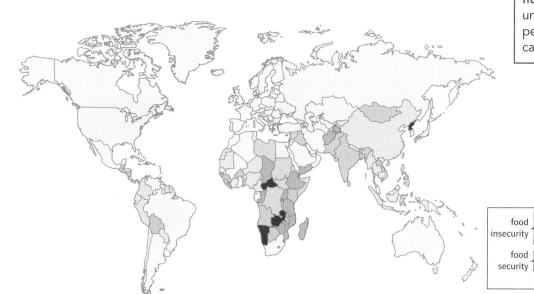

food insecurity	35% and over – Very high
	25%–>34.9% – High
	15%–>24.9% – Moderately high
food security	5%–>14.9% – Moderately low
	<5% Very low
	Missing or insufficient data

Figure 15.1 A map of world hunger (*Source*: FAO).

Future challenges

Global food issues remain an important issue for now and in the future. The United Nations **Sustainable Development Goals** aim to 'end hunger, achieve food security and improved nutrition and promote sustainable agriculture' by 2030. In real terms, this means dealing with not only the traditional problems of food insecurity in LICs (Figure 15.2), but also the emerging issue of malnutrition and obesity in HICs and MICs (Figure 15.7).

ACTIVITY 15.1

1 Look at Table 15.1. Draw pie charts to show:
  a the percentage of rice produced in LICs and HICs
  b the percentage of meat produced in LICs and HICs
  c the percentage of milk and milk products produced in LICs and HICs.
2 What do you notice about the amount of food produced in LICs and HICs?
3 Suggest how the following enabled food to be exported more easily:
  a container ships   b air travel   c temperature controlled storage.
4 Using Figure 15.1:
  a Find three countries that have food security.
  b Find three countries that have food insecurity.
  c Suggest why there is no data for some countries.

Figure 15.2 Food insecurity has been an ongoing issue in LICs.

> ### Key term
>
> **Sustainable Development Goals:** United Nations goals that came into force at the end of 2015; they are international development targets to be achieved by 2030

15.3 Where are the world's food resources consumed?

Of the 50 000 edible plants in the world, just 15 species provide 90 per cent of our global energy intake from food. Rice, maize and wheat are the main staple foods, providing 60 per cent of the global intake of food energy. Table 15.2 shows where the majority of food energy comes from for two different parts of the world. It shows how the staple food for many African nations is cereals, such as rice, wheat and corn. The staple food for many Western European countries is animal products, such as meat, milk and eggs.

The United Nations measure how much food we consume using our calorie intake. The measure of calories is in kcal. Different foods contain different amounts of calories. Rice is a cereal crop; an average portion of boiled white rice contains around 82 kcal (Figure 15.3). Beefburgers are animal products; a burger from a fast food restaurant contains around 508 kcal (Figure 15.4). Although both may be used as part of a meal, one contains significantly more calories than the other.

An average man needs to consume around 2500 calories a day and an average woman needs to consume around 2000 calories per day. People can survive on 1500 calories a day but this depends on the quality of food in the diet. The actual number of calories a person needs depends on things such as the vitamins, minerals and proteins in the food as well as age, height and level of activity. Consuming too few calories causes a lack of energy and weight loss. Consuming too many calories causes weight gain.

Where is the most food consumed?

The amount of calories consumed throughout the world is not even. There are inequalities between different nations. Figure 15.5 shows historic trends in calorie consumption for the 50-year period between 1960 and 2010. All of the countries shown have increased their calorie intake during this time. China has experienced the greatest increase in calorie consumption. Sub-Saharan Africa, which includes countries like Mali, Niger and Chad, lags behind the rest of the world. People in the United States and European Union countries are consuming far more calories than they need. Too few calories causes malnutrition and too many causes obesity. Figure 15.6 projects how the number of calories consumed by different parts of the world is likely to change by 2030.

HICs are consuming the most calories but they are seeing the smallest growth in calorie intake (a 1.7 per cent increase). LICs are consuming the fewest calories, but they are seeing the biggest growth in calorie intake (a 4.6 per cent increase). Within these averages, there are many variations:

Figure 15.3 A portion of rice contains around 82 kcal.

Figure 15.4
A burger contains around 508 kcal.

	Africa	Western Europe
Cereals	46%	26%
Roots and tubers	20%	4%
Animal products	7%	33%

Table 15.2 The energy obtained (as a percentage of total food intake) from cereals, roots and tubers, and animal products in Africa and Western Europe.

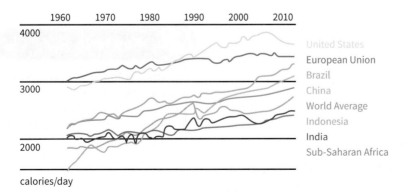

Figure 15.5 Historic trends in calorie consumption.

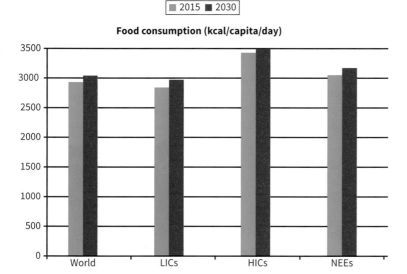

Figure 15.6 Calorie intake for 2015 and 2030 (projection).

Discussion point

How useful is 'average calorie intake' when looking at global food consumption? Can you think of a better way to measure food consumption?

Further research

How many of the 15 edible plants that provide 90 per cent of global energy can you name?

- The LICs Cambodia, Chad, The Democratic Republic of Congo, Haiti, Mozambique and Somalia all have calorie intakes far below 2850. They consumer fewer calories than many other developing nations.
- The HICs USA, Austria, Greece, Belgium, Luxembourg and Italy all have calorie intakes far in excess of 3440. They consume more calories than many other HICs.

World entering era of global food insecurity with malnutrition and obesity side by side within countries, says leading food expert

The Independent, Sunday, 12 July 2015

The world is entering an era of global food insecurity which is already leading to the 'double burden' of both obesity and malnutrition occurring side by side within countries and even within the same families, a leading food expert has warned.

It will become increasingly common to see obese parents in some LICs raising underweight and stunted children because high-calorie food is cheaper and more readily available than the nutritious food needed for healthy growth, said Alan Dangour of the London School of Hygiene and Tropical Medicine.

'We are certainly looking at a period of increased instability in the supply of food, and also the diversity and types of food that are available are going to change.'

Download Worksheet 15.1 from Cambridge Elevate for help with Activity 15.2, questions 2.

ACTIVITY 15.2

1 Use Table 15.2 and Figures 15.3 and 15.4 to explain why people in HICs consume more calories than people in LICs.
2 a On a blank map of the World, plot the countries with the highest and lowest calorie consumption.
  b What patterns do you notice?
3 Looking ahead to 2030 and beyond, why might governments need to manage overconsumption of food?
4 Use the *Independent* article Suggest how a 'double burden' of both obesity and malnutrition' could occur side by side within countries.

15.4 Is there enough food?

Globally, 17 per cent more food is produced per person than 30 years ago. Improvements in farming could lead to us growing up to 70 per cent more crops than current levels by the year 2030.

The world produces enough to feed everyone, but 1 billion people are still hungry. This is because food is not evenly distributed. Some places have a food surplus and people eat more than they need. Other places have a food deficit and people go hungry.

There are two main reasons why food consumption is increasing globally: population growth and economic development.

Population growth

The world's population is growing at a rate of 1.14 per cent per year (Figure 15.7). This means that there are an additional 80 million people on the planet every year. The global population is predicted to exceed 8 billion people by 2030 and 9 billion by 2050. A larger population is leading to a greater consumption of food.

The most rapid population growth is taking place in NEEs and LICs. The greatest increases in population are predicted to occur in Asia. It is thought that China, India and Southeast Asia will make up 60 per cent of the world's population by 2050. Expanding populations lead to a greater demand for food.

Economic development

Food surpluses are common in HICs and NEEs. Increasing wealth often increases both overall demand and the range of food available (Figure 15.8). Wealthier nations can afford to import food and **subsidise** agriculture. Lifestyles here are more sedentary, which means they include little or no physical activity, and many people eat more calories than they need, leading to obesity problems.

Economic development also leads to higher rates of **urbanisation**. City dwellers have easy access to supermarkets, selling convenience food. They also have many fast-food outlets close by (Figure 15.9) and are exposed to more food advertising. The types of jobs in cities mean that people sit down more and have less of a connection with the landscape and natural food sources. People in developed nations eat a greater proportion of processed foods that are higher in calories; many do not think about where their food comes from.

Did you know?

Hunger kills more people every year than AIDS, malaria and TB combined.

Go to the **Worldometers website** (www.cambridge.org/links/gase40144) to see real-time information on births, deaths and global population.

Key terms

subsidise: when the government pays towards the cost of producing something

urbanisation: an increase in the proportion of people living in towns and cities

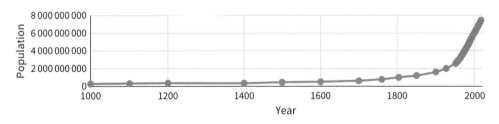

Figure 15.7 Global population over time.

Figure 15.8 A fast food restaurant in China.

Figure 15.9 Fast food is convenient, but often high in calories.

Discussion point

When did you last eat fast food? How often do you have a takeaway? Do you think your grandparents would have eaten similar food when they were your age?

ACTIVITY 15.3

1 a Is there enough food for everyone on the planet?
  b Why do some people not have enough to eat?
  c What could people in the UK do to reduce food waste?
2 Create two flow diagrams to show why food consumption is increasing as a result of:
  a population
  b economic development

Did you know?

The steel containers used on ships, lorries and trains are exactly the same size in every country.

Further research

Around 40 per cent of the ingredients that we use in the UK are imported. Collect some food labels from fruit and vegetables to find out where your food has come from.

15.5 What factors affect food supply?

Most of the people who go hungry (98 per cent) live in rural parts of LICs. The highest prevalence of hunger (that is the highest percentage of the population affected by it) is in Africa but the greatest number of hungry people are found in Asia. There are many reasons for food deficits:

Climate

The global climate is warming by around 0.2 °C per decade. Rainfall levels are increasing in some parts of the world, but decreasing in others. This affects farmers across the world. People who farm **marginal land** already struggle to survive. Even small changes in temperature or rainfall may threaten their existence. Extreme weather events, such as **drought** (Figure 15.10), tropical storms and flooding destroy crops in many places.

Poverty

When people are poor, they cannot afford enough food. This makes them weaker and unable to work effectively. Families in LICs spend 80 per cent of their income on food. Even small increases in the price of food will severely affect them.

Fertiliser contain nutrients like nitrogen, phosphorus and potassium. In HICs they are sprayed onto crops to help them to grow. Many farmers in LICs cannot afford fertilisers. Some use animal manure, but others do not have access to this resource.

Technology

Improvements in transportation and storage have increased the year-round availability of food. People in wealthier nations can now import food from any country, during any season. Transnational corporations (TNCs) have used technology to process food, which helps make it available.

New fuel technologies are also affecting food supplies. Using food crops to fuel cars (biofuels) means there will be less land on which to grow food to eat (Figure 15.11). This increases food prices.

Pests and disease

Pesticides are chemicals that can be applied to crops to kill insects, diseases and weeds. Pesticides have made it possible to grow more crops. Farmers in the USA spend an estimated $4.1 billion on pesticides, whereas farmers in other countries cannot afford to buy them. The UN FAO say that investment in farming is the single best way to reduce poverty and hunger.

Water stress

Some countries do not have sufficient rainfall to grow enough crops. **Irrigation** systems can provide water in drier areas. Crop yields can double when farmland is irrigated. In parts of south Asia, up to 75 per cent of farmers rely on pumped groundwater to water their crops.

Farmers in HICs often have large, computerised irrigation systems. Large-scale irrigation can also be found in some LICs. However, groundwater supplies are diminishing in some parts of the world so even with irrigation technology available, water stress can have a devastating effect on food supplies.

Figure 15.10 Crops like maize can be killed off by drought.

Visit the **Intergovernmental Panel on Climate Change website** (www.cambridge.org/links/gase40145) to see the latest data on global climate change.

Key terms

marginal land: areas which can only be farmed when conditions (e.g. rainfall) are very good

drought: long periods of time without rainfall

irrigation: taking water from a store such as an aquifer or river and distributing it across areas of landscape to make the land suitable for growing crops

Figure 15.11 A factory using corn to make biofuel in the USA.

Conflict

War affects food supply in many ways. Farmers may leave their land to fight in conflict. Sometimes farmers and their families are forced to flee as refugees. Crops can be destroyed in battle. Food can also be used as a weapon, with soldiers trying to starve their enemies into surrender.

Food shortages have occurred in the south of Sudan, where there is an ongoing civil war between the government and armed rebels (Figure 15.12). Millions of people have been forced to flee farmland and food aid workers have been kidnapped.

Figure 15.12 An army tank near farmland in the south of Sudan.

ACTIVITY 15.4

1 Create a table like this one:

	Increase food supplies	Decrease food supplies
Climate		
Technology		
Pests and diseases		
Water stress		
Conflict		
Poverty		

Complete the table to show how each of these factors can affect food supply:

Low income families cannot afford enough food.	Too hot or too cold. Drought or flooding.	Transportation and processing facilities.
Small-scale, basic irrigation systems.	Biofuel production.	Chemical pesticides to kill insects, diseases and weeds.
War and unrest.	Little investment in farming.	Farmers can afford fertilisers to boost crop yields.
Computerised irrigation systems.	A stable government and a strong army.	Moderate temperatures and rainfall.

2 a Do you think it is important to help those countries who can't produce enough food?

b What can be done to help LICs to produce more food?

3 Use the information on conflict and Figure 15.12.
Produce a front page report for a newspaper to show how conflict can create food insecurity problems in countries like Sudan.

 Further research

Farmers in the UK have limits to the amount of fertiliser that they can use. Read the government's guidelines on the use of fertiliser to find out why.

 Download Worksheet 15.2 from Cambridge Elevate for help with Activity 15.4.

 Fieldwork

Open Farm Sunday takes place once a year to encourage the public to visit farms. Go to their website (www.cambridge.org/links/gase40146) to find out if there are any local farms that you can visit to find out more about the use of technology, irrigation and pest control on UK farms.

15.6 What are the impacts of food insecurity?

Food insecurity is a concern for lots of countries, particularly those in sub-Saharan Africa and parts of Asia. There are several impacts of food insecurity: hunger, soil erosion, rising prices and social unrest.

Hunger

Hunger takes two main forms:

- **Undernutrition:** not having enough food to live an active life. Globally, 805 million people are chronically undernourished, most of whom live in poorer countries. When many people starve from undernutrition it is called a famine. Famines can last for months or even years. In 2012, Somalia experienced a famine where 35 000 people died of starvation.
- **Malnutrition:** not having enough protein or nutrients. Malnourished people may consume enough calories, but they don't eat the right foods to keep them healthy. Kwashiorkor is a disease caused by malnutrition whereby children have enlarged stomachs. It prevents children from growing and it can be fatal. Kwashiorkor is mainly found in poorer countries, but there have been isolated cases elsewhere.

Soil erosion

Soil erosion is the removal of soil, often by the wind or rain (Figure 15.13). During times of food insecurity, farmers can cause soil erosion as they try to get more out of their land. They can do this through deforestation, overgrazing and over-cultivation:

- **Deforestation:** trees protect the soil from the wind and rain and their roots bind it together. Deforestation leaves soil exposed.
- **Overgrazing:** vegetation is eaten by animals, which has a similar effect to deforestation. Animals trample the soil and so new vegetation struggles to regrow.
- **Over-cultivation:** this removes nutrients and the soil becomes infertile. Infertile soil is eroded more easily, which makes it more difficult to grow crops.

Rising prices

When the demand for a commodity is greater than the supply of it, the price of the commodity increases (Figure 15.14). The same applies to food. Food shortages lead to higher food prices. Everyone pays more, but those with the least amount of money suffer the most.

Between 2000 and 2015, global food prices increased (Table 15.3). This table shows price changes over time. Prices for the year 2002 have been given the value '100'. The 2015 values are higher than 100, showing that prices have increased.

	Meat	Dairy	Cereals	Vegetable oils	Sugar
2000	96.5	95.3	85.8	69.5	116.1
2015	171.4	167.5	160.8	154.1	189.3

Table 15.3 Food price index for the years 2000 and 2015 (*Source:* FAO)

Visit the **UN FAO website** (www.cambridge.org/links/gase40147) to see an interactive hunger map.

Figure 15.13 Soil erosion in Kenya leaves deep channels (gulleys) in the landscape.

Figure 15.14 The theory of supply and demand.

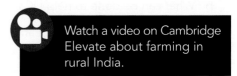
Watch a video on Cambridge Elevate about farming in rural India.

Social unrest

Food shortages can lead to social unrest. Food is vital for survival and so when supplies become scarce, people are willing to fight in order to survive. Food riots have taken place since the 17th century. The potential for food riots is increasing with greater food insecurity.

In 2008 the Côte d'Ivoire and Burkina Faso on the west coast of Africa (Figure 15.15) experienced food riots. The protests were aimed at the government. People felt that governments should have done more to ease food insecurity. Anger and frustration led to violent clashes. Higher fuel prices, an increased demand for food in Asia, the use of farmland for biofuel crops and a poor growing season were all causes of food insecurity in this region.

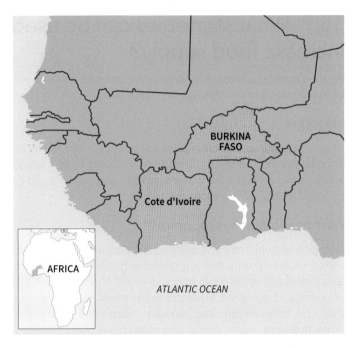

Figure 15.15 The Côte d'Ivoire and Burkina Faso are in the west of Africa.

ACTIVITY 15.5

1 a Which of the two types of hunger is the most severe?
  b Do either of these types of hunger affect HICs or NEEs?
  c Look at Figure 15.13. Describe the causes and effects of the problem shown in this photo.
2 Write the words 'hunger', 'soil erosion', 'food prices' and 'social unrest' in four corners of your page.
  a Draw lines between the words to show how the four impacts of food insecurity are linked. An example might look like this:

HUNGER $\xrightarrow[\text{causes soil erosion}]{\text{Hunger leads to over-cultivation, which}}$ SOIL EROSION

3 Make a copy of Table 15.3.
  a Add two more rows to the bottom of the table. The first row should have the heading 'difference' and the second row should have the heading 'percentage increase'.
  b In the difference row, do the following sum: 2015 index – 2000 index, e.g. for the 'meat' column, this sum will be 171.4 – 96.5, which gives a difference of 74.9.
  c In the percentage increase row, do the following sum: (difference / 2000 index) × 100, e.g. for the 'meat' column, this sum will be (74.9 / 96.5) × 100.
  d Which food type saw the greatest percentage increase in price between 2000 and 2015? What factors affect production, and therefore price?

Download Worksheet 15.3 from Cambridge Elevate to help with Activity 15.5 questions 2 and 3.

15.7 What strategies can be used to increase food supply?

As the global population increases, we will need to find ways to grow more food or make better use of the food that we have.

Irrigation

Irrigation can improve crop production in LICs by 100 to 400 per cent. In Africa, only 10 per cent of food comes from irrigated land. Introducing irrigation to dry areas or areas where the water supply is unreliable could provide enough grain to feed up to 2 billion more people.

Figure 15.16 An aeroponic rice plantation.

Aeroponics and hydroponics

Aeroponics (which means working air) and hydroponics (which means working water) refer to plants that are grown without soil. With aeroponics, plants are supported with their roots hanging in air. This air is then sprayed with a fine mist of water and nutrients. Hydroponics is when plants are grown in a material other than soil, for example sand, sawdust, fibre mats or pebbles. As these contain few or no nutrients, water containing nutrients is used.

Aeroponics (Figure 15.16) and hydroponics (Figure 15.17) are efficient ways for plants to grow as it is easier for the plant to find and absorb nutrients. Plants grown in this way do not need soil, which makes it a more efficient and reliable technique for many countries.

Did you know?

Rice is so important in Asian culture that many things are named in its honour. For example, Toyota translates as 'bountiful rice field' and Honda means 'main rice field'.

The New Green Revolution

The Green Revolution is the transfer of farming methods from HICs to LICs countries, for example the use of fertilisers and pesticides, irrigation and high-yield crops.

The first Green Revolution began in the 1940s, when farming methods from the USA were used to increase food production in Mexico. Between the 1950s and the 1990s, crop yields in LICs almost doubled. However, population continued to increase too and so the gains levelled off.

The New Green Revolution involves developing seeds to help specific areas that are suffering the effects of global warming, such as drought (Figure 15.18), flooding or **salinity ingress**. There is also a focus on improving the nutritional value of crops, rather than just the calorific value. The New Green Revolution should help the poorest areas of the world.

Key term

salinity ingress: when salt water from the sea invades water supplies on the land; this can happen due to sea level rise

Figure 15.17 A hydroponic greenhouse.

Figure 15.18 New varieties of rice can withstand drought.

Biotechnology

Biotechnology is the use of science to produce plants and animals that have specific characteristics. Selective breeding of animals has been happening for a long time. This is where an animal with one beneficial characteristic, such as higher milk production or shorter calving period, is bred with another of that species. Crops can be **genetically modified (GM)** to improve them, e.g. a gene from scorpions which creates poison has been implanted into a type of cabbage (Figure 15.19). The cabbages poison caterpillars but do not harm humans.

> **Key term**
>
> **genetically modified (GM):** specific changes made to DNA to improve them, e.g. to create disease resistance

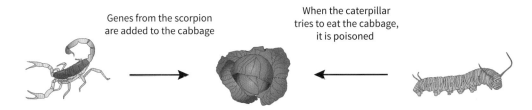

Genes from the scorpion are added to the cabbage

When the caterpillar tries to eat the cabbage, it is poisoned

Figure 15.19 Genetic modification takes the properties of one species and transfers it to another.

Appropriate technology

Appropriate technology is when techniques are introduced to an area to match the level of economic development and local resources and skills found there. These should be affordable, easy to use and maintain and suitable for local needs, e.g. solar powered lighting in areas that have no electricity. Appropriate technology is usually small scale, e.g. mobile solar-powered water pumps for remote areas where there is no electricity, but it can improve yields for many communities.

ACTIVITY 15.6

1 All of the strategies featured on this page increase food supplies. Can you think of any challenges that need to be overcome when using these strategies?
2 a Sketch Figures 15.16 and 15.17.
  b Annotate your sketches to show how aeroponic and hydroponic systems work.
3 Read the statements below, which explain arguments for and against GM crops:

'GM crops could provide enough food for our expanding population. GM crops can be designed to be the perfect crop. They are resistant to drought, insect attack and disease. The UK should be investing more in GM.'

'GM crops are dangerous and we are playing with nature! The crops could cross-pollinate and their seeds can travel to other species. Many of these crops have not been properly tested. There could be unpredicted side-effects or damage to human health.'

  a How might GM crops increase food supply?
  b Why are some people unhappy about the use of GM crops?

> **Discussion point**
>
> Can you think of any useful genes that could help crops?

> Go to the **Full Belly project website** (www.cambridge.org/links/gase40148) to see lots of examples of appropriate technology.

Large-scale agricultural development in Kilombero Valley, Tanzania

Tanzania is a low income country (LIC) in East Africa. The Kilombero Valley is a fertile strip of land in Tanzania (Figure 15.20). It is one of the best rainfed sites in eastern Africa. The Kilombero Valley lies to the south of the capital Dodoma and to the south-west of the main city of Dar es Salaam.

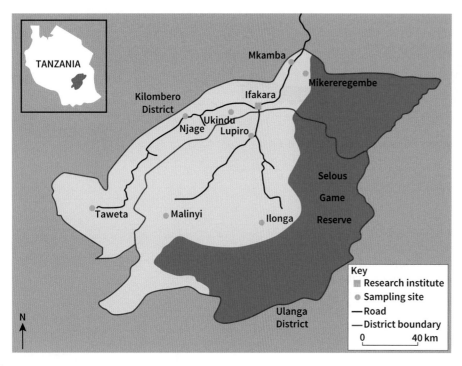

Figure 15.20 The Kilombero Plantations lie in the fertile Kilombero Valley in Tanzania.

What large-scale agricultural development has taken place?

Agribusiness describes the range of processes involved in modern food production, e.g. seed supply, agricultural chemicals, processing and sales. Agribusiness involves profitable, large-scale farming systems. Large-scale farming has increased in the UK and is growing in the developing world.

Kilombero Plantations Limited (KPL) is managed by the UK company, Agrica. In 2008 it acquired a 5818 hectare farm located near Mngeta in the Kilombero Valley. This land was cleared to grow rice crops. Processing and storage facilities were built alongside. Tractors, planters and combine harvesters are in use, and the company has invested in a **biomass** plant and has renovated a small **hydro-electricity** station.

KPL is the largest rice producer in East Africa. Approximately 32000 tonnes tons of rice and 4500 tonnes of beans and pulses are produced by the company every year, improving food security in the region.

Advantages of large-scale agriculture

Large-scale agriculture in Tanzania has brought advantages: KPL supplied local smallholder farmers with training and seeds to help them to improve rice yields. They called this project SRI, which stands for 'System of Rice Intensification'.

Discussion point

Why would a UK company invest in Tanzania?

Key terms

biomass: organic material which comes from living or recently living organisms; it can be used to generate electricity

hydro-electricity: electricity that is generated from running water

Visit the **Agrica website** (www.cambridge.org/links/gase40149) to show how Agrica operates in Africa.

This has improved harvests for over 4000 families, which has helped people to improve their lives.

Agrica have also provided work for the young adults in the area. The company employ local people and they focus on younger people who often have difficulties finding employment.

KPL built 82 houses for families that have been displaced as a result of the new farm. They also constructed school rooms (Figure 15.21), healthcare facilities, water pipes and new roads. KPL spend an estimated $700000 a year on wages and healthcare for locals.

Finally, KPL are contributing towards sustainable energy supplies in the area. Their biomass plant has a 500 KW capacity and their mini hydro-electric power station has a 320 KW capacity.

Disadvantages of large-scale agriculture

Large-scale agriculture in Tanzania has also brought some disadvantages: many farmers are now in debt. The SRI scheme improved yields, but led to farmers borrowing money to pay for new seeds and equipment. In some cases, farmers have had to sell their possessions to meet the repayments on the loans.

The KPL farm covers land that was previously farmed by 230 households. These displaced families were moved from their homes, given alternative land and paid money as compensation. Locals say that the alternative land is less than families originally had (Figure 15.22) and that the cost of renting farmland far exceeds the compensation paid. The amount of land available has decreased as a result of KPL and this has further increased rental prices.

Local farmers claim that chemical runoff from the KPL farm pollutes their land. Some farmers say that their harvests have been ruined, which has meant that their families go hungry. Chemical runoff is also bad for wildlife and habitats.

Figure 15.21 KPL has built school rooms for local children.

Discussion point

Overall, have KPL been beneficial for Tanzania?

Figure 15.22 Displaced farmers are often given less land than they previously had.

ACTIVITY 15.7

1 Create a fact file about large-scale farming in Tanzania. Include:
   a   a location map of the KPL farm
   b   information about the size of the farm
   c   an explanation of how the farming methods differ from traditional Tanzanian farming.
2 Draw a table to show the advantages and disadvantages of large-scale farming in Tanzania:
   a   colour social impacts in blue
   b   colour economic impacts in red
   c   colour environmental impacts in yellow
3 Describe the pattern shown by the colours in your table.
4 Has the KPL farm been good for the Kilombero Valley?
5 Figure 15.23 shows rice being sold at a market in Tanzania.
   a   What might happen to the price of rice if:
      i   More African countries set up large-scale rice plantations?
      ii   A new disease wipes out rice crops?
   b   Is it wise to focus on one crop, like the SRI programme does?

Figure 15.23 Rice sold at a market in Tanzania.

15.8 How can food supply be increased sustainably?

Food supplies are increasing, but the farming techniques that we use are not always good for the environment. We must look for ways to produce sustainable food supplies: farming that balances the use and conservation of natural resources for present and future generations.

Permaculture

Permaculture is the development of agricultural ecosystems intended to be sustainable and self-sufficient. It embraces a range of ideas and techniques whereby food is produced without damaging the environment or causing harm to animals or people. Examples of permaculture are organic farming, urban farming and sustainable fish and meat production.

Organic farming

Organic farming is agriculture which uses natural methods to grow crops and rear animals. Some of these include:

- using natural fertilisers, such as animal manure, known as slurry, and green manure, which are plants that take nitrogen from the air and put it back into the soil, e.g. red clover
- biological pest control, which uses natural predators to control pests (Figure 15.24)
- crop rotation, to break cycles of pest breeding
- high standards of animal welfare to reduce the need for medicines

When farmland becomes organic it takes time for the chemicals to leave the soil. There can be an initial decrease in crop yields, but eventually organic yields are similar to non-organic crops. Organic farms are sustainable in that they use natural products, rather than manufactured chemicals. This uses fewer resources and leads to less environmental damage.

Urban farming

Urbanisation leads to a loss of farmland. Urban farming involves using space in and around cities to grow food (Figure 15.25).

Figure 15.24 One ladybird can eat more than 5 000 aphids over its lifetime.

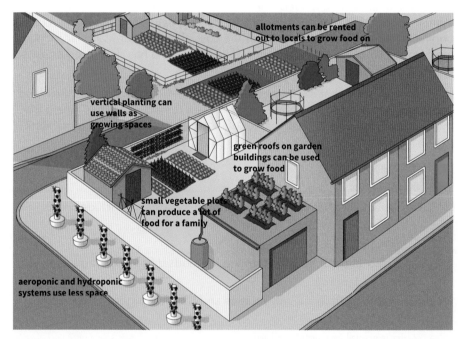

allotments can be rented out to locals to grow food on

vertical planting can use walls as growing spaces

green roofs on garden buildings can be used to grow food

small vegetable plots can produce a lot of food for a family

aeroponic and hydroponic systems use less space

Figure 15.25 Urban farming uses spaces within cities.

Gardens and allotments in urban areas can produce up to 15 times more food than the equivalent area of farmland. Urban farming is small scale, but very productive. Growing food within cities also reduces the need for transport and storage. Urban farming encourages people to eat healthily and think about where their food comes from.

Sustainable fish and meat production

Globally, 77 billion kilograms of wildlife is taken from the sea each year. Sustainable fishing involves catching fewer fish so that there are enough left to repopulate our seas. While this does not immediately increase food supplies, it

 Discussion point

Where could we grow our own food? Is there space in the school grounds or in our gardens at home?

does ensure that there will be fish left for future generations to eat. Sustainable fishing also avoids bycatches, when unwanted fish become accidentally trapped in nets. Better net designs and more careful fishing can reduce these losses.

Rearing meat uses more resources than growing crops. Each kilogram of beef produced uses around 6.5 kg of grain, 36 kg of roughage and 15 500 m³ of water. Many cattle farmers feed their animals on grain, which means that there is less for humans to eat. They also give the cattle special hormones to encourage them to eat more and grow quickly. Sustainable meat production involves animals grazing on grass (Figure 15.26). They do not fatten up so quickly and so less meat is produced.

Seasonal food consumption

Eating food that is 'in season' is more sustainable. Food that is in season is grown locally without the need for heated greenhouses. Food that is out of season is transported from other countries. Food miles is a measure of the distance that our food travels from field to plate. Seasonal produce has lower food miles (Figure 15.27).

Figure 15.26 Sustainable meat production involves animals grazing on grass.

UK products that are imported 'out of season'

Asparagus – from Peru – 6 033 miles

Strawberries – from Mexico – 5 225 miles

Apples – from South Africa – 8 872 miles

Blackberries – from Argentina – 7 378 miles

Sugarsnap peas – from Zambia – 5 068 miles

Broccoli – from Zimbabwe – 7 851 miles

Green beans – from Kenya – 6 896 miles

Figure 15.27 Eating food in season reduces food miles.

Reducing food waste and losses

Food wastage is high in developed nations; an estimated 7 million tonnes of food and drink is thrown away in UK homes every year. Although this food is not eaten, it still counts as having been consumed because other people cannot eat it. Buying only what is needed is more sustainable as it means there is more food to feed others.

Food loss is when products get spilled or spoiled and so are not able to be sold, e.g. cracked eggs. More careful procedures during the stages of food production, e.g. harvesting, storage, packing and transportation, can reduce food losses.

ACTIVITY 15.8

1 a Design a simple logo or icon for each of the sustainable strategies found on these two pages.
  b Briefly explain what your logos show.
  c How could you use these logos to advertise sustainable food consumption? Create an advertising campaign idea to promote one of the examples.
2 How could countries work together to achieve food sustainability?

Did you know?

Flexitarianism is a new trend where people eat meat, but in a sustainable way. The restaurant chain Pizza Express has a 'Meat Free Monday' menu.

Further research

Make a food waste diary. How could you reduce your food wastage? Visit the **LOVE FOOD hate waste website** (www.cambridge.org/links/gase40150) to find out more.

Sustainable food production in Kinshasa

The Democratic Republic of the Congo (DRC) is a low income country in Central Africa (Figure 15.28). The country gained independence from Belgium in 1960. Since then it has struggled to improve its economy and many people face food insecurity.

The population of the DRC has almost doubled between 1990 and 2015, from 34.9 million to 71.2 million. Up to 40 million Congolese will be living in cities by 2025. Most people in cities live below the **poverty line** and only eat one meal a day.

Kinshasa is the capital city of the DRC (Figure 15.29). Between 1995 and 2011, Kinshasa's population grew from 4.5 to 8.8 million people. People came from the countryside looking for work or fled from conflict in the east of the country. With over 1.5 million people, N'djili and Kimbanseke are the city's most populated areas. Many of the migrants are skilled farmers, but they have no land or money to buy seeds. This is a problem because Kinshasa needs to produce 500 tonnes of fruit and vegetables daily in order to feed everyone.

Key term

poverty line: the minimum amount of money needed to be able to live

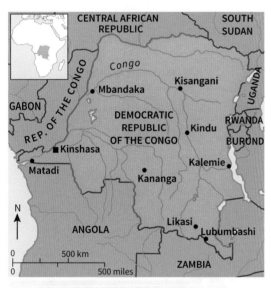

Figure 15.28 The Democratic Republic of the Congo, in Central Africa.

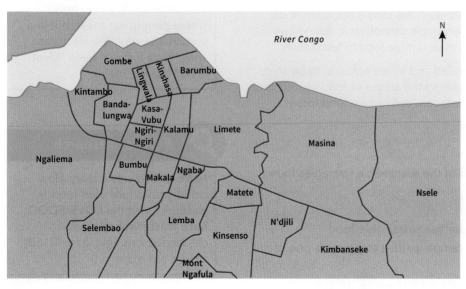

Figure 15.29 The city of Kinshasa.

Urban farming in Kinshasa

Urban and Peri-Urban Horticulture (UPH) is a way to grow food sustainably. It involves people in and around cities (peri-urban areas) growing their own food. UPH has been supported by the Ministry of Rural Development (part of the DRC government) for decades:

- In 1996, the government and the UN FAO set up the National Support Service for Urban and Peri-urban Horticulture (SENAHUP).
- From 2000 the government and the UN FAO have used funding received from Belgium to officially mark out farming areas in Kinshasa.

UPH has been sustainable as it has provided secure areas of land for urban farming, set up an irrigation system (Figure 15.30) and built centres to support farmers. 15 500 UPH farmers have borrowed money from a $1 million fund of **micro-loans**. Women and the unemployed were first offered plots in the 1950s. Today, the area used for UPH has grown to more than 1 000 hectares. 15 000 people produce 65 per cent of the city's food supply.

Figure 15.30 Irrigation allows urban farmers to grow more crops.

Advantages and disadvantages of UPH in Kinshasa

Fruit and vegetable consumption in Kinshasa has increased and so people are healthier. The programme has created 45 000 jobs and wages have increased fourfold between 2000 and 2010. UPH farmers can earn up to US $200 per month, which is more than the average wage. 25 schools in the city now have market gardens, which have taught over 9 500 students and 1 600 adults to grow food. The schools teach about fertilisers and how to choose the best seeds, which helps to improve yields so surplus crops can be sold (Figure 15.31).

In Kinshasa UPH is under threat from new housing. The use of chemicals and fertilisers has caused some water pollution. Many farmers in the N'djili valley were able to access the training or money that they needed. The UPH programmes have been more effective in smaller cities, like Lubumbashi, which has a population of 1.5 million people. Each hectare of UPH land in Lubumbashi provides jobs for 20 producers, 70 workers and 150 traders. The DRC government plans to develop urban farming across the country, but almost all of the money invested so far has come from other countries.

Figure 15.31 Surplus crops, such as cassavas and bananas, can be sold.

 Key term

micro-loans: very small loans with low interest payment

 Discussion point

A recent study in Kimbanseke found that housing developers were selling vegetable plots during the weekend when gardeners were not there. Would this happen in the UK? Why/Why not?

 Assess to progress

1 Suggest why Belgium has given money to the DRC's urban farming schemes. **2 MARKS**

2 Explain the impact of rural to urban migration on food supplies in the DRC. **4 MARKS**

3 Using Figures 15.31 and 15.32 and your own knowledge, explain how Urban and Peri-Urban agriculture in the Democratic Republic of Congo is a sustainable way to produce food. **9 MARKS + 3 SPaG MARKS**

 Tip

Question 3 asks you to refer to the figures so make sure that you talk about each in your answer. It is important also to check things like spellings and the use of paragraphs to structure your answer.

In this chapter you will learn about...

- where water surpluses and deficits are
- where the world's water resources are consumed
- the human and physical impacts on water availability
- the factors that affect water consumption
- the different impacts of water insecurity
- the advantages and disadvantages of a water transfer project
- how we can move towards a sustainable water resource future
- the advantages of a local water conservation scheme.

Key terms

aquifers: bodies of permeable rock that can store water and through which water can easily move

groundwater: water found underground in soil, sand or rock

16.1 Where do our water resources come from?

Global freshwater supplies

Water continuously moves around the Earth, in a cycle known as the hydrological cycle (Figure 16.1). The majority of the world's water (97 per cent) is stored in the oceans and 2 per cent is stored in ice. That leaves just 1 per cent of the world's water to flow through the stages of the hydrological cycle and supply our water needs.

Global freshwater supplies are linked to three physical factors:

- **Climate:** influences rainfall distribution, snowfall and evaporation rates.
- **Geology:** determines the distribution of **aquifers** and **groundwater**.
- **Rivers:** transfer water across river basins.

The flows of the hydrological cycle vary over time, especially in places that experience a fluctuating climate:

- Seasonal rainfall variations result in some places having distinct wet and dry seasons (Figure 16.2).
- Natural climate cycles bring clusters of drier or wetter weather. This variability might be related to solar activity and the jet stream.
- Climate change is increasing global temperatures, increasing evaporation rates, leading to a stormier world.

Rank	Country	Freshwater supply
1	Brazil	13%
2	Russian Federation	10%
3	China	7%
4	Canada	7%
5	Indonesia	7%
6	Colombia	6%
7	United States	5%
8	India	4%
9	Peru	4%
10	Dem. Rep. Congo	3%

Table 16.1 Percentage of global freshwater supplies in different countries (Data: FAO, 2012).

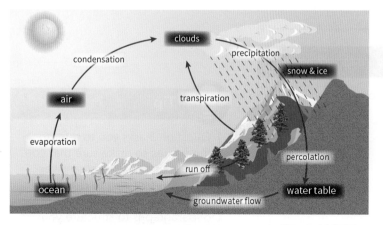

Figure 16.1 The global hydrological cycle.

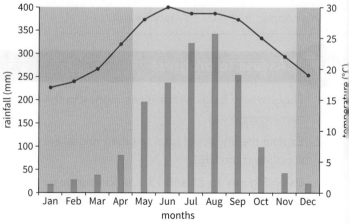

Figure 16.2 Climate graph for Hanoi, Vietnam.

16.2 Global patterns of water surplus and deficit

In many parts of the world, the water needed to meet our increasing water demand comes from three main sources:

Rivers and lakes: lakes are a natural store of surface water, however they are not always located within access of large populations. The world's glaciers feed many rivers. Rivers supply water for **irrigation**, domestic and industrial use, but they are prone to pollution.

Reservoirs: these are artificial lakes created by building a dam across a valley and allowing it to flood. The water collected and stored behind the dam can become an important water supply.

Aquifers and wells: over 94 per cent of the world's freshwater supply lies underground, stored in porous rocks known as aquifers. This groundwater can be extracted by drilling wells or boreholes down to the aquifer. Groundwater in arid and semi-arid regions provides vital water resources in areas of very low rainfall.

Water surplus and deficit

The water cycle is a closed system, so the amount of water circulating in it is fixed. Unfortunately, water is not always available when and where people need it. A place receives water from rainfall but loses water though **evaporation**. Look at the global pattern of water surplus and deficits; in some areas:

- **water demand exceeds supply.** These are known as water-deficit areas. They are located at low latitudes across South America, North Africa and the Middle East, Southern Asia and Australia. Many of these areas are in deficit due to low precipitation. Some places with high rainfall are in deficit due to high evaporation rates. Others are in deficit due to large populations.
- **water supply exceeds demand.** These are known as water-surplus areas. These are located in temperate climate zones with high rainfall and low populations. Located in North America, Northern and Central Europe and Russia. Over 60 per cent of the Earth's freshwater supply is found in just ten countries (Table 16.1).

There are ways of transferring water from surplus to deficit areas. The most widely used way is by long-distance pipelines. In the UK, Welsh water stored in reservoirs in the Elan Valley is transferred 73 miles via aqueduct to supply water to the city of Birmingham.

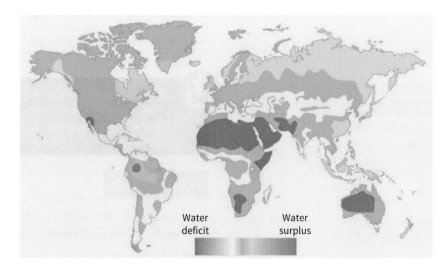

Figure 16.3 A global map of water surplus and deficits

Key terms

irrigation: taking water from a store such as an aquifer or river and distributing it across areas of landscape to make the land suitable for growing crops

evaporation: the effect of the Sun heating liquid water on the Earth's surface and converting it into a gas in the atmosphere

ACTIVITY 16.1

1 Explain why much of the world's water resources are of little use to people.

2 a Describe where areas of water surplus and deficit are found worldwide.

   b Explain why areas of high precipitation can still be areas of water deficit.

3 Study the climate data for Hanoi, Vietnam (Figure 16.2).

   a Which three months are the wettest and the driest?

   b Why might floodwater in the city of Hanoi not be suitable for drinking?

   c Explain why water shortages are most likely in September in London and April in Hanoi.

Download Worksheet 16.1 from Cambridge Elevate for help with Activity 16.1, question 3.

16.3 What is water used for?

Worldwide, agriculture accounts for 70 per cent of all water consumption, compared with 22 per cent for industry and 8 per cent for domestic use (Table 16.2). In developed nations, industries consume more than half of the water available for human use.

Agricultural use

Water is used to irrigate crops and as drinking water for livestock. Irrigation is essential in many regions, such as the American Southwest and Central Asia, where rainfall alone is insufficient to support agriculture. The use of automatic sprinklers can supply water at over 60 litres per second. Where access to modern technology is limited, irrigation can be rain-fed, gravity-fed from local rivers or bucket-fed from wells. These irrigation methods are labour intensive and use less water than automatic systems; therefore, they are more sustainable. Around three-quarters of food in Pakistan and China and over 50 per cent of India's food is now grown on irrigated land.

	Domestic (%)	Industrial (%)	Agricultural (%)
World	8	22	70
Higher income countries (HICs)	11	59	30
Lower income countries (LICs)	8	10	82

Table 16.2 Percentage of water consumption by sector in 2014.

Industrial use

Industrial demand for water increases as a country's economy develops, rising from 10 per cent in LICs to 59 per cent for HICs (Table 16.2). This is occurring particularly rapidly in the BRICS (the five emerging economies of Brazil, Russia, India, China and South Africa) due to the growth of manufacturing and power generation. China builds at least one new power station a week. The Chinese province Jiangsu plans to add more power capacity by 2030 than exists in the whole of the US and the EU today (Figure 16.4). Power generation is the single largest industrial user of water, requiring water for cooling, steam generation and other water-intensive processes.

Domestic use

Water is used in the home for drinking, cooking and cleaning (Figure 16.5). Domestic water demand is higher in HICs because water infrastructure and piped water is available and people own many water-demanding white goods such as dishwashers, washing machines and even sprinkler systems for their lawns.

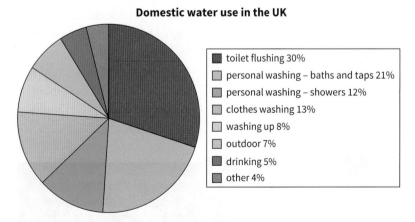

Domestic water use in the UK

- toilet flushing 30%
- personal washing – baths and taps 21%
- personal washing – showers 12%
- clothes washing 13%
- washing up 8%
- outdoor 7%
- drinking 5%
- other 4%

Figure 16.5 Domestic water use in the UK.

Further research

Carry out a water survey in your household. Work out what you use water for and what activities are the most wasteful of water. Compare your results with your class to find out what the most wasteful activities are.

There are plenty of learning resources about water available online, for example **The Pod website** (www.cambridge.org/links/gase40151).

Figure 16.4 A coal-fired power station recently opened in Jiangsu, China.

Download Worksheet 16.2 from Cambridge Elevate for help with Activity 16.2, question 1.

ACTIVITY 16.2

1 Compare the amount of water used in industry between HICs and LICs.

16.4 What are the reasons for increasing water consumption?

Our demand for water has increased dramatically over the last century. This demand is driven by global population growth, economic development and rising standards of living.

Population growth

Between 1990 and 2015, the global population increased from 5.3 billion to 7.3 billion. Most of this growth was concentrated in LICs, particularly in Asia where 61 per cent of the global population live.

Considering the UN advice that each person needs at least 20 litres of clean water a day for drinking, cooking and washing, rapid population growth is putting a strain on global water supplies. In China, population growth rates may have slowed with the One Child Policy, but there are still 1.4 billion people who need water. India has just 4 per cent of the world's fresh water shared among 17.5 per cent of its people. Currently, there are 354 million people in India without access to safe, clean water (Figure 16.6). With only $200\,m^3$ of water storage capacity per person, compared with $6000\,m^3$ per person in the United States, India's water crisis can only get worse as projected water demand for industry, agriculture and energy production all continue to rise.

Discussion point

27 million children are born in India every year, how will this affect the country's water demand?

Figure 16.6 Slum dwellers without access to water in Damu Nagar, Mumbai illegally tap into water pipes.

Economic development

The process of industrialisation associated with economic development creates a greater demand for water. The industries that produce energy and petroleum products, chemicals, metals, wood and paper products are major users of water. Water is used during the production process for washing, diluting, processing, cooling and transporting products, as well as for the personal use of the workers.

Virtual water refers to water that is embedded in agricultural and manufactured products, as well as the water used in the growing or manufacturing process. When a country exports goods, it also exports virtual water. It takes up 4000 litres to produce a cotton T-shirt and as much as 450 000 litres to manufacture a car. The number of cars on the road is expected to rise from 1.2 billion today to 2 billion in 2030. This will drive up demand for water, particularly in China, a newly emerging economy (NEE) with the largest automotive industry in the world (Figure 16.7).

Figure 16.7 Car production in China required 10.8 km³ of water to produce 24 million vehicles in 2014.

Improving standards of living

As the standard of living for many people in LICs improves, people are choosing to eat more meat. On average, it takes 1800 litres of water to grow 1 kg of wheat compared with 9500 litres of water for 1 kg of beef. If this trend continues, up to 24 per cent more water will be needed to grow the world's food in 20 years' time. With increasing wealth, people buy more domestic goods. A washing machine requires around 50 litres of water. Two billion people now have access to a washing machine. This has transformed the way people live, as they no longer have to wash clothes by hand.

Future trends

Freshwater withdrawals have tripled over the last 50 years. With the world population expected to rise to 9 billion by 2050 this trend is set to continue. Increasing demand coupled with the effects of climate change mean that future water supplies are not secure unless we invest heavily in water management and water efficiency strategies (Figure 16.8).

The Chinese government has outlined three national goals for water, called the 'Three Red Lines', which sets targets to cap total water use at 700

Further research

The thorny issue of exporting cut flowers from Kenya regularly hits the headlines. Find out about the impact of the flower-growing industry on Lake Naivasha and the people of Kenya. To start your research type 'Kenya's flower industry' into the BBC website's search box.

Key term

virtual water: the 'hidden' volume of water that is used in the agriculture and industry.

billion m³, increase irrigation use efficiency to 60 per cent and protect water quality by 2030.

Discussion point

Should a water tax be paid by people who export virtual water from water stressed countries?

Using water more efficiently can reduce the amount of water consumed. You can:

 Take a short shower rather than a bath to use 30 litres not 80 litres to wash with. Avoid power showers as they use much more water than a bath

 Turn off the tap while brushing your teeth – and save 12 litres if you clean your teeth for 2 minutes

 Wait for a full load before starting your washing machine or dishwasher

 Reuse grey water such as bath water or washing-up water to clean your car or water the garden

 Have a water meter fitted. Metered customers use approximately 10% less water than unmetered customers

Figure 16.8 Strategies to use water more efficiently in the home.

 Find out more about water footprints online.

ACTIVITY 16.3

1 Why is the global demand for water increasing?
2 To what extent can rainfall variability cause water supply problems?
3 India receives over 80 per cent of the annual rainfall in the four months from June to September.
   a How might India attempt to manage this water resource, so that it can be used year round?
4 Look at the Environmental Agency's monthly water situation report for your region.
   a Investigate whether rainfall, river flow and reservoir levels are normal or not for the time of year.
   b What problems might there be this year?

16.5 Water stress and deficit

Water is essential to life. The problem is that water demands are growing and water supplies are diminishing. Globally, **water stress** occurs when annual water supply drops below 1700 m³ per person per year. This already affects around 2.8 billion people worldwide. With the onset of climate change and rapid economic development in Asia the number of people experiencing water stress is likely to rise to 4 billion by 2050.

Water scarcity, which leads to water deficit (also referred to as water insecurity), is either the lack of enough water (quantity) or lack of access to safe water (quality). It occurs when the annual supply of water per person drops below 1000 m³. There are two types of water scarcity:

- **Economic water scarcity** occurs when the development of water sources is limited by lack of capital and technology; this affects 1.6 billion people worldwide.
- **Physical water scarcity** affects 1.2 billion people and occurs when more than 75 per cent of a region's river flows are being used (Figure 16.9).

16.6 Is there enough water for everyone?

In theory, there is enough fresh water worldwide to supply the needs of over 7 billion people, but uneven distribution and deterioration in water quality means not everyone receives enough safe drinking water.

Factors affecting water availability

About 80 per cent of the world's population lives in areas where the freshwater supply is threatened by deficit and pollution. Water insecurity occurs when there is not enough water available to ensure the population of an area enjoys good health, livelihood and earnings.

Climate

In regions that usually receive a surplus of water, such as the South East of England, seasonal variability can reduce rainfall and increase vulnerability to water shortages; this happened between 2010 and 2012. In Asia, monsoon rainfall is an important source of fresh water; a poor or late monsoon creates all

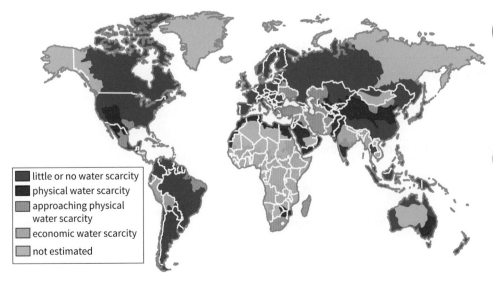

Figure 16.9 Global water scarcity.

- little or no water scarcity
- physical water scarcity
- approaching physical water scarcity
- economic water scarcity
- not estimated

Key term

water stress: water stress occurs when the demand for water exceeds the available amount during a certain period or when poor quality restricts its use

Discussion point

Why is it not possible to supply everyone in the world with clean, safe water?

Discussion point

Every £1 invested in water and sanitation generates an £8 return in the form of time saved, increased productivity and reduced healthcare costs. So why is water infrastructure not in place everywhere in the world?

sorts of water supply issues. Excessive rainfall can also create water quality issues as tube wells and other water stores are easily contaminated. Climate change is warming the Earth, causing changes to the amount of water stored as ice and the timing of annual snowmelt. This is cause for concern as a third of the global population rely on snowmelt-fed rivers for their water supplies.

Geology

The type of rock underlying a river basin can affect water supply. Where rocks are impermeable, water remains on the surface in rivers. Rivers can transfer and supply huge amounts of water over thousands of miles. Where rocks are permeable, surface water quickly disappears into underground drainage systems.

Pollution of supply

Human action can pollute both surface and groundwater supplies. Industrial waste can include heavy metals such as lead, cadmium and nickel and is often dumped directly into local rivers. As many people are forced to drink unsafe river water, they also ingest metals which accumulate in the soft tissues of their body and slowly poison them. Sewage disposed directly into rivers in lower-income countries causes waterborne diseases such as cholera, hepatitis and typhoid to spread easily. Water pollution kills 2000 children every day and it reduces the capacity of rivers to provide water for human use. Chemical fertilisers used on the farm run off the land, contaminate rivers and cause eutrophication, which leads to oxygen depletion and the death of aquatic animals. There are now thought to be hundreds of biologically dead rivers in Asia, including large stretches of the mighty Ganges and Yangtze (Figure 16.10).

Over-abstraction

When too much water is abstracted from an aquifer this can limit future water supplies. In Mexico City, over-abstraction has caused the city to sink more than 9 metres in the last century. The Angel of Independence monument, built at ground level in 1910 to celebrate the 100th anniversary of Mexico's War of Independence, now requires 23 additional steps to reach its base as the ground has sunk below it. Ground subsidence can also break water pipes, causing fresh water to escape. In Chennai, India, over-extraction of aquifers has caused seawater to seep over 10 km inland. Saltwater intrusion is also a concern in Los Angeles, where most of the water used by the 10 million residents comes from groundwater stores.

Water infrastructure and poverty

HICs can afford to pay for water infrastructure (drinking water, wastewater and stormwater) and therefore have water security. Around 2.6 billion people worldwide live in poverty without access to a toilet and 27 per cent of the urban poor do not have access to piped water. Slum dwellers in Nairobi (Kenya), Lusaka (Zambia) and Manila (Philippines) pay ten times more for water than Londoners (from water vendors, as shown in Figure 16.11). Some 15 million people lack access to running water in Mexico City and spend $18 a month on bottled water, a level only exceeded in the US.

As the world's population continues to grow, the number of people affected by water deficit is expected to rise sharply.

ACTIVITY 16.4

1   What is the difference between water stress and water deficit?
2   How might climate change impact on future water supplies across Southeast Asia?
3   What is meant by saltwater intrusion and why is it such a problem to coastal cities?

Figure 16.10 Every day, over 600 tonnes of rubbish are pulled from the Yangtze River, China.

Key terms

eutrophication: a form of water pollution, caused when nutrients from sewage or agricultural fertiliser run off into waterways and cause algal blooms to grow. These use up all the available oxygen. Fish and other marine life then suffocate

over-abstraction: when water is taken from rivers and other sources more quickly than it is being replaced

Figure 16.11 Water vendors are common in areas of water deficit such as Lusaka in Zambia.

Explore the **World Resources Institute GIS Aqueduct Project** online.

16.7 What are the impacts of water insecurity?

Globally, there are 785 million people without access to enough safe water, a situation defined as water insecurity. This is affecting the well-being of people, the environment and the economy.

Lack of clean water

With limited availability and access to water the global poor are trapped in a cycle of extreme poverty. Although the United Nations (UN) Millennium Development Goal to halve the number of people without access to safe drinking water was met five years before the 2015 deadline, it still leaves 11 per cent of the global population water insecure. Table 16.3 shows the global variation in access to safe water. These water statistics only deal with water quantity; **water quality** is yet to be monitored.

Waterborne diseases

2.6 billion people lack access to improved sanitation such as a toilet or latrine, of whom 70 per cent live in Asia. If not carefully managed, water sources located in areas without sanitation can become contaminated with parasites that can cause malaria and infectious diseases such as cholera and dysentery.

Malaria is spread to humans via the bite of an infected mosquito. It causes headaches, fever, sweats and vomiting and can result in death if not diagnosed and treated promptly. Malaria is found in 100 countries and affected over 200 million people in 2014. Cholera is a bacterial infection caused by drinking contaminated water or eating food washed in contaminated water. It causes diarrhoea, vomiting and stomach cramps. The resulting dehydration can be fatal. There are over 3 million cases of cholera worldwide every year. Dysentery is an infection of the intestines caused by bacteria or a single-celled parasite. It causes a high temperature, diarrhoea, vomiting and stomach cramps.

Water pollution

The biggest global polluter is agriculture. The use of fertiliser and pesticides contaminates both groundwater and surface water supplies (Figure 16.12). Industrial pollutants can be even more dangerous to human health. Mercury from mining in the Niger Delta spills into waterways, people eat the contaminated fish and mercury accumulates in their bodies causing rates of miscarriages and deformities in babies to quadruple. Oil refining by-products,

Did you know?

Every day, 1.8 million tonnes of sewage and industrial and agricultural waste are discharged into the world's waterways.

Country	% population with access to clean water	Life expectancy
Canada	100	81
Colombia	92	75
Ethiopia	44	61
Iraq	79	68
Jordan	97	78
Malaysia	100	74
Morocco	81	73
Rwanda	65	64
Somalia	29	57
Tanzania	53	61

Table 16.3 Access to safe drinking water (UN MDG Data, 2010)

Did you know?

The number of people in the world that have a mobile phone is higher than the number of people who have access to a toilet.

Key term

water quality: a measure of the chemical, physical, and biological content of water; high levels of bacteria or suspended material can result in poor quality water and pose a health risk for people

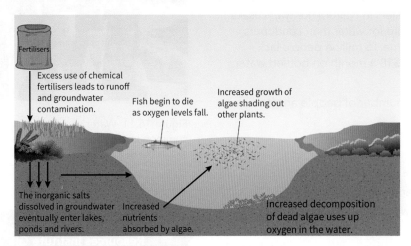

Figure 16.12 The process of eutrophication.

such as Benzopyrene, seep into the Athabasca River from the vast Canadian tar sands in Alberta (Figure 16.13). Locals drink the contaminated water resulting in 30 per cent higher than average blood and lymphatic cancer rates.

Food production

In 2015, there were 795 million undernourished people in the world. Hunger is caused by a wide range of factors common in overpopulated urban slums. These include poor hygiene and sanitation, limited access to clean water, and calorie or protein deficiency. Global food prices have been rising since 2012 and since one third of food production occurs where water is becoming scarce, it is likely that food price rises will continue. Wealthier countries are less susceptible to food insecurity as they can import produce from other countries (Figure 16.14).

Industrial output

In developing countries, 70 per cent of industrial wastes are dumped untreated into waters, polluting the water supply. Industrial waste such as nickel, lead and mercury from precious metals refining, used lead acid battery recycling and jewellery making in Manila has resulted in the Pasig River being declared biologically dead.

Potential for conflict

Where water supply does not meet demand, competition among water users intensifies and there is a risk of conflict. Potential water disputes can be handled diplomatically. A total of 150 water treaties have been signed in the last 50 years. **Water conflict** on the Nile was prevented in 2015 when Egypt, Ethiopia and Sudan signed an agreement or 'Declaration of Principles' to work together and promote shared use of the river.

The last 50 years have seen 37 violent disputes over water. This risk is greatest in arid areas and in the 276 river basins that cross national boundaries. The Grand Ethiopian Renaissance Dam (GERD) is currently under construction on the Blue Nile River in Ethiopia and is the source of a potential conflict with the two countries located downstream – Sudan and Egypt – over rights to water.

ACTIVITY 16.5

1 Outline how waterborne diseases can affect human health.
2 How have the people of the Niger Delta been affected by polluted water?
3 Using data from Table 16.3:
   a Draw a scatter graph to show life expectancy against access to safe drinking water.
   b Decide whether there is a correlation between the two sets of data; if so, add a line of best fit.
   c Explain the correlation between life expectancy and access to safe drinking water.

 Download Worksheet 16.3 from Cambridge Elevate for help with Activity 16.5, question 3.

Figure 16.13 Pollutants leak from the tailing ponds, which store mining waste, into the Athabasca River, Canada.

 Key term

water conflict: disputes between different regions or countries about the distribution and use of fresh water; water conflicts arise from the gap between growing demands and diminishing supplies

Figure 16.14 UK supermarkets import food from abroad; food worth £10 billion is thrown away every year.

 Further research

Investigate the causes and effects of the shrinking Aral Sea, once the world's fourth largest inland sea, on the **Earth Observatory website** (www.cambridge.org/ links/gase40152).

16.8 What strategies can be used to increase water supply?

The world is not running out of water, but it is not always available when and where people need it. There are a range of strategies that can be used to increase water supply. Engineering megaprojects such as water storage, transfers and **desalination** are large in scale, time consuming to construct and often come with high economic, social and environmental costs. Some of these are outlined below. More sustainable solutions are smaller in scale, less expensive and can be easily maintained by the local community (see 16.4 What factors affect water consumption?).

Around 5000 km³ of water is stored in facilities, either in underground aquifers or in reservoirs. This represents a sevenfold increase since 1950.

Aquifer recharging

In places with high evaporation rates, or with plenty of storage capacity underground, recharging aquifers with water can be a sustainable way of storing excess water.

Dams and reservoirs

Over 50 000 large dams have been built worldwide, 90 per cent of them since 1950. In the USA, the Glen Canyon and Hoover dams built by the US Army Corps of Engineers store water from the River Colorado in Lake Mead (Figure 16.15) and Lake Powell. They also control flooding and create recreation areas for boating, camping and whitewater rafting in the Grand Canyon. Initially seen as a success, low rainfall and rapid urbanisation in the last few decades has resulted in a worrying decline in the water levels and some dire environmental problems (see Figure 16.16).

Building more dams can only solve water supply problem in areas with secure rainfall. Where rainfall is low or increasingly variable due to climate change, alternatives include **water transfer schemes** and desalination.

Figure 16.15 Lake Mead, with the Arizona intake towers of the Hoover Dam, has shrunk to its lowest level since the 1930s. The light 'bathtub ring' shows the high water mark.

ℹ Did you know?

Lake Mead is the largest reservoir in the USA and contains enough water to flood the entire state of New York with 30 cm of water.

🔑 Key terms

desalination: the process of creating fresh water by removing salts and minerals from seawater

water transfer schemes: systems of pipelines and aqueducts to move water long distances from areas with plentiful supply to those of high demand

Sediment is deposited into the reservoir as flow slows.

Clear water is released below the dam and quickly scours the bedrock, causing high rates of erosion.

Water pollution from farm runoff as well as industrial and domestic effluent have all been deposited into the reservoirs.

The local climate is altered as a large body of water increases local evaporation and cloud formation.

Water flow is reduced, affecting plants and animals. For example, migratory fish cannot travel upstream and reach spawning grounds.

Figure 16.16 Environmental impacts of dams and reservoirs.

Water transfer

Water supplies can be increased by transferring water from one place to another. Water transfer schemes can vary in scale from huge engineering schemes to small-scale community schemes. Large-scale water transfer schemes divert water from river basins with a **water surplus** to those with a **water deficit** via pipes, canals and tunnels. They are often multi-use schemes combining water storage with water-based leisure activities and can provide a supply of **hydroelectric power**. However, they are very expensive to construct, can displace communities and put control of water into the hands of a few key players.

Desalination

In many water-scarce countries removing salts and minerals from seawater, known as desalination, is the only way to supply much needed water. This is particularly true of the Gulf States including Saudi Arabia, the United Arab Emirates and Kuwait (Figure 16.17). Currently, 1 per cent of the world's population are dependent on desalinated water. This figure is set to soar as more areas face water deficit. Desalination requires large amounts of energy to extract the salts from the water and is therefore a very expensive technology (the process is shown in Figure 16.18), costing just under £2 per cubic metre of water. The environmental cost is also high, as desalination plants emit high levels of carbon dioxide by burning fossil fuels and the brine wastewater damages marine ecosystems.

Figure 16.17 A desalination plant.

Did you know?

China plans to increase its use of desalination technology. With 400 cities facing serious water shortages, Beijing aims to quadruple its seawater desalination capacity to 3.6 billion litres a day by 2020.

Further research

Find out more about the process of desalination, how it works and the impacts on the environment. Start by using the **US Geological Society website** (www.cambridge.org/links/gase40153).

Key terms

water surplus: water surplus exists where water supply is greater than demand

water deficit: this exists where water demand is greater than supply

hydroelectric power (HEP): electricity generated by turbines that are driven by moving water

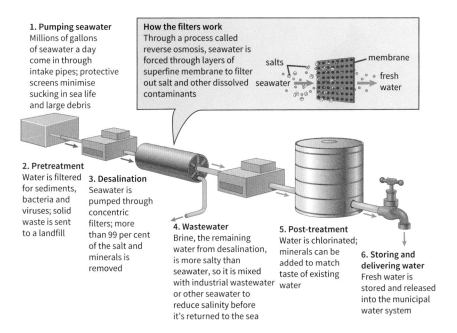

1. Pumping seawater
Millions of gallons of seawater a day come in through intake pipes; protective screens minimise sucking in sea life and large debris

How the filters work
Through a process called reverse osmosis, seawater is forced through layers of superfine membrane to filter out salt and other dissolved contaminants

salts
membrane
seawater
fresh water

2. Pretreatment
Water is filtered for sediments, bacteria and viruses; solid waste is sent to a landfill

3. Desalination
Seawater is pumped through concentric filters; more than 99 per cent of the salt and minerals is removed

4. Wastewater
Brine, the remaining water from desalination, is more salty than seawater, so it is mixed with industrial wastewater or other seawater to reduce salinity before it's returned to the sea

5. Post-treatment
Water is chlorinated; minerals can be added to match taste of existing water

6. Storing and delivering water
Fresh water is stored and released into the municipal water system

Figure 16.18 Desalination using the multi-stage flash distillation process.

ACTIVITY 16.6

1 Why is water storage necessary?

2 Describe three strategies that can be used to increase water supply.

3 Draw a labelled sketch of Lake Mead (Figure 16.15) to show the characteristics of the dam and reservoir.

4 What problems are caused to people and the environment by:
   a dam construction
   b desalination?

The South-to-North Water Transfer Project, China

The South-to-North Water Transfer Project in China is the largest interbasin water transfer scheme ever to be undertaken anywhere in the world (see Figure 16.19). Mao Zedong first proposed it in 1952, but construction did not start until 2002. At twice the cost of the controversial Three Gorges Dam, the $80 billion project aims to divert 45 billion cubic metres of water each year from the Yangtze to the Yellow River basin and beyond to the north. The aim is to ease the growing water shortages in Beijing, Tianjin and the northern provinces (Figure 16.20).

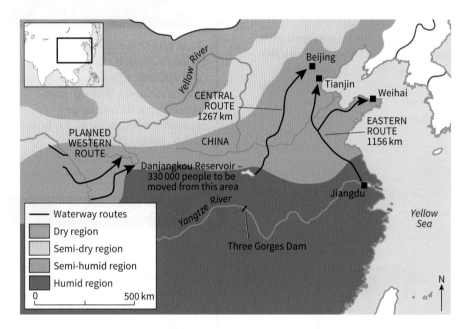

Figure 16.19 The South-to-North Water Transfer Project.

Why is such a mega-project needed in China?

China has a fifth of the world's population but only 7 per cent of its freshwater resources. Shared amongst a population of 1.4 billion people this gives each citizen 2000 cubic metres each year (less than a third of the global average). It is expected that by 2030 China's water supply will not be able to meet demand.

Southern China has 77 per cent of China's water resources, but the densely populated north demands huge amounts of water. Farming in the rural north has become ever more difficult as water resources are diverted to supply the manufacturing and urban areas, leaving farmers without enough water to grow crops and turning the sandy loess soil to dust (Figure 16.21). With an estimated 134 million people facing hunger, the greatest number in any single country globally, China must manage water resources carefully to ensure future prosperity.

Figure 16.20 The South-to-North Water Transfer Project moves water over 1200 km across China.

Figure 16.21 Drought and soil erosion will increasingly affect farmers in the rural south of China.

Advantages

- Over 45 billion m³ of water will be diverted from the Yangtze River Basin to supply the people and industry in the north.
- When completed, the Chinese government claim the project will benefit up to 500 million people, with increased availability of water resources in Beijing.
- The project is seen as a key to unlocking the development potential in the north of China.

Disadvantages

Many of the unintended impacts (negative externalities) of the project will affect the environment and the people of China.

- In Hubei and Henan provinces, around 350 000 people were resettled to allow the Hanjiang Valley to flood for the Danjiangkou Reservoir.
- Water diverted from the lower basin of the Yangtze River is of a low water quality and contains heavy metals, agricultural chemicals and human waste.
- Over £2 billion has been spent on pollution prevention projects and much more will be required in the future.
- There will be uncertain water futures due to external shocks to the system such as climate change. In recent years rainfall has been higher in the north than the south, and people have questioned whether water diversion is necessary.
- In times of drought, such as in 2011, there was not enough usable water to satisfy the demand in the south, let alone transfer water from the Yangtze to the north, making people question the decision to go ahead with this project.

 Discussion point

Why might the people and industry in the north be more supportive of the transfer scheme than the farmer shown in Figure 16.21?

 Visit the **International Rivers website** (www.cambridge.org/links/gase40154) for more information about the south–north water transfer scheme.

16.9 How can we move towards a sustainable water resource future?

With 1.1 billion people in the developing world without access to safe water and global water demand continuing to rise, it is essential that water resources are used sustainably and efficiently.

Water conservation

In areas of high water consumption, **water conservation** strategies can be employed to use water more efficiently. Figure 16.22 shows a variety of water conservation strategies to reduce water usage in the home. The state of Nevada has reduced its water consumption by 23 per cent over the past decade, despite the population of Las Vegas growing by half a million new residents. This has been achieved by limiting sprinkler use and the 'Cash for Grass' rebate in which the Southern Nevada Water Authority has paid out $200 million to homeowners to replace their lawns with desert landscaping.

> **Key terms**
>
> **water conservation:** strategies that use water more efficiently
>
> **hydroponics:** a method of growing plants using mineral nutrient solutions, in water, without soil

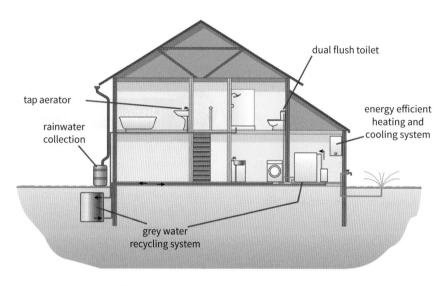

Figure 16.22 A water efficient house.

Industrial output can be limited without access to sufficient water resources. For this reason it is important for industry, particularly food and beverage companies who use large quantities of water, to use water efficiently and ensure future sustainability by taking the following steps:

- fit push taps and aerators in sinks so water cannot be left running and flow rates are reduced
- replace workplace grass with an artificial lawn to save water and reduce maintenance costs
- harvest rainwater to flush toilets and use in industrial processes

Farmers can use water more efficiently by using automated sprinkler irrigation with pipes rather than inefficient central-pivot irrigation (Figure 16.23). This would reduce water loss by evaporation. Drip-feed irrigation releases water from a pipe directly to individual plants and uses even less water. Growing crops in containers full of water and nutrients in greenhouses, known as **hydroponics** (Figure 16.24), is an alternative choice where soil and water resources are limited.

Figure 16.23 Central-pivot irrigation is an inefficient use of water, due to high evaporation rates.

Figure 16.24 Hydroponics used to grow salad leaves in the UK.

Recycling wastewater

Homes and industries can recycle wastewater (**grey water**) to reduce water consumption. Water from showers, baths and washbasins can be stored and used for other purposes such as to water lawns. Coca-Cola has improved water efficiency across 300 bottling partners by 25 per cent by harvesting rainwater. This is then used for washing vehicles and for flushing staff toilets. Coca-Cola return all their wastewater to the environment after treatment.

Groundwater management

In areas of water deficit, groundwater levels are falling. Water companies need to manage the amount of water removed from underground aquifers to maintain healthy water supplies. In the UK, the Environment Agency is tasked with balancing abstraction for water supply with the needs of the environment. Farmers providing Coca-Cola with sugar have switched from flood irrigation to automated sprinkler systems to reduce water need to 50 per cent. From 2005 to 2013, Coca-Cola replenished 109 billion litres of water, benefiting over 1.9 million people. One project – the Replenish Africa Initiative (RAIN) – replenished nearly 2 billion litres of water, providing access to safe water for 2 million people in 35 African countries, sparing women the chore of fetching water from rivers far away and allowing girls more time in school.

Appropriate technology

In lower income countries, in areas of water deficit, appropriate technology that is small-scale, locally controlled and sustainable can be used to increase the quality and quantity of water. Appropriate technology available to communities includes:

- hand-dug wells – these are up to 15 m deep to reach underlying aquifers
- rainwater catchment – rooftop gutters collect water into a holding tank
- water purification systems – treatment systems remove contaminants from existing water stores
- subsurface or sand dams – sand and water collects behind a small concrete dam; filtered water can be dug out when required
- gravity-fed systems – use the force of gravity to move water downhill into a community from a river or spring
- latrines (Figure 16.25) and toilets – prevent water pollution, reduce the spread of disease and improve sanitation for 2.5 billion people.

ACTIVITY 16.7

1 Study Figure 16.22.
  a Outline the different ways water can be conserved in the home.
2 Study the text on appropriate technology.
  a Produce a diagram to summarise the different methods used to increase the quality and quantity of water available to communities.
3 Use a website such as piktochart to create an infographic about the benefits of one **sustainable water supply** solution. There are nine to find out about online at the **Charity Water website** (www.cambridge.org/links/gase40155).

Key terms

grey water: wastewater from people's homes can be recycled and put to good use; treated grey water can also be used to irrigate both food and non-food producing plants

groundwater management: regulation and control of water levels, pollution, ownership and use of groundwater

Watch a video on Cambridge Elevate about urban communities gaining access to clean water and sanitation in Bangalore, India.

Figure 16.25 A latrine in Kenya is built where there is currently no wastewater treatment infrastructure.

Key term

sustainable water supply: meeting the present-day need for safe, reliable and affordable water, which minimises adverse effects on the environment, while enabling future generations to meet their requirements

The Kyeni Kya Thwake water conservation scheme in Kenya

In many regions of the world annual rainfall is sufficient for life. Unfortunately, rain does not fall regularly, leading to periods of water shortages. Climate change is exacerbating the problem by increasing rainfall variability and drought frequency. If you also consider that global water demand is projected to increase, then it is essential for people living in dry LICs such as Kenya to harvest and store rainwater when it does fall for later use when it is dry.

A **sand dam** is a concrete wall built across a seasonal sandy riverbed (Figure 16.26). During the rainy season, a river forms and deposits soil, sand and water behind the dam. The water, filtered by the sand, can then be hand-dug or pumped out from the reservoir for use by the community for up to 50 years (Figure 16.27). Over 1500 sand dams have been built in Kenya over the last 40 years. In 2012, a sand dam was constructed for the Kyeni Kya Thwake community with help from the water charity African Sand Dam Foundation.

Why was this project needed?

Before 2012, the Kyeni Kya Thwake community were dependent on selling crops, goats and baskets at market to provide an income to pay school fees and hospital fees. When the rains failed, and they did regularly, their crops failed resulting in a hungry community without access to healthcare or an education.

New opportunities

The sand dam is a good example of sustainable development and has allowed the community to farm independently from erratic rainfall. It also means women are spared the chore of walking for hours to collect water. Instead they can tend crops on the farm while their children are at school. Vegetables such as spinach, kale and tomatoes can now be grown to eat and highly profitable onions can be sold at market.

Figure 16.26 A sand dam in Kenya.

Key term

sand dam: a sand dam is a concrete wall built across a seasonal sandy riverbed

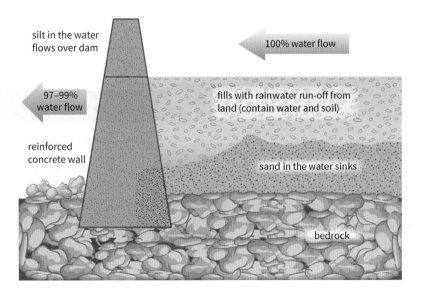

silt in the water flows over dam

100% water flow

97–99% water flow

fills with rainwater run-off from land (contain water and soil)

reinforced concrete wall

sand in the water sinks

bedrock

Figure 16.27 Cross-section through a sand dam.

Extra funding from The Water Project, a UK-based water charity, was used to develop sustainable farming activities including terracing, tree planting and the creation of a tree nursery and seed bank. Terraces are steps cut into slopes. The flat areas are used to grow crops and trees stabilise the soil; this stops valuable soil from being washed or blown away.

Factfile

- **Location:** Kenya
- **Community:** Kyeni Kya Thwake
- **Community size:** 500
- **Cost:** £8000
- **Stakeholders:** African Sand Dam Foundation, The Water Project

Benefits for the community

The Kyeni Kya Thwake water conservation scheme offers the following benefits to the community:

- up to 20 million litres of water can be stored, supporting 1000 people
- increasing food security throughout the year
- children can access full-time education
- sand filters the water clean
- with no surface water as breeding ground for mosquitoes, the risk of malaria is reduced.

Assess to progress

1 Outline why water is essential for life. **4 MARKS**
2 Describe the global pattern of water surplus and deficit. **4 MARKS**
3 Explain the impacts of water insecurity for people and the environment. **9 MARKS**
4 Explain the social and economic benefits of a local water conservation scheme. **4 MARKS**
5 What are the problems associated with over-abstraction of water from an aquifer? **4 MARKS**
6 Explain why water transfer schemes and large dams are often described as unsustainable. **6 MARKS**
7 Explain the advantages and disadvantages of a named water transfer scheme. **6 MARKS**

Discussion point

The UN reports that by 2030, the world will need at least 50 per cent more food, 45 per cent more energy, and 30 per cent more water. How could sand dams help meet these demands?

Did you know?

Dripping taps in rich countries lose more water than is available to the poorest billion people on Earth.

Drylands cover approximately 40 per cent of the world's land area and support 80 per cent of the world's poorest people.

Further research

Read more about local water conservation schemes at **The Water Project website** (www.cambridge.org/links/gase40156).

Tip

Preparing your revision notes as you go will be helpful for practising exam questions and revising subject knowledge.

17 Energy resources

In this chapter you will learn about...

- where global energy supplies come from
- how global energy consumption is changing over time
- how a country's location affects its energy supplies
- why energy supplies change over time
- how energy insecurity affects different countries
- what countries could do to improve their energy supply
- the advantages and disadvantages of oil exploitation in the North Sea
- measures that can be taken to reduce energy use.

17.1 Where does our energy come from? (1)

Energy powers homes, industries and transport networks. **Non-renewable** energy refers to fossil fuels, such as coal, oil and gas, and nuclear power. **Renewable** fuels come from sources such as solar, hydro or wind power. Fossil fuels are the most important energy resource, making up a large part of the world's **energy mix**.

Energy resources can be used directly, e.g. burning wood to produce heat. They can also be used indirectly, e.g. burning coal in power stations to generate electricity. Most electricity in the UK is generated using gas, coal and nuclear fuel. Global energy production and consumption affects every country.

 Key terms

non-renewable: a resource that is limited in supply and cannot be replaced

renewable: a resource that can be replaced or replenished over time

energy mix: the combination of energy sources that make up the total supply

Oil

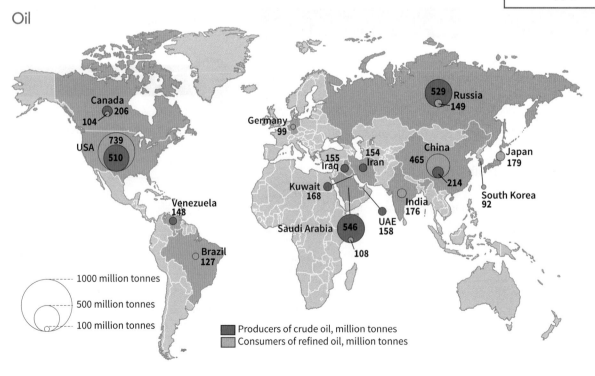

Figure 17.1 The top ten oil producing countries and oil consuming countries in 2014. Figures are in million tonnes.

Figure 17.1 shows the ten largest suppliers and consumers of oil products. Five of the oil producing countries are in the Middle East, where 66 per cent of the world's total oil reserves are located. All five countries are members of the Organization of the Petroleum Exporting Countries (OPEC). This group of countries work together to control global oil supplies and prices. Five of the biggest consumers of oil have to import supplies. Oil is a non-renewable resource (Figure 17.2), so although new reserves may be discovered, it will eventually become increasingly scarce. Oil is not only an energy resource, it is also used in the manufacture of products such as paints and plastics.

Coal

Figure 17.4 shows the largest suppliers and consumers of coal. China, the USA and India both supply and consume the most coal. There are around 70 countries worldwide with coal reserves that are profitable enough to extract (Figure 17.3). Coal is used directly to create heat (both domestically and industrially) and in coal fired power stations to generate electricity.

Figure 17.2 Oil production.

Figure 17.3 Coal mining.

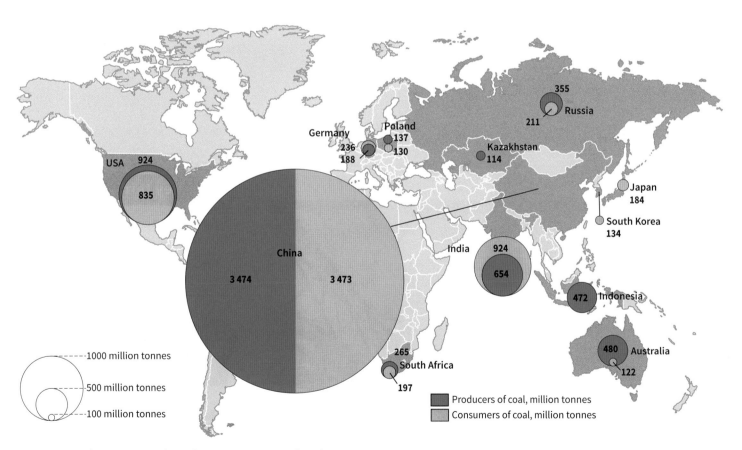

Figure 17.4 The top ten coal producing countries and coal consuming countries in 2014. Figures are in million tonnes.

17.1 Where does our energy come from? (2)

Gas

Figure 17.5 shows the largest suppliers and consumers of gas. The USA and Russia both produce and consume the most gas. Natural gas is used for electricity generation, heating and for powering manufacturing processes. In 2015, approximately 40 per cent of the UK's gas supply came from the North Sea (Figure 17.6). The country also relied on European gas pipelines and gas tankers to import gas. Some supplies are stored for winter months when the demand is higher. The UK has the capacity to store 3 months' supply.

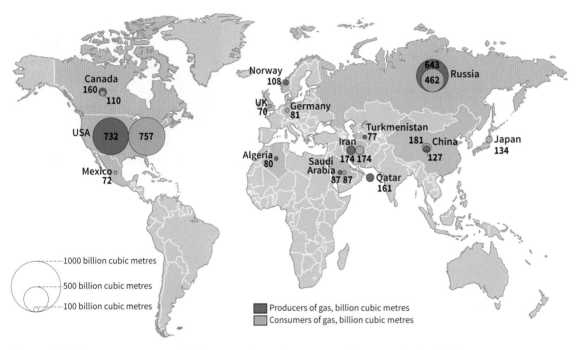

Figure 17.5 The top ten gas producing countries and gas consuming countries in 2014. Figures are in billion cubic metres.

Figure 17.6 Gas production.

Key terms

power stations: places that produce electricity

decommissioned: closed down and safely dismantled

Watch a video on Cambridge Elevate about biogas production in India.

Discussion point

How might the UK's energy mix change in the future? Why?

Nuclear power

Nuclear power comes from the release of energy when atoms of uranium are split (Figure 17.7). Uranium is a heavy metal found in rocks. It is non-renewable as it cannot be created and will eventually run out. Kazakhstan, Canada and Australia produce the most uranium and the USA, France and China are the biggest consumers (Figure 17.8). 18 per cent of electricity generated in the UK comes from nuclear power. In the past this figure has been higher, but some older nuclear **power stations** have reached the end of their lifespan and have been **decommissioned**. All but one of the UK's current reactors will have to be decommissioned by 2023, although new nuclear power stations are being commissioned, such as Hinkley Point C.

Figure 17.7 A nuclear power plant.

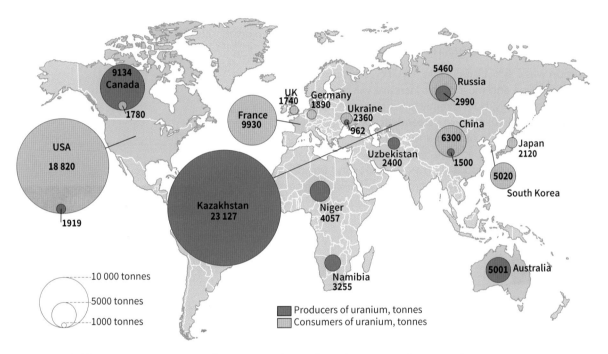

Figure 17.8 The top ten uranium producing countries and uranium consuming countries in 2014. Figures are in tonnes.

Renewable sources

Renewable energy sources, such as wind and solar power, currently make up less than 10 per cent of the UK's energy use. The government has a target to increase this figure to 30 per cent by 2020. In order to meet this target (and to avoid shortfalls as other resources become scarce and therefore more expensive) the use of new technologies must increase. Solar panels and wind turbines are examples of renewable energy (Figure 17.9). **Biogas** could be generated using organic waste products, creating recycled energy. As renewable energy technologies become cheaper, the 2020 goal may become more achievable.

Key term

biogas: energy derived from decaying organic matter

Figure 17.9 Solar panels and wind turbines are renewable sources of energy.

ACTIVITY 17.1

1 What might happen as non-renewable energy resources begin to run out?
2 Create a table to show the social, environmental and economic advantages and disadvantages of renewable energy sources. Include solar, wind, tidal and HEP (see later sections).

17.2 How are energy consumption and economic development linked?

Global energy **consumption** has been increasing and is expected to grow by over 50 per cent between 2010 and 2040. Much of this growth will come from lower income countries (LICS). Figure 17.10 shows the projected increase:

- Over a 59-year period the consumption of most types of energy will increase.
- Gas and the use of other resources (renewables) will increase the most.
- Nuclear power will decline.

Projected future demand

Billion tonnes oil equivalent

Figure 17.10 Projected future global demand in energy.

Why is global energy consumption increasing?

In 1950, the global population was 2.5 billion people. By 2050, this number is predicted to rise to more than 9 billion and most of this increase will be seen in LICs. Growing populations require more energy, both directly in terms of individual usage, and indirectly. For example, individuals use energy directly to power household appliances and indirectly in the production of food and to power industries. Economic development and energy consumption are linked. Greater wealth leads to increased personal and public spending on goods such as electricity and technology, for example smartphones, laptops and cars. **Newly emerging economies** are seeing the biggest growth in these areas as they continue to develop (Figure 17.11). In China, demand for energy could increase by 75 per cent by 2035.

💬 Discussion point

What could your school do to reduce its energy consumption?

ⓘ Did you know?

Switching a single computer and monitor off overnight will save enough electricity to make 30 teachers six cups of coffee each!

The largest consumers of non-renewable fuel

Table 17.1 shows the top five consuming countries of oil, coal and gas for 2000 and 2014. Some patterns can be found within the data:

- The USA and China are big consumers of fossil fuels.
- China has increased its consumption of each type of fuel.
- Coal has seen the biggest growth in consumption within the table. China and India's consumption have more than doubled.

Figure 17.11 China has seen a rapid growth in industry.

	Oil (million tonnes)				Coal (million tonnes)				Gas (billion cubic metres)			
Rank	**2000**		**2014**		**2000**		**2014**		**2000**		**2013**	
1	USA	837	USA	739	China	1365	China	3473	USA	661	USA	757
2	Japan	232	China	465	USA	983	India	924	Russia	391	Russia	462
3	China	204	Japan	179	India	359	USA	835	UK	103	China	181
4	Germany	121	India	176	Germany	244	Germany	236	Canada	92	Iran	174
5	Russia	119	Russia	149	Russia	232	Russia	211	Germany	88	Japan	134
Total	**1513**		**1708**		**3183**		**5679**		**1335**		**1708**	

Table 17.1 The top five consuming countries of oil, coal and gas in 2000 and 2014.

The largest consumers of renewable fuel

Table 17.2 shows the top five countries that have the highest percentage of renewable energy use in their electricity energy mix. This means that renewable energy is used to generate electricity. The same five countries appear in both columns, so these countries have used renewable energy for some time. This could be because their physical geography gives them a natural advantage when generating energy, e.g. they have sufficient rainfall to generate **hydroelectric power (HEP)**.

Rank	Renewables (as a % of total electricity)			
	2000		**2014**	
1	Norway	99.7	Norway	98
2	Brazil	89.5	New Zealand	79.0
3	Colombia	75.5	Brazil	73.4
4	Venezuela	73.7	Colombia	70.0
5	New Zealand	71.5	Venezuela	62.8

Table 17.2 The top five countries that have the highest percentage of renewable energy use in their electricity energy mix.

The majority of Norway's renewable energy comes from HEP. The country is also developing onshore and offshore wind farms. Portugal (62.6 per cent), Canada (62.5 per cent) Sweden (58.5 per cent), Chile (42.8 per cent) and Italy (42.1 per cent) all appeared in the top 10 countries in 2014.

ACTIVITY 17.2

1 a On a blank map of the world, mark on the top five countries for oil, coal, gas and renewable energy consumption.

   b Describe and explain the links between energy consumption and economic development.

Download Worksheet 17.1 from Cambridge Elevate for help with Activity 17.2, question 1a.

2 Read the following text written by a factory worker in India (Figure 17.12):

'India is a great place to live right now. There are more jobs available and wages have increased. Our standard of living has improved as we can afford to buy new technology, like mobile phones and tablet computers. I am saving up to buy a car so that I can drive to work.' Nimesh, Mumbai

   a Why is the biggest growth in energy consumption being seen in emerging economies like India?

   b Do you think the energy consumption of the emerging economies will continue to rise? Explain your answer.

3 How could we persuade governments to invest money in renewable energy?

   a Create a tweet (no more than 140 characters) encouraging your followers to reduce energy consumption.

   b Explain what you think the UK government should do about supporting renewable energy. Explain where the money might come from and what benefits will result from the changes.

Visit the **National Grid website** (www.cambridge.org/links/gase40157) for real-time information about electricity production and consumption in the UK, to see how demand changes over the course of the day, and how electricity is being generated.

Figure 17.12 Nimesh works at a factory in Mumbai, India.

17.3 Why are there areas of energy surplus and deficit?

Some parts of the world have a **surplus** of energy. This is an advantage as other countries rely on them and so it gives them more power and money. As we saw previously, Russia and the USA both have supplies of more than one energy source and the Middle East has 66 per cent of the world's oil.

Fossil fuels

Figure 17.16 shows the countries that have the largest reserves of fossil fuels. Most of these areas have a surplus of fossil fuels. This means that they produce more than they consume. Other countries have to import energy, e.g. the USA and China both consume more oil than they produce.

Fossil fuels are found in specific parts of the world. Oil formed beneath seas, when plants and animals died and fell to the seabed. It is found in geological formations where rocks have trapped the oil. Natural gas is often found above oil reserves as it was created from the decaying remains of plants and animals. Coal formed beneath swamp forests, where decaying trees were compressed into coal.

Renewable energy

Some countries are able to generate lots of renewable energy. Iceland takes advantage of its position on a plate boundary to generate **geothermal energy**. Lesotho and Paraguay both have a suitable landscape and climate to generate hydroelectric power (Figure 17.13).

Once generated, surplus amounts of renewable energy can be transferred to other countries via electricity interconnectors – cables that allow electricity to be transported from one country to another. When electricity demand is low and wind speeds are high, Denmark produces more electricity through wind power than it needs which it exports to neighbouring Germany. Similarly, Canada exports its surplus hydroelectric power across the border to the USA.

Areas of deficit

People in some parts of the world do not have access to electricity. South Sudan, Chad and Burundi are all African countries where more than 90 per cent of the population live without electricity (Figure 17.14). In the region of sub-Saharan Africa, two-thirds of the population have no access to electricity. Fuelwood and charcoal is the main fuel used for cooking. Several factors have created these areas of **deficit**.

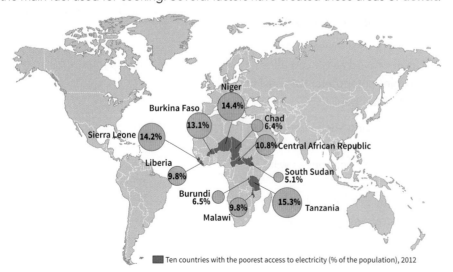

Figure 17.14 The top ten countries with poorest access to electricity in 2012.

Key terms

surplus: having more of something than is needed.

geothermal energy: where the heat from within the Earth is harnessed to generate electricity; geothermal power stations are usually found near areas of volcanic activity

deficit: not having enough of something

Figure 17.13 A hydroelectric power station in Paraguay.

Discussion point

What are the advantages of having surplus energy supplies?

Did you know?

The use of fuelwood and charcoal in LICs causes health problems, such as breathing difficulties, burns and eye diseases. Women and children spend the most time cooking and so suffer these illnesses more.

Many African countries have supplies of fossil fuels and suitable conditions for renewable energy, e.g. sunlight to generate solar power (Figure 17.15). However, the extreme poverty faced in some parts of the continent means that there is no money to spend on an electricity infrastructure. Some countries are landlocked or have an inhospitable environment, which can make importing fuel more difficult. Other countries have experienced issues such as rapid population growth, unstable governments, conflict and disease.

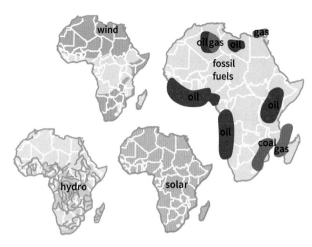

Figure 17.15 Africa has fossil fuels and the potential for renewable energy.

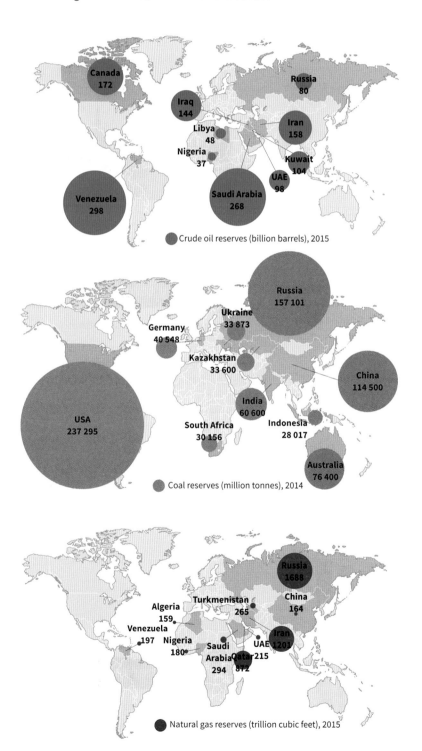

Figure 17.16 The top ten countries with proven global reserves of fossil fuels in 2015 (the coal reserves are from end 2014).

ACTIVITY 17.3

1 a What is a fuel surplus?
  b Describe the patterns of fossil fuel reserves shown in Figure 17.16.
  c How might global patterns of energy surplus and deficit change over time?
2 How do the following factors affect energy supply?
  a geology
  b poverty
  c physical geography
3 Read this diary entry, written by Aretta who lives in Niger (West Africa).

'Each morning I get up early and walk to collect the wood for the stove. I have to walk further and further each day. When I have collected enough wood, I carry it home. I then get changed and spend the afternoon studying in school.'

  a Why does Aretta have to walk further each day to find fuelwood?
  b How does fuelwood put Aretta's health at risk?
  c If Aretta had a reliable source of energy (electricity) how would her life be improved?

17.4 What factors affect energy supply?

Many factors affect global energy supplies. Over time, changes in these factors result in surpluses and deficits. Prices fall when there is a surplus and rise when there is a deficit.

Physical factors affecting energy supply

Fossil fuels are non-renewable resources that took millions of years to form. They are only found in certain locations on the planet. They will eventually run out, although new supplies of fossil fuels are sometimes found. A new oilfield, thought to hold 100 billion barrels of oil, was discovered in the south of England near Gatwick Airport, Sussex in 2015.

Climate affects the availability of energy supplies, e.g. warm countries are able to generate more solar power. In other countries, rivers and seas freeze for part of the year, which makes it difficult to generate HEP or tidal power. Global climates are experiencing a period of warming, which may allow new fossil fuel resources to be found and **exploited** (Figure 17.17). A warmer planet will mean changes in sunlight hours, wind speeds and rainfall.

The cost of exploitation and production

The cost of energy is partly dependent upon how much people are paid to produce it. When wages increase, fuel costs more money to buy. Production may move to parts of the world where both the resource is available and where people are willing to work for less money.

As non-renewable fuels run out, less profitable resources will be used. The UK has over 200 million tonnes of coal. Much of this coal has been too expensive to extract. As coal becomes scarce, UK coal mines are becoming profitable again. New opencast coal mines are being planned and opening across the UK, e.g. Druridge Bay in Northumberland.

Energy prices can also fall. In 2016 oil prices fell to 2009 levels. This was due to overproduction (partly due to the lifting of sanctions that had prevented Iran from selling oil), a fall in demand (partly due to the slowing Chinese economy), and the production of shale oil in the USA.

Changes in technology

Improvements in technology can alter the availability of energy resources. The method used to extract gas from shale is known as **fracking**. The USA has used new drilling and mapping techniques to extract increasing amounts of shale gas

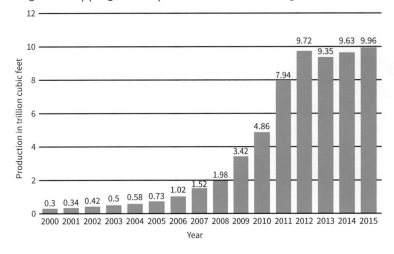

Figure 17.18 Shale production in the USA between 2000 and 2015.

Did you know?

In 1962 a fire spread to an underground coal seam in the town of Centralia, in Pennsylvania, USA. The fire is still slowly burning to this day. The town has been demolished and all residents evacuated.

Figure 17.17 An oil drilling platform in the Arctic Ocean.

Key terms

exploited: to use something

fracking: extracting gas from rocks by injecting water, sand and chemicals into them at high pressure

Further research

Fracking is a controversial technique that extracts natural gas trapped within rocks. Investigate the arguments for and against fracking.

Find out more:
Go to the 'Drill, Baby, Drill' map on the **Shale Bubble website** (www.cambridge.org/links/gase40158) to find out how much shale gas is being extracted in the USA.

(Figure 17.18). Natural gas makes up over a quarter of the energy mix of the USA and 64 per cent of this gas will come from shale by 2020. This increase in the supply of energy is lowering energy prices. Fracking is a controversial, heavily debated process as it uses chemicals, which may contaminate water supplies. It could also cause **seismic activity** and damage to property.

New generation wind turbines and solar panels are making renewable energy cheaper and more efficient. Taller wind turbines with longer, lighter rotor blades can 'capture' more wind. Solar panels (Figure 17.19) are becoming cheaper and 'black solar cell' technology can generate electricity on cloudier days.

Political factors affecting energy supply

War and corruption both affect the supply of energy. War can make energy resources difficult to access. In 1991, Iraqi troops set fire to oil wells as they withdrew from their invasion of Kuwait (Figure 17.20). Corruption can mean that money is diverted from energy projects.

Key terms

seismic activity: movement within the Earth's crust

Sustainable Development Goals: United Nations goals that came into force at the end of 2015. They are international development targets to be achieved by 2030

Further research

How is the the UK Government encouraging renewable energy use? Research the 'Green Deal' to find out how homeowners are being encouraged to reduce their greenhouse gas emissions.

Use the Government's energy simulator to plan an energy strategy for the future.

Figure 17.19 New-build housing with solar panels.

Figure 17.20 Troops leaving Kuwait set fire to oilfields.

The 2015 UN **Sustainable Development Goals** aim to achieve sustainable energy by 2030. The EU aims for its member states to cut greenhouse gas emissions by 40 per cent between 1990 and 2030. These targets put pressure on governments to reduce fossil fuel use and develop more renewable energy.

ACTIVITY 17.4

1 What four factors affect global energy supplies?
2 How can changing global temperatures alter the supply of energy?
3 How has new technology increased the supply of both renewable and non-renewable energy?

17.5 What are the impacts of energy insecurity?

The International Energy Agency (IEA) defines energy security as 'uninterrupted availability of energy sources at an affordable price'.Energy insecurity is therefore when countries have an interrupted supply of energy, or cannot afford energy.

Energy insecurity, or energy supply problems, can have a number of impacts:

The exploration of difficult and environmentally sensitive areas

When countries cannot obtain enough energy, they are forced to find new sources. Fracking extracts oil and gas deposits that were once locked away beyond reach within shale rocks.

Energy insecurity sometimes means that hostile landscapes or environmentally sensitive areas are opened up for development. The USA drilling for oil in the Arctic is a good example. The extreme temperatures create icebergs that can pose a threat to drilling platforms. The fragile ecosystem is put at risk through noise and the potential for oil spills.

Economic and environmental costs

Economic costs of energy insecurity stem from a reliance on imported fuel. Countries exporting the fuel set prices and so they have power over those countries importing. This means that some countries pay much more to access fuel supplies than others.

Environmental costs occur when a country relies on fuels that cause pollution, or when fuel is over-exploited in order to meet demand. In 2016, a report by the Royal College of Physicians and the Royal College of Paediatrics and Child Health was the first to prove a link between air pollution and health. The report found that 40 000 people in the UK die every year as a result of air pollution.

China has a surplus of coal on which it relies on to reduce the money it has to spend on importing oil. The use of coal-fired power stations has created problems with **smog** in the cities (Figure 17.21) and large **opencast** coal mines scar the landscape.

Food production

Growing biofuels take up valuable farmland. This reduces the amount of food that can be grown and pushes food prices up. If energy is in short supply, then it also costs more to produce and transport food.

Thailand has an Alternative Energy Development Plan (AEDP). The 15-year programme aims for renewable energy sources to make up 20.4 per cent of its energy needs by 2022. Over this time, biofuel production will increase fivefold, using sugar cane (Figure 17.22), cassava and palm oil. Rice production will decrease and so food prices will rise.

Industrial output

If energy is in short supply, **manufacturing** becomes more expensive or unachievable. Some types of industry use more energy than others. 75 per cent of all energy used in manufacturing is for industries such as petrol refineries, aluminium plants and glass production.

Figure 17.21 Smog in Beijing.

Key terms

smog: a mixture of smoke and fog that can be dangerous to human health

opencast: mining that takes place on the surface of the Earth by scraping away soil and rock

manufacturing: the process of making a product

Visit the **Biofuel website** (www.cambridge.org/links/gase40159) to find out about the different types of biofuel. What are the advantages and disadvantages of developing biofuels?

Figure 17.22 Sugar cane fields in Thailand.

In Iceland, almost 70 per cent of the energy generated is used for the aluminium industry. Iceland can only do this because they have an abundance of renewable energy.

Potential for conflict where demand exceeds supply

A recent surge in global demand for energy is now increasing conflicts between countries. In 2015, Russia threatened to cut gas supplies to Europe due to its conflict with Ukraine. Ukraine is a transit country, which means that Russia's gas supply to Europe travels by pipelines running through the country. A dispute between Russia and Ukraine threatens gas supplies to Europe.

In 2015 Egypt, Sudan and Ethiopia held talks about a dam that Ethiopia wants to build on the River Nile. The Grand Ethiopian Renaissance Dam would generate hydroelectric power for Ethiopia, but is likely to reduce the amount of water reaching Sudan and Egypt (Figure 17.23). The countries remain in disagreement about whether the dam should go ahead.

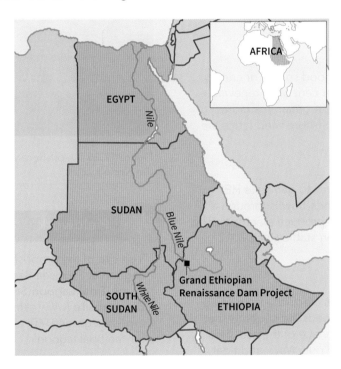

Figure 17.23 The River Nile runs through Ethiopia, Sudan and Egypt.

> ### Discussion point
>
> What problems are being caused through illegal charcoal trading in Somalia?

ACTIVITY 17.5

1 Explain five impacts of energy insecurity.
2 Using the Thailand Alternative Energy Development Plan (AEDP), explain how energy insecurity may be linked to food insecurity.
3 Countries are interrelated so when one place faces energy insecurity other countries are affected. Draw flow diagrams to show some of these positive and negative effects. Present your work like this:

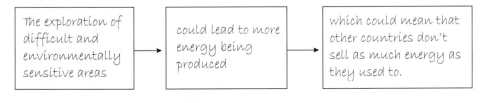

311

17.6 What strategies can be used to increase energy supply?

Energy demand is increasing over time. If we are to maintain our current lifestyles we must find new ways to increase our energy supplies or use existing supplies more efficiently.

Renewable solutions

Biomass

Biomass is organic matter used as a fuel. It can be used in a power station for generating electricity. It differs from fossil fuels (which also come from living things) as it has formed recently, e.g. in the UK, chicken droppings are used to produce biomass. Waste plants burn non-recyclable rubbish to generate electricity (Figure 17.24). This a way to deal with waste and generate heat and electricity. The South East London Combined Heat & Power (SELCHP) plant generates enough electricity to power 48 000 homes.

Wind power

Wind power converts wind energy into electricity. It is a good source of energy for the UK, as it is often windy. By the end of 2014, 54 per cent of all renewable energy generated in the UK came from wind farms. The UK generates more wind energy using onshore turbines, but the number of offshore wind farms are increasing (Figure 17.25).

Hydroelectric power

Hydroelectric power (HEP) uses water and gravity. One way to generate HEP is to build a dam across a river to trap water. The water then travels through a penstock (a channel in the dam) and turns a generator to produce electricity. The amount of UK electricity generated through HEP has been stable since 2012. Building more dams could increase the capacity of HEP.

Tidal power

It is thought that the UK could generate 20 per cent of its electricity from using waves and tides. Less than 0.01 per cent of our energy needs are currently met this way. Wave energy uses smaller movements on the surface of the sea. Tidal power uses larger movements of the tides coming in and out. Wave and tidal power generation takes many forms and the UK is researching into which is best. There are plans for six tidal **lagoons** in the UK, e.g. the Swansea Tidal Lagoon in South West Wales.

Geothermal power

Geothermal power taps into the Earth's buried heat and converts it into steam, which is then used to generate electricity. This means that other forms of energy do not have to be used. This process is easier where sub-surface temperatures are higher, e.g. Iceland. The British Geological survey thinks that the UK could generate geothermal energy too and there is already a geothermal borehole in Southampton. In addition to this, **ground source heat pumps** have been developed that extract heat from the ground (Figure 17.26).

Solar power

The UK Government estimates that 4 per cent of our electricity could be generated using solar power by 2020. Solar panels use photovoltaic (PV) cells to turn sunlight into electricity. They can be installed as solar farms or on individual buildings. The amount of solar power generated in the UK is increasing as technology becomes cheaper and more effective.

Figure 17.24 Waste plants generate electricity by burning non-recyclable rubbish.

Figure 17.25 An offshore wind farm off the Norfolk coast.

Fieldwork

Visit site of the proposed Tidal Lagoon at Swansea Bay or go to the **Tidal Lagoon Swansea Bay website** (www.cambridge. org/links/gase40160) to see plans for the tidal lagoon.

Key terms

lagoons: stretches of salt water separated from the sea by a barrage or beach

ground source heat pump: a system for converting heat from the ground into heating systems within homes

Visit the **Wavehub website** (www.cambridge.org/links/ gase40161) to see the latest research into tidal power in Cornwall, UK.

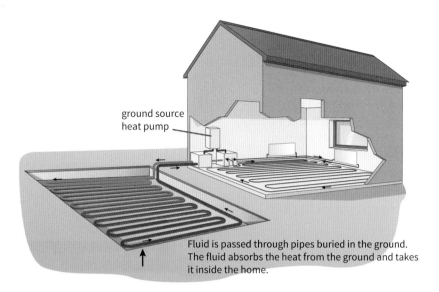

ground source
heat pump

Fluid is passed through pipes buried in the ground.
The fluid absorbs the heat from the ground and takes
it inside the home.

Figure 17.26 Ground source heat pumps take heat from the ground and use it to heat homes.

Non-renewable solutions

The UK is still dependent on non-renewable energy, like fossil fuels and nuclear power. Here are some of the things that can be done to use both types of energy more efficiently:

- **Combined-cycle systems:** power stations that are powered by fossil fuels such as coal or natural gas can reuse heat that would otherwise be wasted. A combined-cycle system produces up to 50 per cent more electricity from the same amount of fuel than a traditional power station. Less fuel is used to produce the same amount of electricity.
- **Co-firing:** all major UK power stations now burn small amounts of biomass, like wood and agricultural waste, alongside fossil fuels (usually around 3 per cent biomass to 97 per cent fossil fuel). Burning biomass reduces the amount of fossil fuels that are needed to generate electricity.
- **Nuclear fuel reprocessing:** fuel for nuclear power stations comes from uranium fuel rods, which remain an efficient fuel for about four years. Reprocessing is a chemical process that separates useful uranium from nuclear waste. This allows it to be reused and not wasted. The UK, France, Japan, Russia and India all have nuclear fuel reprocessing plants.

ACTIVITY 17.6

1 Should we focus on developing renewable energy or increasing the efficiency of our non-renewable energy generation?
2 a Which renewable strategy do you think is the best for the UK? Give reasons.
  b Which renewable strategy to you think is the worst for the UK? Give reasons.
  c Which renewable strategies would work well in other countries? Try to include some specific countries in your answer.
3 How can energy consumption be reduced?

Download Worksheet 17.2 from Cambridge Elevate for help with Activity 17.6, question 3.

The Gannet oilfield in the North Sea

Oil is a fossil fuel that is found beneath the North Sea. The UK is allowed to drill for the oil that lies around its coastline. Norway, Denmark, Germany and the Netherlands can also drill for oil in other parts of the North Sea. The UK and Norway have the largest reserves of North Sea oil, although it is estimated that less than 50 per cent of it now remains. Gas is also found in the North Sea.

The UK Department of Energy and Climate Change (DECC) grants licences for companies to drill for the oil and sell it. There are over 100 oil installations in North Sea oilfields (Figure 17.27). These are operated by companies like BP, Shell and Total.

The Gannet oilfield

The Gannet oilfield lies 180 km (112 miles) to the east of Aberdeen in Scotland. The sea is 312 ft (95 m) deep here. Shell have been **extracting** oil and gas from the Gannet Alpha Platform since November 1993. They have a manned platform, which means that people live and work on the **drilling rig** for periods of time (Figure 17.28). 88 000 billion barrels of oil per day is taken from the Gannet and there is enough oil for this to continue beyond 2019. Once extracted, the oil is transported by pipeline to Teesside.

> ### 🔑 Key terms
>
> **extracting:** removing something, e.g. oil
>
> **drilling rig:** a floating or fixed structure for removing oil from deep within the seabed

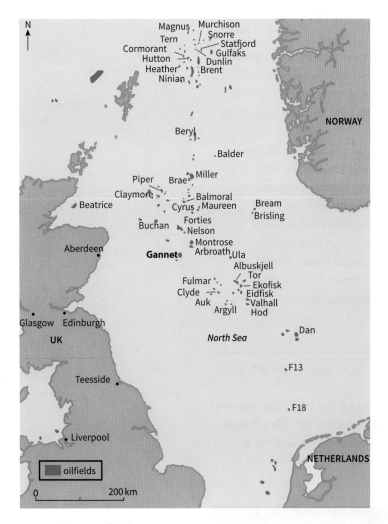

Figure 17.27 North Sea oilfields.

Figure 17.28 A drilling rig off the coast of Scotland.

Advantages of drilling for North Sea Oil

There are a number of advantages of drilling for oil in the North Sea:

- Drilling for oil at the Gannet and other North Sea oilfields means that the UK can produce its own oil. This means the UK is less reliant on other countries and is less vulnerable to changes in global oil prices. North Sea oil is good for the country's economy too as the UK can sell oil and make money from it.
- There are many jobs created through oil extraction. Aberdeen has benefited from this as the increased employment has created a **multiplier effect** in the city. Unemployment in the city is just 1 per cent and on average people in Aberdeen earn £150 more per week than the rest of Scotland.
- Oil is a raw material used by many other industries in the area. A local and reliable supply of oil could create more local employment opportunities.

Disadvantages of drilling for North Sea oil

There are also many disadvantages of drilling for oil in the North Sea:

Supplies of oil in the North Sea will eventually run out. There is also not enough oil to supply the whole of the UK and so the country is still reliant on importing oil from other countries. Exploring the North Sea for new oilfields is very expensive.

Oil prices fell in 2015, which made North Sea oil less profitable. As a result of this, Shell announced that they may sell their stake in the Gannet oilfield. In February 2016, the UK's offshore oil association, Oil & Gas UK, asked the government to reduce the amount of tax that they pay so that they can still make a profit. If North Sea oil becomes unprofitable, then production may have to stop.

Although oil rigs are largely safe, there is potential for oil spills and accidents. In August 2011, a leak in a flow line between the Gannet Alpha platform and Aberdeen led to an estimated 1300 barrels of oil being spilled into the sea. Oil leaks and spills create many environmental problems (Figure 17.29).

Finally, drilling for oil in the North Sea encourages the UK's use of fossil fuels, which release carbon dioxide emissions. The UK has a commitment to lowering carbon dioxide emissions and tackling climate change.

ACTIVITY 17.7

1 Create a fact file about the Gannet Alpha Platform in the North Sea.
2 a Make two spider diagrams to show the advantages and disadvantages of North Sea oil extraction.
  b Are there more advantages or more disadvantages?
  c How might these advantages and disadvantages change in the future?
3 Work in pairs to produce a short news report about the impact of North Sea oil on the city of Aberdeen. Half the class must focus on the advantages of oil extraction and half must focus on the disadvantages.

Key term

multiplier effect: the 'snowballing' of economic activity, for example, if new jobs are created this gives people more money to spend which means that more workers are needed to supply the goods and work in the shops

Figure 17.29 Oil leaks and spills create problems for wildlife and habitats.

Further research

How many years are fossil fuel reserves predicted to last? Why do estimates vary?

Discussion point

What types of jobs will be available to the people of Aberdeen? Why are these jobs so well paid?

Discussion point

On the whole is the extraction of North Sea oil good or bad?

17.7 How can we use energy more sustainably?

Encouraging each person to use less energy will reduce overall demand. We can measure our energy use and impact on the planet by calculating our **carbon footprint** (Figure 17.30). The size of our carbon footprint depends how much carbon we produce in our lives. Calculating our footprint considers things such as:

- Where our food comes from – buying locally uses less energy.
- How many new products we buy – energy is used to produce new goods.
- Whether we walk or travel by car – travelling by car uses petrol or diesel.
- How many foreign holidays we take – planes use more fuel than other forms of transport.
- Whether we switch off electronic devices and lights when we're not using them – electronic devices still use electricity when on standby.

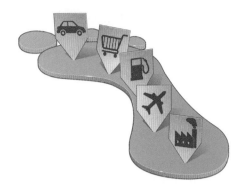

Figure 17.30 Carbon footprints measure our energy use.

Homes and buildings

Homes in the UK that are built, sold or rented out are rated according to their energy use (A is the most efficient type of house and G is the least efficient). Energy Performance Certificates (EPCs) also recommend ways to reduce energy use at the property, such as fitting insulation to reduce heat loss, using radiator thermostats to ensure that a home is not overheated and using smart meters to monitor energy use.

New houses are often designed to minimise energy use. Measures include:

- effective insulation and sealing to retain heat
- screening out light to prevent overheating without the need for air conditioning
- using materials such as concrete, which store heat; these can help to cool the building during the day and heat it at night.

20 per cent of the UK's carbon emissions are generated in workplaces. The government encourages businesses and organisations to become more energy efficient. Public authority buildings, like council-run schools, must have a Display Energy Certificate (DEC) to show how much energy they use. New workplaces can be designed to conserve energy by making full use of natural lighting and motion sensors to reduce any wasteful use of electric lights, incorporating natural ventilation to reduce the need for air conditioning, fitting solar panels to harness renewable energy and using durable materials to avoid the need for unnecessary manufacturing. Existing workplaces can conserve energy by switching off lights and computers when they're not being used and by improving loft and wall insulation.

Sustainable transport

Much of the energy that we consume is used for transport. Cars are now designed to use less fuel and produce smaller amounts of harmful emissions. Electric and hybrid cars run on electricity, which can be produced from renewable sources. Other cars can run on biofuels instead of fossil fuels. The government's system of taxing vehicles encourages people to buy cars that use less fuel; drivers of fuel-efficient cars pay no tax, whereas drivers of cars that use the most fuel pay over £500 a year. Some cities have introduced measures such as charging people to drive into certain areas (Figure 17.31) or have 2+ lanes, which are special lanes where only cars with two or more people can drive. This reduces car use and conserves energy as people are reluctant to pay congestion charges and are more likely to share their journeys.

Go to the footprint calculator on the **WWF website** (www.cambridge.org/links/gase40162) to calculate your carbon footprint.

Figure 17.31 The congestion charge in London was introduced in 2003.

Key terms

carbon footprint: the amount of carbon dioxide released into the atmosphere by an economic activity (e.g. person, business or event)

public transport: shared methods of travelling, such as buses, trams and trains

The government is looking at ways to make transport more sustainable. Many cities have improved their **public transport** or provided cycling facilities. These measures encourage more people to cycle, which reduces car use and conserves energy.

Improved technology

All electrical items use energy. Table 17.3 shows how much electricity (measured in kilowatt hours) household appliances use in a year. People generally own more electrical items nowadays, but improvements in technology has meant that these items are using less electricity than they used to. If less electricity is used, then fossil fuels are used more efficiently and overall energy consumption decreases.

Device	Annual energy use (kWh)
DAB radio	3
Laptop	11
HD-DVD/Blu-Ray player	16
Vacuum cleaner	26
Satellite TV set top box	51
42" smart LED TV	71
Games console	120
Fridge-freezer	149
Washing machine	153
Dishwasher	291

Table 17.3 The annual electricity use of household items.

The European Union (EU) has put a system in a place where all major household appliances, such as washing machines, dishwashers and fridge-freezers, are rated according to the amount of energy that they use (Figure 17.32). The rating ranges from A+++, which are the most efficient appliances, to G which are the least efficient. This helps consumers to make sensible choices about how much energy they are using.

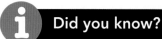
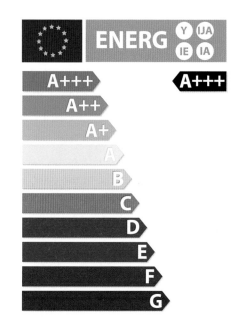

Figure 17.32 A European Union energy label.

ACTIVITY 17.8

1 Look at Table 17.4, showing electricity use by dishwashers.
  a 2015 average electricity prices were 12.5p per kWh. Calculate the cost of running each type of dishwasher over the course of that year.
  b Create a graph to show your results.
2 Use Table 17.3.
  a Which devices do you use in your house?
  b What other devices would contribute to your electricity usage?
  c Describe three ways to reduce your household energy use.
3 Here is a quote from Joanna Ward, Project Manager at Sustrans.
  'Sustrans work with families, communities, policy-makers and partner organisations so that people are able to choose healthier, cleaner and cheaper journeys, with better places and spaces to move through and live in.'
  a What sustainable forms of transport are found in your nearest town or city?
  b How can transport be improved to reduce energy use?

Dishwasher rating	Annual electricity use (kWh)
A+++	237
A++	266
A+	291
A	327
B	344

Table 17.4 Annual energy use for a range of dishwashers.

The Solar Mini Grid Scheme in Melela, Tanzania

Tanzania is a lower income country (LIC) in East Africa (Figure 17.33). It is one of the world's poorest nations. The World Bank identifies LICs as those countries with a **GNI per head** of less than $1045 in 2015. This figure changes over time and the most up-to-date value can be seen on the **World Bank website** (www.cambridge.org/links/gase40163). Tanzania's GNI per head is less than $1000 which means that 9 out of 10 Tanzanians live on less than $3 a day.

Tanzania is a hot country with between 2800 and 3500 hours of sunshine per year. It is ideal for solar power and it also has the capacity to generate electricity using wind, geothermal heat and biomass.

Many people in Tanzania cannot afford to pay for electricity. Local, renewable energy schemes help people to access energy.

The Melela Solar Mini Grid

Melela is a village that lies approximately 100 miles to the south-east of the capital, Dodoma. In the past, locals have had to rely on burning **kerosene** oil to provide power for their homes. Kerosene can be bad for people's health and the environment, but the villagers had no choice as they were not connected to the national **electricity grid** in Tanzania.

In 2013 a solar mini grid was installed to provide a renewable and sustainable source of fuel. The grid uses **photovoltaic (PV) panels** to generate electricity (Figure 17.34), which is then fed into a small grid. The solar grid is owned and managed by a small company called Devergy, who specialise in renewable solar energy solutions in Ghana and Tanzania. Locals in Melela only pay for the energy that they use, but they must pay in advance.

Impacts of the Solar Mini Grid Scheme

Around 200 households in Melela are connected to the mini grid. Each home receives enough electricity to power two small lamps, a mobile phone charger and a TV. Local shops and services have also bought into the scheme.

Figure 17.33 Tanzania, East Africa.

Figure 17.34 A Devergy solar panel in Melela.

Figure 17.35 An electric-powered welding machine; reliable electricity supply helps local businesses.

There are many benefits to this local renewable scheme:

- Businesses stay open and trade for longer as they have lighting. This creates more employment in the village (Figure 17.35).
- People watch TV and use mobile phones so they feel more connected with the outside world.
- Less kerosene is used, which is better for health and the environment.
- Energy costs are lower so people have money to spend on other essentials.
- The technology is relatively cheap and easy to fit and maintain (Figure 17.36).

While the solar mini grid is better than using kerosene, there are some drawbacks. Some villagers are still not able to afford the electricity. Those who can afford it often need more than they are given. Finally, the power generated using the PV panels can decrease during the seasonal rains.

Figure 17.36 A Devergy worker fitting a battery pack.

Discussion point

The solar mini grid is a small-scale solution to energy issues in Tanzania. What are the advantages and disadvantages of small-scale solutions?

The rapper Akon started a project in 2014 to introduce solar powered street lighting across Africa. Read the latest news about Akon's 'Lighting Africa' project at the **Akon Lighting Africa website**. (www.cambridge.org/links/gase40164)

Tip

Question 2 asks you to use Figures 17.35 and 17.36. Make sure that you refer to these in your answer. Explain means to give a reason for something. This question is asking you to give reasons why small-scale solutions are suitable for Tanzania.

Tip

It is important to use connectives, such as 'because' and 'therefore', to show that you understand the relationships between ideas. For each benefit you should name the improvement and then explain your ideas fully.

Assess to progress

1 Why is it more difficult to access electricity in rural parts of Tanzania?　2 MARKS

2 Use Figures 17.35 and 17.36. Explain the suitability of small-scale solutions to the energy problems faced in Tanzania.　3 MARKS

3 The following are benefits linked to improved access to electricity in rural Tanzania.
   - energy costs are lower
   - fewer people are using kerosene.

For each of the benefits explain how the lives of the people living in rural Tanzania will be improved.　4 MARKS

Fieldwork, skills and assessment preparation

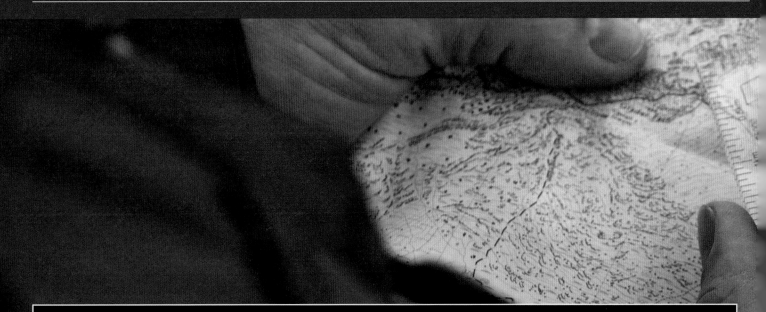

18 Fieldwork

In this chapter you will cover:

- choosing an enquiry
- methods of collecting, processing and presenting data
- analysis and conclusions; evaluation

19 Geographical skills

In this chapter you will cover:

- graphical skills (graphs and charts, population pyramids, choropleth maps)
- numerical skills (number, area, scale, ratio, proportion, sampling)
- statistical skills (central tendency, % increase and decrease, relationships)
- map skills (gradients, contours, cross-sections, transects, co-ordinates, GIS)

20 Assessment preparation

In this chapter you will cover:

- structure of the exam (number of papers, sections, types of question)
- issue evaluation questions
- how you will be assessed
- the grading system (point and level marking)
- exam technique (understanding the question, key exam words, how to use case studies and named examples)

Fieldwork, skills and assessment preparation – an overview

This section explores the use of fieldwork, geographical skills and techniques that will be helpful when preparing for assessments.

Fieldwork is an important part of geography because it offers the opportunity of investigating a local geographical issue by using newly collected information.

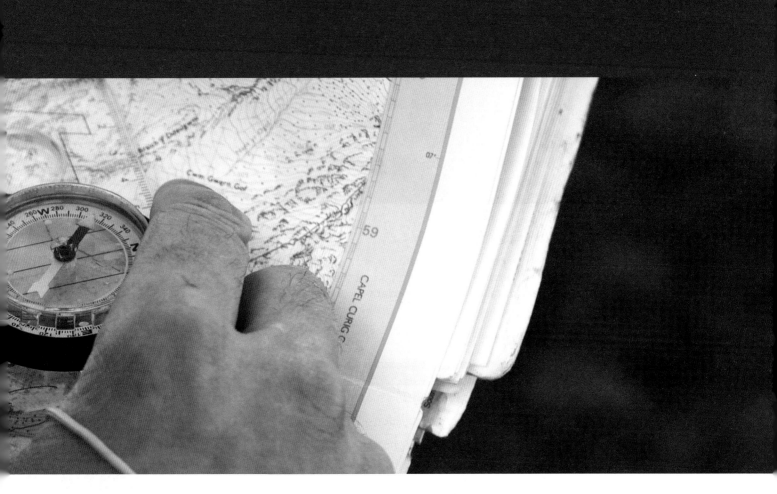

Completing a successful fieldwork enquiry involves a number of stages. Starting with a clear aim is important because it will then be easy to identify the research and fieldwork data needed to address the aim. After that, it is really about collecting and presenting the data and then using it to reach a conclusion which is clearly related to the original aim.

Geography provides an opportunity to develop a wide range of skills. These include cartographic (mapping), graphical, visual and mathematical skills. All of these skills can be seen by looking through this textbook. This shows the range of skills that can be used when collecting, illustrating and using geographical information in order to develop an understanding of geographical topics.

Planning and preparing is important when preparing for examinations. There are a number of revision and preparation techniques that can be used in order to help you. Being comfortable with the style of questions you might find in the examination and how answers are going to be assessed is an important part of your examination preparation.

Figure S7.1 The study of geography lets you explore the world around you and find explanations for what you discover. Keep an enquiring mind and enjoy the course.

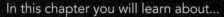

18 Fieldwork

In this chapter you will learn about...

- selecting a topic for enquiry
- planning an enquiry
- methods of collecting, processing and presenting data
- analysing results and reaching conclusions
- evaluating the enquiry process.

18.1 What is a fieldwork enquiry?

A fieldwork enquiry is where a particular topic is investigated by gathering data. These data are then interpreted, providing information about the topic. A fieldwork enquiry must use **primary data**, but **secondary data** can also be used (Table 18.1).

What are primary and secondary data?

Type of data	Definition	Examples
Primary	Original data collected first hand by fieldwork. Raw data that has not been manipulated.	Counting and measuring Asking questions Sketching/photographs Raw census data
Secondary	Information from published sources which was collected/manipulated by someone else.	Textbooks and newspapers Television reports Maps Planning documents Websites Processed census data

Table 18.1 Types of data.

Tip

Many people use the terms 'data' and 'information' interchangeably. However, these are not the same thing. Data are the facts from which information is derived.

Tip

It can be beneficial to use secondary data to show a general understanding of the topic and primary data to link it to the local area.

The process of enquiry

Completing a geographical enquiry is like a journey. There are a number of stages to go through if the journey is going to be completed successfully. The stages are:

1 selecting a suitable topic for enquiry and identifying the types of data required to carry it out
2 identifying the precise data required, and then collecting and recording it
3 presenting the data
4 describing and analysing the data and explaining what it shows
5 writing a conclusion which addresses the original aim of the enquiry
6 evaluating the whole enquiry process.

Selecting a suitable topic for enquiry

A successful enquiry needs to have a clear aim so that it is easy to understand what it is about and how it is going to be carried out. Here are some ideas about choosing a topic to investigate.

What makes a successful enquiry?

A successful enquiry should be SMART: what does this mean?

S – Simple It should be a single question, simple **hypothesis** or have a clear link to geographical **theory**.

M – Measurable Can it be measured? Is it possible to get adequate data?

A – Achievable Is the location easily accessible? Can it be done in the time?

R – Realistic Can it be done? Will it be possible to reach a viable conclusion?

T – Timed Does it require data over a long period of time to draw reasonable conclusions?

What about safety – risk assessment?

There are always risks associated with collecting data. In urban areas there are obvious risks of traffic-related accidents. When carrying out a physical geography enquiry there may be a number of potential hazards such as falling rocks or steep slopes. Some enquiries involve talking to strangers so there is a need to be aware of personal safety.

Before carrying out any practical fieldwork you need to identify any potential risks and take appropriate steps to reduce these risks.

ACTIVITY 18.1

1   Explain how using SMART will help to produce a successful enquiry.
2   Consider the potential risks associated with:
    a   a town centre traffic enquiry (Figure 18.1)
    b   a river flow enquiry (Figure 18.2).
3   What could be done to reduce the potential risks?

Tip

It is important that you can write about all of the stages of the enquiry process.

Key terms

hypothesis: an idea or explanation that is tested through study and investigation

theory: one or more statements that explain a situation or a course of action. It has to be proved by evidence from an investigation

Tip

An enquiry can be used to complete a part of the GCSE course. It might then also be useful as an example in an examination.

Tip

If the potential risks cannot be reduced to an acceptable level the enquiry should not be carried out.

Figure 18.1 A busy high street in Whitby, North Yorkshire.

Figure 18.2 River Fechlin, Fort Augustus, Highland Scotland.

18.2 Methods of collecting data

In order to complete a geographical enquiry, it is important that the data is clearly linked to the title and the geographical ideas expressed in the enquiry and not just collected for the sake of it! Data can be:

- **quantitative** – data that can be measured or counted
- **qualitative** – data that is descriptive.

Asking the following two questions is a useful starting point:

1 What data are required to address the title of the enquiry?
2 Why are the data important to the enquiry?

A 'thought shower' exercise like Figure 18.3 below will help to show if the aim of the enquiry is clear and also help to identify the range of data collection opportunities. Put the title of the enquiry at the centre and then identify any appropriate methods of data collection.

Use the 'thought shower' exercise to:

1 identify the most important data for the enquiry
2 explain (justify) why the data are important
3 develop a data collection plan.

Tip

Remember that qualitative data can give:

- objective information: facts
- subjective information: opinions

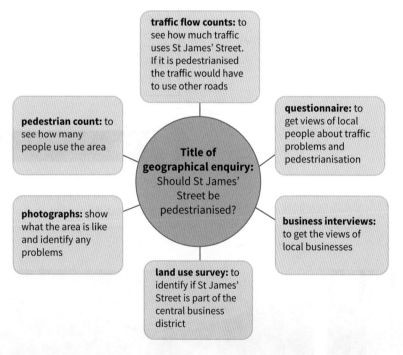

Figure 18.3 An example of a thought shower.

Tip

A geographical enquiry must use primary data. When collecting primary data always make a note of:

- when and where it was collected
- what the weather conditions were like
- if there were any problems

Further research

Find out what data the following pieces of equipment can be used to collect. Are any of them appropriate to your enquiry?

- barometer
- max–min thermometer
- hygrometer
- anemometer
- clinometer
- infiltrometer

Using questionnaires

Questionnaires are an excellent source of primary data and can be used to obtain information about people's views, habits and opinions.

Constructing a questionnaire

Always start by asking yourself, 'What do I need to find out?' and then construct questions around these ideas. There should be a justifiable reason for every question in relation to the aims of the enquiry.

There are basically two types of question:

- **closed questions** – these are questions that allow only a limited set of responses, e.g. short answer, yes/no or tick box questions. Closed questions are a good way of getting a lot of data quickly.
- **open questions** – these are questions that allow unlimited responses (they can be like mini interviews). Open questions can provide more detail and individual opinions, but they are harder to process and collate.

Piloting a questionnaire

It is a good idea to 'test' a questionnaire. This can be done by trying it out on a small number of people to make sure that the questions work.

Tip

Interviews can be a good source of data. Be prepared – have a list of questions ready for an interview.

ACTIVITY 18.2

1 a Why should you avoid questions like 'How old are you?' or 'How much do you earn?'?

   b How else might you get this type of data?

2 a Why is 'What do you think of the coastal protection methods?' not a very useful question?

   b How might you get more accurate data about people's opinions of coastal protection methods?

Sampling

Sampling is about selecting the people or places where data is going to be collected. For example, it would not be possible to give everyone a questionnaire or complete hundreds of river measurements, so a sampling method needs to be used.

There are three main types of sampling:

1 **random sampling**: where every person or place has an equal chance of being selected

2 **stratified sampling**: where people or places are chosen according to the topic; for example, a question about the condition of cycle tracks might be given to a higher proportion of cycle users

3 **systematic sampling:** where a regular sample is taken, for example, every tenth house or person, every 50 metres along a river.

Tip

Remember to learn about the different types of sampling and why a particular sampling method was used.

18.3 Methods of presenting data

There are a wide range of data presentation methods. It is important to select the most appropriate presentation method for each individual set of data.

Starting point

All enquiries are set in a place so a map showing the **situation** and **site** is useful to locate the enquiry (Figure 18.4). Photographs can be added to a site map to identify points that are important to the enquiry.

> **Key terms**
>
> **situation:** the general area or surroundings
>
> **site:** the precise area where the enquiry is taking place

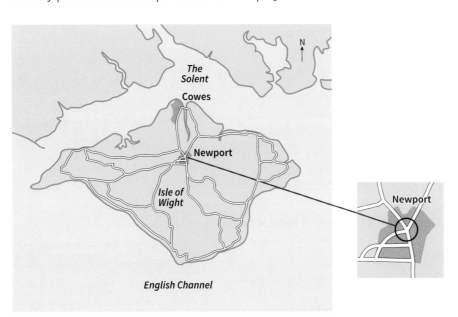

Figure 18.4 Situation and site map indicating the location of Newport, Isle of Wight.

Photographs can be both a good source of data and also used to present data or identify important points. To use photographs effectively:

- it needs to be clear where they were taken (label or locate on a map)
- each photograph needs to be labelled or annotated (Figure 18.5) with each point clearly linked to the enquiry.

Presenting quantitative data

There are a wide range of techniques available for presenting quantitative data, including:

- **tables**: a good way of showing raw data
- **graphs**: useful for showing data visually so that changes or patterns can be identified
- **maps (cartography)**: useful for showing spatial data.

Table 18.2 suggests suitable presentation techniques for different data.

Figure 18.5 An effectively labelled photograph.

Data	Presentation technique
temperature recordings over the course of a day	line graph
ages of people entering a public library between 9 am and 11 am	bar chart or pie chart
the radius of limpets in two beach locations	dispersion graph
the location of flooding events in the UK over the last ten years	annotated map

Table 18.2 Examples of appropriate presentation techniques for different types of data.

Skills link

Refer to Chapter 19 Skills practice for guidance on creating graphs, charts and maps.

18.4 Describing, analysing and explaining data (1)

You will need to describe, analyse and explain your collected data. To do this successfully, you will need to link the data back to the original aim of the enquiry.

The key parts to this section of your enquiry are:

- **Describing and analysing the data** – what does it actually show? Identify any patterns in the data. Are there any **anomalies**? Can you use any statistical techniques to identify relationships?
- **Explaining the data** – why does the data show the patterns or relationships that you have described? Suggest reasons for any anomalies.
- **Drawing out links** – how are different data sets linked to each other?

Using statistics

Data is often collected in the form of statistics. Statistical techniques can be used to:

- describe data
- identify **spatial** patterns
- measures of central tendency (Table 18.3)
- examine relationships between different data sets.

Tip

Use specific data to express points, rather than words like 'larger' or 'smaller'.

Key terms

anomaly: a data point that does not follow the general pattern or that seems slightly odd

spatial: linked to a location or place

Skills link

Refer to Chapter 19 Skills practice for guidance on using statistical techniques.

Measures of central tendency (used to calculate average)		
Measure	**How is it calculated?**	**Limitations**
Mean (arithmetic average)	The total divided by the number of items	Can be distorted by extreme values
Median	Middle value of a ranked set of data	Gives no idea of other values or extremes
Mode	The most frequently occurring value in a set of data	Gives no idea of other values or relationship with other values

Table 18.3 Measures of central tendency.

Using dispersion diagrams

A dispersion diagram plots each value against a vertical scale. Figure 18.6 shows how often each person visits a shopping centre each month. Each dot represents one person.

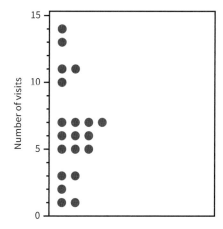

Figure 18.6 A dispersion diagram.

18.4 Describing, analysing and explaining data (2)

Using scatter graphs

A scatter graph is a visual way to describe the relationship between two sets of data. Figure 18.7 plots pedestrian count data against distance from a town centre and clearly shows that pedestrian numbers decrease with distance from the town centre. It then plots a line of best fit on this data.

Development of statistical techniques

The strength of the relationship between two sets of data can be calculated by using a statistical calculation called the Spearman's Rank Correlation Coefficient.

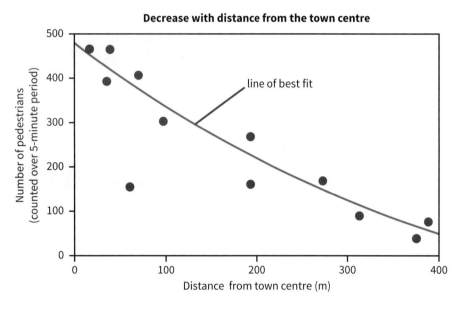

Figure 18.7 A scatter graph.

ACTIVITY 18.3

1 In a beach enquiry a student measured the long axis of ten randomly selected pebbles. The pebble sizes were: 6, 12, 9, 6, 8, 8, 7, 10, 8, 9 cm.

a What were the mean, median and mode values for the pebble sample?

Drawing conclusions

A conclusion is an opportunity to use evidence from the data to reflect on the original aim of the enquiry. Consequently a useful starting point will be to have a look at the original aim of the enquiry and read it through so that the 'journey' of the enquiry is clear.

A successful conclusion will:

- return to the original aim of the enquiry
- use the most important evidence from the data collection to make specific points in relation to the original aim
- identify any important links between different sets of data
- provide an overall conclusion.

Evaluation of geographical enquiry

Evaluation is about reflecting upon how effectively the enquiry satisfied the original aim and making observations about how the enquiry could be improved or developed. The three elements of the evaluation are shown below. It is always useful to start by looking at the limitations of the data collection methods because it is usually these that affect the results and conclusions.

Tip

It is important to know that a statistical relationship is not the same as a causal relationship. For example, a Spearman's Rank calculation might show a relationship between the location of pebbles in a river and pebble size. This does not imply that the location of the pebble influences the pebble size.

If the enquiry is based on:

➡ a question – make sure that the question has been answered

➡ a hypothesis – make it clear whether it has been proven or not

➡ an issue – make sure that both sides of the argument have been considered

➡ a theory – make sure that a clear comparison with the original theory has been made

 AND

always back up any points made with evidence from the data collected

| Limitations of data collected | What other data might have been useful? | How reliable were the conclusions? |

Questions that might be considered

- How accurate were the data collection methods?
- How reliable were the data collection methods?
- Was the data collection programme affected by any particular problems?

- Should the data have been collected at different times?
- Was there sufficient data collected from each method?

- Was there sufficient evidence to draw reliable conclusions?
- How well did the data collection methods fit the aim of the enquiry?
- How did the accuracy and reliability of the data affect the conclusion?

 Assess to progress

1 a Identify two potential risks when carrying out a geographical enquiry about coastal processes. **2 MARKS**

 b Suggest how one of the risks in part **a** might be reduced. **4 MARKS**

2 Describe and justify one data collection method that might be used in a physical geography enquiry. **6 MARKS**

3 Assess the usefulness of cartographic and visual presentation methods to geographical enquiries. **6 MARKS**

4 For either a physical or a human geographical enquiry that you have carried out:

 a State the title of the enquiry.

 b Consider how effectively the enquiry satisfied its original aims. **9 MARKS**

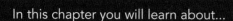

19 Skills practice

In this chapter you will learn about...

- the types of graphs and charts that can show data
- population pyramids and choropleth maps
- different types of numbers and how area and scale are used
- ratios, proportions and sampling techniques
- comparing different sets of data
- using lines of best fit to interpret scatter graphs
- using contour lines to see gradient
- drawing cross-sections and transects
- four and six-figure grid references.

19.1 Graphical skills – graphs and charts (1)

Graphs and charts are a useful way to show numerical data. They help to show patterns and to reach conclusions. Choosing the right type of graph or chart to present data is very important. Data that has been presented badly is very difficult to understand.

There are some common graph and chart types that you may have come across already. Some are more complex than others.

Charts and graphs

There are many ways to show data through charts and graphs:

- **Bar graphs** (Figure 19.1) – show categories of data as rectangular bars, e.g. the rates of deforestation for different continents. Divided bar graphs (Figure 19.2) are when the bars are divided up to break down the information further. A divided bar chart could be used to show the breakdown of energy types used.

> **i** **Did you know?**
>
> The terms 'graph' and 'chart' are often used interchangeably. However, they are different:
>
> - graphs are usually used to represent data over a period of time
> - charts are better for showing frequency or spread at a single point in time.

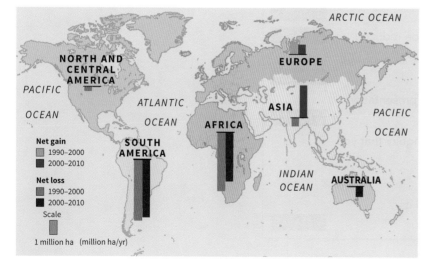

Figure 19.1 Bar graphs showing rates of rainforest deforestation.

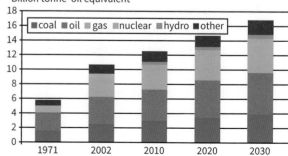

Figure 19.2 Divided bar graph to show projected future demand in energy.

- **Line graphs** (Figure 19.3) – show how data changes over time. A line chart could be used to show changes in global calorie consumption over time.

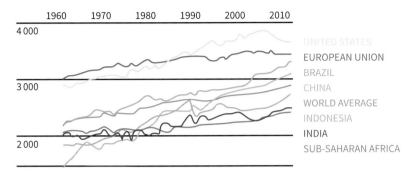

Figure 19.3 Line graph to show trends in calorie consumption.

- **Pie charts** (Figure 19.4) – show percentages as a circle, divided into segments. A pie chart could be used to show the different uses of water in UK homes.

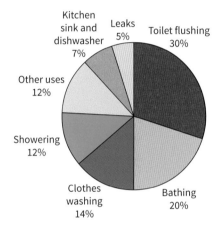

Figure 19.4 Pie chart showing how water is used in UK homes.

- **Scatter graphs** (Figure 19.5) – shows relationships between two sets of data. A scatter graph could be used to show the GDP and carbon dioxide (CO_2) emissions of a country.

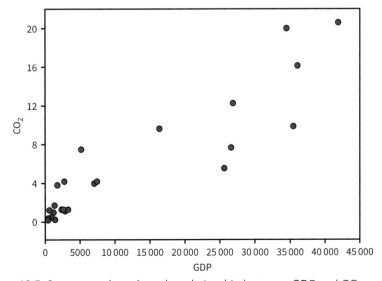

Figure 19.5 Scatter graph to show the relationship between GDP and CO_2 emissions.

19.1 Graphical skills – graphs and charts (2)

- **Pictograms** (Figure 19.6) – pictograms use small pictures or icons to compare data. Pictograms could be used to show levels of recycling.

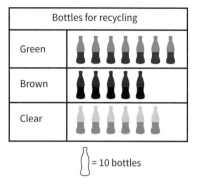

Figure 19.6 Pictogram showing the number of bottles recycled by a restaurant in one week.

- **Histograms** (Figure 19.7) – These are similar to bar graphs, but they show frequencies rather than categories. A histogram could be used to show the temperatures for each day in a month.

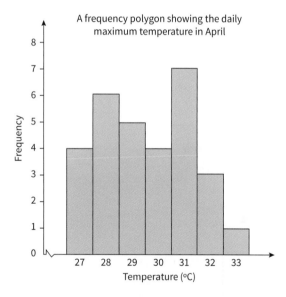

Figure 19.7 A histogram showing the temperatures for each day in a month.

 Go to the **gap minder website** (www.cambridge.org/links/gase40165) to see a unique way of presenting geographical data.

 Discussion point

Think of data that you could collect to fit into each chart or graph type. What are the weaknesses of each type of graph?

ACTIVITY 19.1

1 a Match the data on the left to the type of chart in Table 19.1.

The source country for migrants into the UK, plus the ages of the migrants.	Bar chart
The percentage of people working in each employment sector in the UK.	Divided bar chart
The relationship between the velocity and gradient of a river.	Line chart
The number of wind farms in each country in the EU.	Pie chart
The level of a river over a 24-hour period.	Scatter graph

Table 19.1 Types of charts used in geography.

b What would you write as the axis labels for the scatter graph?

2 Table 19.2 shows the ages of people that participated in a questionnaire carried out in three locations around Oxford.
  a Draw a divided bar graph to show the data.
  b Put the location on the *x*-axis and the total number of people on the *y*-axis.
  c Split each bar into the age categories and colour each age category in.

	City centre	Suburbs	Countryside
0–20 years	4	18	0
21–40 years	32	9	1
41–60 years	21	11	5
61+ years	14	1	6
Total	71	39	12

Table 19.2 Ages of questionnaire participants in three locations around Oxford.

d Describe the patterns that you can see. Think about:
  i Where is the total number of questionnaire participants the highest?
  ii Which location has the greatest number of younger people?
  iii Which location has the greatest number of middle-aged people?
  iv Which location has the greatest number of older people?
e Try to think of some reasons to explain the patterns shown in the data.
f What are the strengths of this type of graph?

Tip

You can draw graphs and charts by hand or on a computer. You must always think about titles, axis labels, keys and the use of colour. When drawing a graph by hand, you must use a pencil with a sharp point and a ruler to keep the lines straight and tidy.

Download Worksheet 19.1 from Cambridge Elevate for help with Activity 19.1 question 2.

19.2 Graphical skills – population pyramids, choropleth maps

There are some types of graphs and maps that are very useful to geographers. Two examples are population pyramids and choropleth maps.

Population pyramids

Population pyramids are graphs that show what the population of a particular place is like. The pyramids are bar graphs, where the bars stretch sideways, rather than upwards.

- The x-axis shows the number of people in a particular place. It can be shown as the total number of people or the percentage of the total population.
- The y-axis shows the ages of people in a particular place. These age ranges are usually grouped into categories, e.g. 0–4 years, 5–9 years.
- The bars on the left-hand side of the graph show the number of males.
- The bars on the right-hand side of the graph show the number of females.

Figure 19.8 is an estimate of what the UK's population pyramid will look like in 2025. We can see that there will be a lot of people in the age brackets of 35–39 and 55–59. The base of the pyramid is smaller than these two age brackets; this means that either the birth rate is falling, or that older migrants are moving into the country. By 2050 there will be many people living beyond the age of 70. This means that we will need to think more about healthcare for the elderly and ways to fund **pensions**.

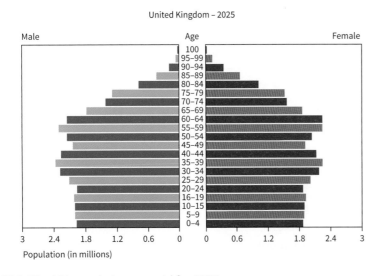

Figure 19.8 The UK population pyramid for 2025.

Population pyramids are useful for comparing different countries. Lower income countries (LICs) usually have a population pyramid that is very wide at the bottom and very narrow at the top. This shows high **birth rates** and low **life expectancies**. Middle income countries (MICs) usually have a population pyramid that is very narrow at the bottom and wider at the top. This shows lower birth rates and longer life expectancies. Population pyramids can also show important events in a country's history. An example of this is the current bulge in the UK's pyramid for the generation of people around 70 years old. These are the **baby-boomers** – they were born when their parents were reunited at the end of the Second World War. One weakness of a population pyramid is that they only show a snapshot of total population. It is not possible to identify whether population changes are as a result of natural change or migration.

 Further research

How and why might the UK's population pyramid change over the next 50 years? Go to the **Office of National Statistics website** (www.cambridge.org/links/gase40166) to see an interactive population pyramid of the UK. Use the slider on the screen to move the pyramid from 1971 to 2085. What changes are expected to take place?

 Did you know?

The UK has been formally collecting census data every 10 years since 1841. The only time that the census was not carried out was in 1941, during the Second World War.

 Key terms

pension: money paid to someone who has retired from work

birth rate: the number of live births per thousand people per year

life expectancy: the average number of years a person is expected to live from birth in a particular society at a certain time

baby-boomer: someone who was born during a 'baby boom', which is a time of unusually high births

WORKED EXAMPLE 19.1

1 Look at the population pyramid for the UK (Figure 19.8):
   a How many males are in the age bracket 0–4?
   There are approximately 2 million males in the 0–4 category.
   b Which gender has the highest infant mortality? How can you tell?
   There are fewer females below the age of 5. Since birth rates tend to be about equal, this suggests fewer female infants are surviving to the age of 5.
   c How might the pyramid change if birth rates increased?
   If the birth rate increased, both of the 0–4 bars would be larger.

ACTIVITY 19.2

1 Look at the population pyramid for the UK (Figure 19.8):
   a How many females are in the age bracket 40–44?
   b Which gender has the highest life expectancy and how can you tell?
   c Which age category will you fit into in 2025?
   d How might the pyramid change if another World War breaks out?
   e How would the pyramids of other countries look different to this one?
2 Draw a population pyramid using the data for Bangladesh in 2015 (Table 19.3).
   a Describe your pyramid.
   b How is it different to the UK's population pyramid?

Download Worksheet 19.2 from Cambridge Elevate for help with Activity 19.2 question 2.

Choropleth maps

Choropleth maps show information as colours. Different areas of a map are shaded to show data. The data shown in choropleth maps is in intervals, which is when numbers follow on from one another.

- Choropleth maps are usually shaded using one colour, but they can sometimes use colour progressions too, e.g. colours like yellow, orange and red that blend together well.
- The darker shades of the colour show higher numbers of something.
- The lighter shades of the colour show lower numbers of something.
- There should be enough shades of the colour to show a pattern.
- A choropleth map should always have a key so that the reader can understand what it shows.

Choropleth maps can be easy to interpret as patterns shown on them are often clearer than using text. Figure 19.9 shows the percentage of people living in London who cycle five times or more a week. If this data was presented in the form of a table, it would be very difficult to understand. Seeing the data as a choropleth map makes it easier to read.

The Office for National Statistics (ONS) stores information about the UK's population. Go to their website (www.cambridge.org/links/gase40167) to see the choropleth maps using the data that was collected in the 2011 UK Census.

Age	% male	% female
0–4	10.6	9.7
5–9	11	10.1
10–14	11.4	10.6
15–19	10.1	10.1
20–24	8.2	9.3
25–29	7.3	8.5
30–34	7	7.7
35–39	7.2	7.2
40–44	6.3	6.4
45–49	4.7	4.9
50–54	4.7	4.6
55–59	3.5	3.4
60–64	2.8	2.5
65–69	2.2	1.9
70–74	1.5	1.4
75–79	0.9	0.9
80–84	0.4	0.5
85–89	0.2	0.2
90–94	0	0.1
95–99	0	0
100+	0	0

Table 19.3 Population data for Bangladesh (2015).

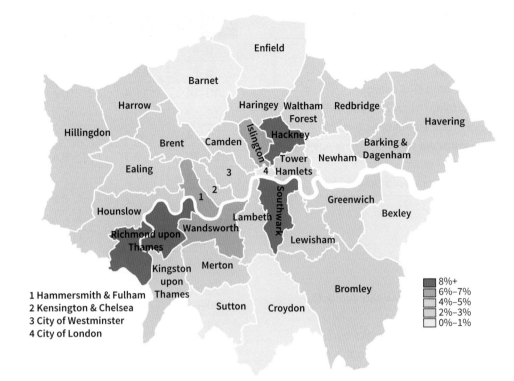

1 Hammersmith & Fulham
2 Kensington & Chelsea
3 City of Westminster
4 City of London

Key:
- 8%+
- 6%–7%
- 4%–5%
- 2%–3%
- 0%–1%

Figure 19.9 Percentage of people who cycle five times or more a week in London.

ACTIVITY 19.3

1 Look at the choropleth map of London (Figure 19.9).
   a Describe the patterns shown by the map. Where do people cycle the most and the least?
   b Suggest reasons for the distribution of people who cycle the most.
   c Suggest reasons for the distribution of people who cycle the least.

19.3 Numerical skills – number, area, scale

Numbers

The use of numbers in geography is often referred to as data. Numbers are important because they allow us to reach conclusions and compare different places. They are useful on fieldwork as a way to take down data about a place.

Numbers come in different forms:

- **nominal data** – data that is in categories, e.g. 1 = male and 2 = female.
- **ordinal data** – data that has an order, e.g. the rank order of countries by level of development. We know that country 1 is more developed than countries 2 and 3, but we don't know how much more developed it is because the spacing between each country is not equal.
- **interval data** – this is similar to ordinal data, but the difference between each number is equally split, e.g. degrees centigrade.
- **ratio data** – this is when there is a relationship between two types of data, e.g. number of people per doctor.

Area

Land area can be measured in square metres, hectares and square kilometres. These are called metric measures as they are all based around the metre. Table 19.4 shows the size of area in m².

Numbers come in different forms:

- The area of a square or rectangle is found by multiplying the length of the two sides.
- The area of a triangle is found by multiplying the base by the height and then dividing by two.

 Key term

interval data: data where the difference between each value is equally split

Unit	Area in m²
hectare	10 000
square kilometre (km²)	1 000 000

Table 19.4 Metric measures of area.

 Did you know?

The UK only started using the metric system of measurement in 1965. Metrication was increasingly applied to weights and measures between then and 2000. However, road distances still use the imperial system of yards and miles.

- The area of an unusual shape requires a more complicated procedure. Either try to find common shapes within the area and calculate them, or split the area down into 1 m² squares and then count how many there are.

WORKED EXAMPLE 19.2

1 Calculate the area of the following shapes:
   a A square with sides that are 5 cm long. 5 cm × 5 cm = 25 cm²
   b A rectangle with one side that is 3 cm and another that is 9 cm.
     3 cm × 9 cm = 27 cm²
   c A triangle that has a base of 8 cm and a height of 9 cm.
     8 cm × 9 cm = 72 cm   72 cm ÷ 2 = 36 cm²

ACTIVITY 19.5

1 Calculate the area of the following shapes:

   a A square with sides that are 3 cm long.
   b A rectangle with one side that is 5 cm and another that is 12 cm.
   c A triangle that has a base of 14 cm and a height of 22 cm.

 A handy tool for calculating the area of different shapes (www.cambridge.org/links/gase40168).

Scale

Scales are a form of ratio data; they show how two measurements are related. Maps, diagrams and some graphs can have scales. This is because we want to show something that is too big to fit onto the page or because we want to provide a formula for working things out. The most common scales on Ordnance Survey maps are:

- **Large scale** – 1 : 1250, 1 : 2500 and 1 : 10000. Features appear larger on the map and you may see individual houses. This may be used for maps of cities, towns and villages.
- **Small scale** – 1 : 25000, 1 : 50000 and 1 : 100000. Features appear smaller on the map, so things are simplified. This may be used for maps of National Parks, remote areas or regions.

WORKED EXAMPLE 19.3

1 For a 1 : 50000 map:
   a How much would 1 cm on the map be worth on the ground?
     A map scale of 1 : 50000 means that 1 cm on the page represents 50000 cm (500 m) in reality.
   b How far would a route of 3 cm be on the ground?
     A route of 3 cm would be 500 m × 3. This is 1500 m or 1.5 km.

 Further research

How could you adapt the method of measuring straight line distances (Activity 19.6) on a map to measure actual routes, or curved lines?

ACTIVITY 19.4

1 Match the measures to the correct data types in Table 19.5:

The ages of people answering a questionnaire, where the categories are: 0–19 20–39 40–59 60+	Nominal data
A 'travelling to school' survey where: Foot = 1 Bike = 2 Bus = 3 Car = 4	Ordinal data
Counting the number of wells in Ugandan villages and comparing it with the number of people living in that village.	Interval data
A bipolar survey collecting information about graffiti in areas of a city, where: 10 = lots of graffiti 0 = no graffiti	Ratio data

Table 19.5 Types of data used in geography.

ACTIVITY 19.6

1 For a 1 : 25000 map:

   a How much would 1 cm on the map be worth on the ground?
   b How far would a route of 23 cm be on the ground?
   c What scale of map would have features that appear larger?
   d What scale of map would have features that appear smaller?

19.4 Numerical skills – ratio, proportion, sampling

Ratio

Ratio data is when there is a meaningful relationship between two types of data, e.g. the number of people per doctor. Ratios are usually expressed as two numbers, with a colon between them.

WORKED EXAMPLE 19.4

Country	Number of people	Number of doctors
UK	64 100 000	180 057
Afghanistan	30 550 000	8 126

Table 19.6 The number of people and doctors in the UK and Afghanistan.

1 Look at Table 19.6 and calculate the number of doctors per 1 000 people for each country.

Divide the number of doctors by the number of people.

As this ratio is per 1 000 people and not per person, multiply the answer by 1 000.

The UK calculation is (180 057 ÷ 64 100 000) × 1 000 = 2.81 doctors per 1 000 people or 2.81 : 1 000

The Afghanistan calculation is (8 126 ÷ 30 550 000) × 1 000 = 0.27 doctors per 1 000 people, or 0.27 : 1 000

ACTIVITY 19.7

Country	Number of people	Number of doctors
Ghana	25 900 000	2 486
Switzerland	8 081 000	32 720
Sierra Leone	6 092 000	134
USA	318 900 000	781 943

Table 19.7 The number of people and doctors in Ghana, Switzerland, Sierra Leone and the USA.

1 Look at Table 19.7 and calculate the number of doctors per 1000 people for each country.

Proportion

Proportion is similar to ratio, but it is expressed differently. For example, in London, 1 in 10 people have a first language other than English or Welsh. This means that out of every 10 people in London, 1 person will not have English or Welsh as their first language. The proportion of people in London with a first language other than English or Welsh is 1 in 10, or 10 per cent; the ratio of people in London with a first language that is not English or Welsh to people whose first language is English or Welsh is 1 : 9.

WORKED EXAMPLE 19.5

Year	% of people working in tertiary jobs
1911	41
2011	80

Table 19.8 % of people working in the tertiary sector in the UK

1 Look at Table 19.8. What was the proportion of people working in tertiary jobs in 1911 and 2011?

In 1911, 41 per cent worked in the tertiary sector. This is 41 out of every 100 workers.

In 2011, 80 per cent worked in the tertiary sector. This is 80 out of every 100 workers.

ACTIVITY 19.8

Year	% of people working in manufacturing jobs
1911	39
1961	38
2011	9

Table 19.9 % of people working in the manufacturing sector in the UK.

1 Look at Table 19.9. What was the proportion of people working in manufacturing jobs in 1911, 1961 and 2011?

Sampling

All data and information collected must be as accurate as possible. Collecting the wrong sort of data or information leads to inaccurate conclusions. For example, if you carry out a travel survey, but you only ask people in your school, it will look like no one drives cars in your town or city. This is unlikely to be true! Sampling techniques help you to avoid making mistakes like this.

There are three main types of sampling technique:

- **random sampling** – where you look at your data and randomly select a person or site from it. For example, you may have ten sites and decide to review the data for site six. Online random number generators and Microsoft Excel® can generate numbers to use to select samples from your data. Typing the formula '=RANDBETWEEN(1,10)' into a cell will produce a random number between 1 and 10. Random sampling is good if you have a large sample of data. It is also unbiased as you are not making assumptions about people or places.
- **systematic sampling** – when you collect data in an ordered or a regular way. You might choose every 10 m or every tenth person that passes by. Systematic sampling is good for covering a whole area or a whole population.
- **stratified sampling** – where you split something into categories. You might choose three sites from each part of a city, or ten people from each age category. You can combine stratified sampling with random and systematic sampling. Stratified random sampling would take random samples from within each category. Stratified systematic sampling would take regular samples from within each category.

Control measures can improve the reliability of results. For example, if different pairs of people are measuring different sites within a city, a control site could be used. Measurements for the control site would be agreed collectively so that all pairs collecting data have a set of standards before they carry out fieldwork independently.

 The Royal Geographical Society has a webpage that explains more about sampling techniques in Geography (see www.cambridge.org/links/gase40169)

 Did you know?

Magnitude and frequency are also examples of numerical data. Magnitude shows the size of something and frequency shows how often something occurs. Earthquakes can be measured using both magnitude and frequency.

ACTIVITY 19.9

1 What sort of sampling is represented in each of the following data collections?
   a Recording family sizes, using ten migrants from each country in the EU.
   b Noting down the number of species found at every 10 m along a sand dune transect.
   c Asking 50 men and 50 women about how much they recycle at home (half of the men and women must be under 40 and half over 40).

19.5 Statistical skills – central tendency, percentage increase and decrease (1)

Measures of central tendency

When we have a wide spread of data, it is often more useful to find averages. Averages are measures of 'central tendency' as we try to find the central piece of data. They commonly take three forms:

1 **mean** – add up all the total of all values that you've collected and then divide by the number of values
2 **median** – write out all of the values that you've collected in numerical order and find the middle number
3 **mode** – the most commonly appearing number in the data that you have collected.

Discussion point

What are the advantages and disadvantages of using measures of central tendency? Where do you think averages are either useful or not so useful?

WORKED EXAMPLE 19.6

1 Use Table 19.10 to work out the mean, median and mode:

Approximate distance travelled to work each day	Number of commuters
10 miles	7
20 miles	5
30 miles	2
40 miles	1

Table 19.10 Data to show how far commuters travel to work each day.

The **mean** distance is 18 miles.
It is calculated in the following way (Table 19.11):

Approximate distance travelled to work each day	Number of commuters	Total distance travelled by all commuters
10 miles	7	10 miles × 7 = 70 miles
20 miles	5	20 miles × 5 = 100 miles
30 miles	2	30 miles × 2 = 60 miles
40 miles	1	40 miles × 1 = 40 miles
Total	15	270

Table 19.11 Calculating the mean distance.

270 miles ÷ 15 commuters = 18 miles

*The **median** distance is 20 miles.*
It is calculated in the following way (Table 19.12):
Number of commuters = 15
Number of commuters ÷ 2 = 7.5
The middle commuter = 8

Commuter number	1	2	3	4	5	6	7	8	9	10	11	12	13	14	15
Distance travelled	10	10	10	10	10	10	10	20	20	20	20	20	30	30	40

Table 19.12 Calculating the median distance.

*The **modal** distance is 10 miles.*
It is the most commonly occurring number in the table.

Further research

Which measure do you think is the best representation of the average commuting distance? Why?

ACTIVITY 19.10

1 Work out the mean, median and mode for the data in Table 19.13:

Average rate of coastal erosion	Number of sites
1 m per year	7
5 m per year	8
10 m per year	2

Table 19.13 Data to show average erosion rates at 17 different sites.

Download Worksheet 19.3 from Cambridge Elevate for help with Activity 19.10 question 1.

Cumulative frequency and quartiles

Cumulative frequencies and quartiles allow us to compare the spread of data. For example, the data found in Table 19.10 has a wide spread, but most of the data is found in the first two categories.

Cumulative frequency is calculated first. It is found by taking a running total of the data. If we did this with the data from Table 19.10 it would look like this (Table 19.14):

Approximate distance travelled to work each day	Number of commuters	Cumulative frequency (cumulative number of people)
10 miles	7	7
20 miles	5	12
30 miles	2	14
40 miles	1	15

Table 19.14 Calculating cumulative frequency.

19.5 Statistical skills – central tendency, percentage increase and decrease (2)

We could then plot this information as a line graph, where the x-axis is the number of miles travelled and the y-axis is the cumulative number of people (Figure 19.10).

Quartiles are found by splitting the values on the y-axis into four equal sections. If we draw a horizontal line across from the joining point of these four sections, we can read down and find the upper, median and lower quartile values. The interquartile range is the difference between the upper and lower quartiles.

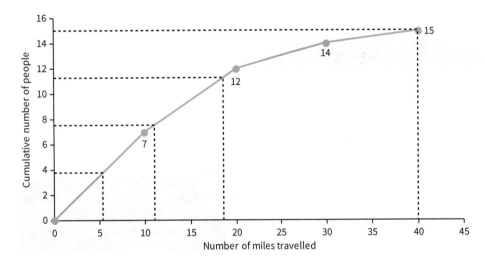

Figure 19.10 Cumulative frequency line graph with quartiles.

ⓘ Did you know?

Cumulative frequency graphs often have an S-shaped line. It is called an ogive.

Percentage increase and decrease

When we are looking at an increase or decrease in something, it can be hard to work out the amount of change over time. An example of this is the UK National Debt (money that the UK owes to other countries). The debt has increased over time, but it is difficult to say when the biggest increases occurred. Calculating a percentage increase helps us to see how the data has changed over time.

WORKED EXAMPLE 19.7

1 Table 19.15 shows the UK National Debt between the years 2008 and 2015.

Year	National Debt total	Difference	Difference ÷ the original number	× 100
2008	£0.53 trillion	–	–	–
2009	£0.62 trillion	£0.09 trillion	0.15	15%
2010	£0.76 trillion	£0.14 trillion	0.18	18%
2011	£0.91 trillion	£0.15 trillion	0.16	16%
2012	£1.10 trillion	£0.19 trillion	0.17	17%
2013	£1.19 trillion	£0.09 trillion	0.08	8%
2014	£1.26 trillion	£0.07 trillion	0.06	6%
2015	£1.36 trillion	£0.1 trillion	0.07	7%

Table 19.15 Calculating percentage increases.

To calculate the percentage increase found in the final column:

* work out the difference (increase) between the two numbers you are comparing (column 3)
* divide the increase by the original number (column 4)
* multiply by 100 (column 5)

The largest increase occurred in 2010. The smallest increase occurred in 2014.

ACTIVITY 19.11

1 Use Table 19.16 to calculate the percentage increase in the energy from onshore wind farms between 2009 and 2014.

Year	Onshore wind power generation (GWh)	Difference	Difference ÷ the original number	× 100
2009	7 529	–	–	–
2010	7 136			
2011	10 346			
2012	12 111			
2013	16 992			
2014	18 333			

Table 19.16 Onshore wind power generation (GWh) in the UK.

Download Worksheet 19.4 from Cambridge Elevate for help with Activity 19.11 question 1.

19.6 Statistical skills – how do we find relationships between data?

Scatter graphs show relationships (or **correlations**) between two sets of data. It can be difficult to see patterns on scatter graphs and so we draw a **line of best fit** to help us. This is a straight line that runs through the middle of all the points on the graph. There is usually an equal number of points on either side of this line.

Strong and weak correlations

A strong correlation is when the points on the scatter graph are very close to the line of best fit. This means that the two variables on the x and y axes are related to one another; as one variable changes, so does the other.

A weak correlation is when the points are far away from the line of best fit. This means that the two variables on the x and y axes are not necessarily related to one another; a change in one variable does not lead to a change in the other.

Outliers are points that are not close to the line of best fit.

Positive and negative correlations

A positive correlation is found when the line of best fit runs from the bottom left to the top right of the graph. This means that as one variable increases, so does the other.

A negative correlation is found when the line of best fit runs from the top left to the bottom right of the graph. This means that as one variable increases, the other decreases.

> **Key terms**
>
> **correlations:** relationships between two sets of data
>
> **line of best fit:** a straight line drawn through the points on a scatter graph

> **Fieldwork**
>
> How could you use scatter graphs and lines of best fit in your geography fieldwork?

WORKED EXAMPLE 19.8

1   Plot the information from Table 19.17 onto a scatter graph.
2   Draw a line of best fit and interpret the correlation.

River cross-section area (m²)	River velocity (m/s)
50	0.1
80	0.2
115	0.25
190	0.4
220	0.45

Table 19.17 River measurements that were collected during GCSE fieldwork.

We can plot this data onto a scatter graph and draw a line of best fit (Figure 19.11):

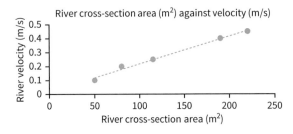

Figure 19.11 A scatter graph showing river cross-section area (m²) against velocity (m/s).

The points are very close to the line, so it is a strong correlation with no outliers. The line of best fit runs from the bottom left to the top right and so this is a positive correlation.

ACTIVITY 19.12

1   Plot the information from Table 19.18 onto a scatter graph.
2   Draw a line of best fit and interpret the correlation.

Sediment size (cm)	Velocity (m/s)
19	0.1
15	0.2
12	2.5
9	0.4
4	0.45

Table 19.18 Data to show sediment size and velocity at five different sites.

 Download Worksheet 19.5 from Cambridge Elevate for help with Activity 19.12 questions 1 and 2.

Making predictions

We can use the line of best fit to make predictions about our data. We may want to estimate data that is missing within our records or predict data that falls outside of the range that we have collected.

Interpolation is finding a value inside the range of data. For example, a scatter graph shows the number of tourists (y-axis) compared with the distance from a seaside resort x-axis. Our line of best fit might show a negative correlation, in that the number of tourists decreases with distance from the resort. Interpolation would allow us to estimate the number of tourists at any given point along that **transect**. To find this value, we would draw a straight line upwards from the chosen point on the x-axis. Where this line meets our line of best fit, we would read across to see the value on the y-axis.

Extrapolation is finding a value outside of our range of data. Using the same tourism scatter graph, we may want to predict the number of tourists further away from the seaside resort than we were able to measure. To find this value, we follow the same procedure, but this time we would need to extend our line of best fit beyond our original data collection.

> **Key term**
>
> **transect:** a straight line on the ground along which measurements are taken

WORKED EXAMPLE 19.9

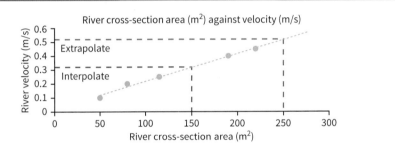

Figure 19.12 A scatter graph showing river cross-section area (m²) against velocity (m/s).

To interpolate a new data point to show velocity for a cross-sectional area of 150 m² (Figure 19.12):

- draw a straight line up from 150 m² until it meets the line of best fit
- read across to the y-axis to find the velocity value. Here it is 0.32 m/s.

To extrapolate a new data point to show velocity for a cross-sectional area of 250 m²:

- extend the line of best fit
- draw a straight line up from 250 m² until it meets the extended line of best fit
- read across to the y-axis to find the velocity value. Here it is 0.52 m/s.

ACTIVITY 19.13

1   Using the scatter graph that you created in Activity 19.12, interpolate a new data point at 10 cm and extrapolate a new data point at 22 cm.

19.7 Cartographic skills – gradients, contours

Ordnance Survey (OS) is the national mapping agency of Great Britain. OS maps are accurate representations of places and are used by government and businesses across the country. You should be able to use and interpret OS maps at a range of scales and use OS maps alongside photographs of places. See Chapters 8, 9 and 10 for activities using OS maps.

Gradient

Gradient is a change in the height of a landscape. An increase in gradient occurs when the land becomes steeper. A decrease in gradient occurs when the land becomes flatter.

Contour lines

Contours are lines drawn on a map to show changes in gradient on the land (Figure 19.13). They are usually shown as thin orange or brown lines that join areas of equal height. Some contour lines have numbers on them to show the exact height of the land along that line. In addition to contour lines, spot heights show information about particular points on the map. Spot heights that have been calculated from the air are usually shown in orange numbers and those that have been calculated from a ground survey are usually shown in black numbers.

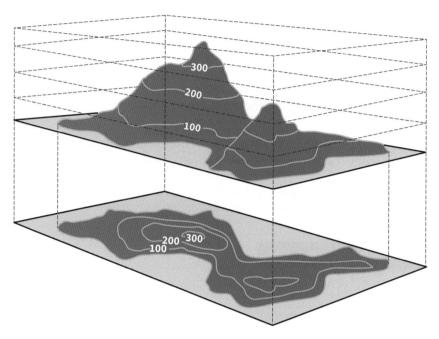

Figure 19.13 Contour lines are used on maps to show changes in gradient.

Contour lines that are close together show land that is steep. Contour lines that are far apart show land that is flat or gently sloping.

WORKED EXAMPLE 19.10

1 Look at Figure 19.14, a map extract of Blithfield Reservoir and Abbots Bromley in Staffordshire.

a Why does the contour line closest to the edge of the reservoir have the number 100 written along it? *This means that it is 100m above sea level.*

b The causeway running across the reservoir has a picnic bench symbol on the north-eastern section. Why is this a good location?
This is a good place for a picnic as it is slightly higher than the surrounding area so it will have good views of the reservoir.

c What do the contour lines to the north of the reservoir show?
The stretch of land that extends out into the reservoir's northern section has contour lines that are close together. This shows that the land here is steeper than the surrounding areas.

d Admaston village on the south-western edge of the reservoir has the number 125 written beside it in black. What does this mean?
This is a spot height that has been calculated from a ground survey. It shows that this point is exactly 125m above sea level.

e What do the contours that run in a line with a river beyond the dam to the south-east of the reservoir show?
This shows a valley.

Further research

Isoline maps show data in the form of lines. The lines join areas that are equal. A good example of this is contour lines, which join areas of equal height on a map. What other types of isoline maps can you think of?

Key terms

gradient: the change in height of a landscape

contours: lines on a map that show places of equal height

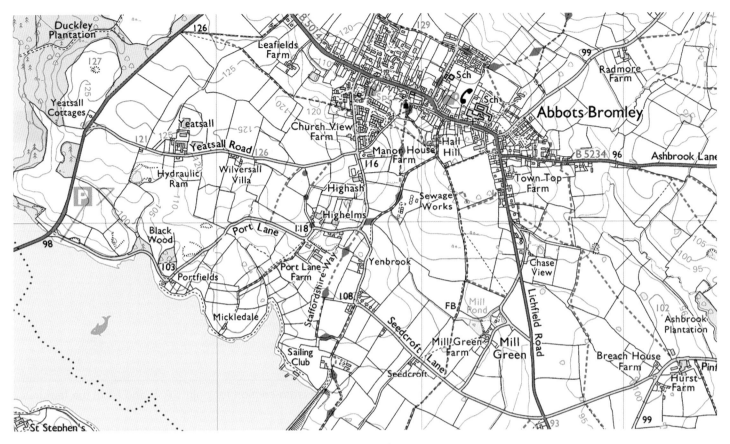

Figure 19.14 OS 1:25 000 map of Blithfield Reservoir and Abbots Bromley in Staffordshire.

ACTIVITY 19.14

1 Locate the village of Abbots Bromley on Figure 19.14.

  a Approximately how high is the police station to the north-west of the village?

  b Suggest a good location for a picnic bench to overlook this picturesque village.

  c Which is steeper, the land to the north-east or the land to the north-west of the village?

  d What number is written on the spot height near to Manor House Farm, which lies to the south of the village?

  e Where is the nearest valley to Abbots Bromley?

Further progress

Sketch the area around your school and try to estimate where contour lines would lie. Where are the steepest areas of your school grounds? Which places are flat?

Visit the Ordnance Survey website for a key to the symbols used on OS maps (www.cambridge.org/links/gase40170)

19.8 Cartographic skills – cross-sections, transects

Cross-sections and transects help us to interpret information about the shape and use of the land on a page. Although the two terms are often used to mean the same thing, in geography the following tends to apply:

- A transect is a line along which samples are taken or observations are made. Transects can consist of one or more straight lines, They tend to be used to record **qualitative** information.
- Cross-sections are line representations of physical landscapes. They show the shape of a feature viewed from the side. Cross-sections tend to show **quantitative** information.

Key terms

qualitative: information that is in the form of words

quantitative: information that is numeric

Cross-sections

Cross-sections can be small scale, e.g. a cross-section from one riverbank to the other, or large scale, e.g. from one side of a valley to another. Cross-sections are labelled to show important information, such as different landforms or the location of features like roads and rivers.

WORKED EXAMPLE 19.11

1 Figure 19.15 shows a section of coastline in north Somerset. Draw a cross-section for points A–B and label the features on your cross-section.

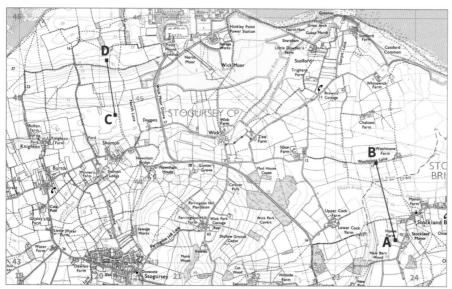

Figure 19.15 OS 1:25 000 map of the north Somerset coastline with cross-sections.

How to draw a cross-section:

- Draw a line on the map. Figure 19.15 has two lines already drawn on it: A–B and C–D.
- Draw an x-axis line on your paper that is the same length as the line on the map.
- Draw a y-axis to show the height above sea level. In the A–B cross-section on Figure 19.15, the highest point is 30 m, so the y-axis goes from 0–30 m.
- Wherever the line crosses a contour, put a mark on the y-axis. At each of these marks, plot the contour height, using the x-axis.
- Join all of the points, without using a ruler.
- Label the cross-section to show features of the landscape, such as valley sides, rivers, hilltops, farms, roads and streams.

ACTIVITY 19.15

1 Draw a cross-section for points C–D on Figure 19.15. Label the features on your cross-section.

Transects

Transects are straight lines that show information about a human landscape. We usually annotate transects to show important information about a place. Transects can be drawn to scale, or they can be sketches carried out while on fieldwork. Transects show differences between places, e.g. from the CBD to the outskirts of the city.

Further research

What are the differences between cross-sections A–B and C–D? What are the similarities and differences between the two? Why do these cross-sections look so steep? How could you make the cross-sections resemble the real landscape (i.e. be less steep)?

WORKED EXAMPLE 19.12

1 Draw a transect for Bristol in the south-west of England. The transect runs from the **CBD** in the centre to the southern edge of the city.

Figure 19.16 A transect from Bristol CBD to the south of the city.

How to draw a transect:

- Draw a line on your map. Make sure that the line is in a useful place, i.e. it will show a change from one place to another.
- Either walk along the transect to collect primary data, or research places along the transect to find secondary data.
- Sketch the buildings along the transect. You won't be able to draw them all, but you can draw one or two that are typical of the area.
- Label the transect with information about each place. Figure 19.16 has information about dates that the buildings were constructed.

ACTIVITY 19.16

1 Draw a transect of your nearest city, town or village.
   a Sketch buildings along the transect.
   b Add notes to show information about each place.

Key term

CBD: central business district (in a town or city)

19.9 Cartographic skills – coordinates

Coordinates on a map are called **grid references**. They help us to accurately locate places on a map. Every Ordnance Survey map has a grid of faint blue lines. The vertical lines across the bottom of the map are called **eastings** as their numbers increase as they travel towards the east. The horizontal lines up the side of the map are called **northings** as their numbers increase as they travel towards the north.

Four-figure grid references

Four-figure grid references locate an object on a map within a grid square. To find a four-figure grid reference:

- Find the bottom left corner of the square that you wish to locate.
- First, write the eastings number (found along the bottom of the map).
- Then, write the northings number (found along the side of the map).

Key terms

grid references: coordinates on a map that are used to locate places

eastings: numbers found along the x-axis of a map

northings: numbers found along the y-axis of a map

WORKED EXAMPLE 19.13

1 Use Figure 19.17 to find the four-figure grid references for:

a Mappleton

The village of Mappleton lies in grid squares 22 43 and 22 44.

b Broom Hill

Broom Hill lies in grid square 20 43.

c A golf course.

There is a golf course (shown using a blue flag) in grid square 20 45.

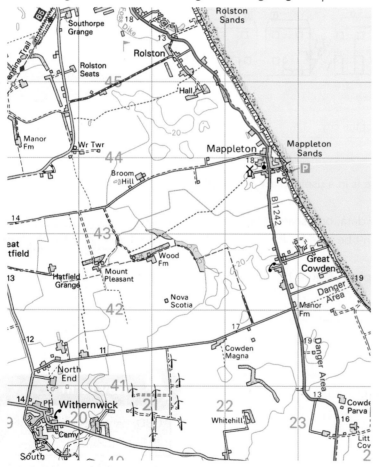

Figure 19.17 OS 1:50000 map of Mappleton on the Holderness Coast.

Six-figure grid references

Six-figure grid references locate an object on a map within a specific part of a grid square. Six-figure grid references are much more precise. To find a six-figure grid reference:

- In your head, split the grid square into ten mini-squares along the bottom and ten mini-squares up the side.
- Write the eastings number as you did for a four-figure grid reference, but then write a third number to estimate how many tenths of the way across the grid square the object lies.
- Write the northings number as you did for a four-figure grid reference, but then write a third number to estimate how many tenths of the way up the grid square the object lies.

Figure 19.18 shows how to work out six-figure grid references:

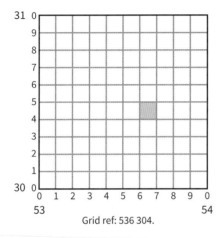

Grid ref: 536 304.

Figure 19.18 Six-figure grid references.

WORKED EXAMPLE 19.14

1 Use Figure 19.17 to find the six-figure grid references for:
  a Mappleton church
   Mappleton church has the six-figure grid reference 225 439.
  b Broom Hill
   Broom Hill has the six-figure grid reference 208 438.
  c A golf course.
   The golf course (shown using a blue flag) has the six-figure grid reference 206 455.

ACTIVITY 19.18

1 Use Figure 19.17 to find six-figure grid references for:
  a Manor Farm at Great Cowden.
  b The car park (shown using a blue P) at Mappleton.
  c Wood Farm to the west of Great Cowden.
2 What features are found at the following locations?
  a 197 407
  b 220 415
  c 216 449

Tip

Some people remember the order of grid references by the phrase 'along the corridor, up the stairs'. This reminds us to use the eastings first (along) and the northings second (up).

ACTIVITY 19.17

1 Use Figure 19.17 to find the four-figure grid references for:
  a Great Cowden.
  b The windmill (shown using a small black cross and building) to the west of Mappleton and to the north of Great Cowden.
  c Wood Farm to the west of Great Cowden.

Discussion point

When might you need to use grid references? When would four-figure references be more useful? When would six-figure references be more useful?

Further research

Find out about the UK's National Grid. Go to the **Ordnance Survey website** (see www.cambridge.org/links/gase40171). Use the National Grid to create eight-figure grid references.

Go to the **grid reference finder website** (see www.cambridge.org/links/gase40172) to find a grid reference for any place on the UK map.

19.10 Cartographic skills – GIS

How can we show information on a GIS?

GIS stands for Geographical Information Systems. GIS are tools for showing **spatial** data **electronically**. A GIS map contains layers of information. These layers can be changed to show information and find patterns.

Uses of GIS

Most websites contain GIS maps; the user can search for places, find routes and add layers, such as the location of the nearest supermarkets or petrol stations. The emergency services rely on GIS data.

Satnav systems also use GIS technology; these are linked to **satellites** and can show exactly where on the map the driver is located. Taxi drivers and bus drivers use satnavs to plan the best route.

GIS maps can be used within fieldwork. **Secondary data** can be found through GIS maps online and **primary data** can be added to GIS maps in the classroom. You can create a GIS map for any fieldwork that you carry out. This can be done through an online system, such as the Ordnance Survey's OpenSpace and Google Earth. Or you can use a specialist piece of software, such as AEGIS or ArcView.

How can we use GIS maps?

GIS maps can show information in a variety of ways (Figures 19.19–19.22):

Key terms

satnav: satellite navigation systems; these link to satellites to provide the best routes for drivers

satellites: space stations that orbit the earth

secondary data: data that someone else has collected, but that you are going to use

primary data: data that you have collected yourself

Area shading

Figure 19.19 Sections of the map are coloured in to show different information. Shading can be small scale, such as individual buildings, or large scale, such as entire countries.

Flow-line and desire-line maps

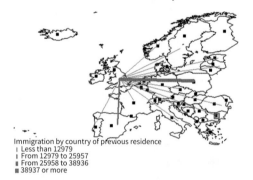

Immigration by country of previous residence
- Less than 12979
- From 12979 to 25957
- From 25958 to 38936
- 38937 or more

Figure 19.20 The movement of something from one place to another is shown using a line. The size of the movement is shown by the thickness of the line. Flow lines follow the exact path of movement, whereas desire lines show a more generalised movement. This map shows that immigration to the UK from Poland is greater than from any other EU country.

Proportional symbols

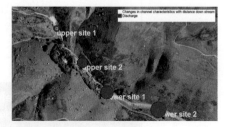

Figure 19.21 Symbols are added to a map to show information about different places. The same symbol is used, but it appears larger or smaller, depending on how much of something has been recorded.

Chart

Figure 19.22 Bar graphs, pie charts and line graphs are superimposed to show information about different places.

GIS systems can perform complex searches to narrow data down or find common characteristics of places. For example, you could search the data on a GIS map of your school to find places where it is both sunny and sheltered from the wind.

 Free GIS software is provided by Google Earth and GE Graph. Plot pins onto Google Earth and then add your data and present as graphs through GE Graph. GE Graph can be downloaded from http://ge-graph.soft32.com/.

ACTIVITY 19.19

1 Use Figure 19.23 and an atlas to discuss access to water in cities and GNI per capita for the following countries:
   a Iceland
   b India
   c Brazil
   d Angola

Discussion point

How could you use a GIS to show river velocity, shop type, pedestrian numbers or amount of litter?

Further research

What would a desire-line map of students' journeys to your school look like? Where would the thickest line be, i.e. where do the most students travel from? How easy would it be to draw a flow-line map using the same information?

WORKED EXAMPLE 19.15

1 Figure 19.23 is area shaded to show the percentage of urban residents with access to safe, clean water in each country. It also has bar graphs to show GNI per capita. Use Figure 19.23 and an atlas to discuss access to water in cities and GNI per capita for the following countries:

a Egypt

   Egypt is in North Africa.
   - *In Egypt, 94–99 per cent of urban residents have access to safe, clean water. It is coloured in the second darkest shade of blue.*
   - *Egypt has a very low GNI per capita.*
   - *We can conclude that, although Egypt appears quite poor, it has good access to safe, clean water within its cities.*

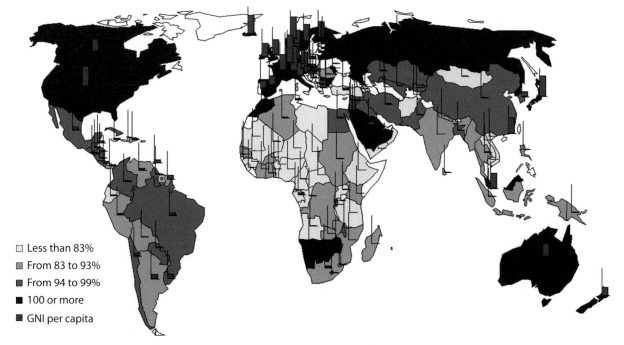

Legend:
- ☐ Less than 83%
- ■ From 83 to 93%
- ■ From 94 to 99%
- ■ 100 or more
- ■ GNI per capita

Figure 19.23 A GIS map showing access to water in urban areas and GNI per capita.

19.11 Atlas skills

An atlas is a collection of maps at various scales. Atlases usually have maps of the world and also maps of individual continents and countries. Atlas maps can either be:

- **Physical** – These maps show natural features of the landscape, such as high land and lowland and areas of water. Physical maps use colours to represent different features; usually high land is shown in brown, low land is shown in green and water is shown in blue.
- **Political** – These maps show country boundaries and large cities. Different countries are usually shaded in using different colours to make them stand out. Capital cities are shown using different symbols or bold font.
- **Thematic** – These maps show different information as choropleth maps. Atlas maps may show information such as global biomes or levels of wealth.

World maps have a grid to show latitude and longitude to help with navigation. As the earth is a sphere, **latitude** and **longitude** are measured as circles, using degrees.

Latitude is the angular distance, measured in degrees, from the equator (Figure 19.24). Lines of latitude run from east to west and are sometimes called **parallels** as they run parallel to the equator. The longest line of latitude is the equator and the lines become shorter as they near the Poles.

Longitude is the angular distance, measured in degrees, from Greenwich Meridian in London (Figure 19.25). They run from north to south and are sometimes called **meridians**. They are all the same length as they all meet at the North and South Pole. They are spaced furthest apart at the equator.

Key terms

latitude: imaginary lines that run from east to west around the globe

longitude: imaginary lines that run from north to south around the globe

parallels: another name for lines of latitude

meridians: another name for lines of longitude

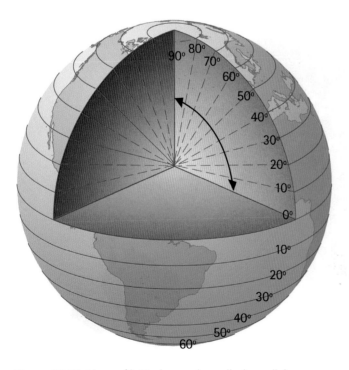

Figure 19.24 Lines of latitude are also called parallels.

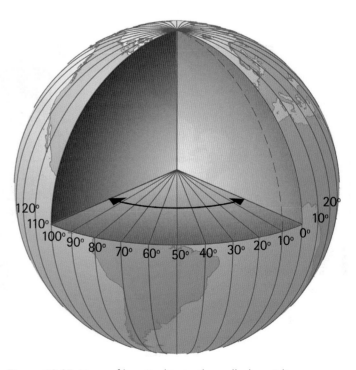

Figure 19.25 Lines of longitude are also called meridians.

WORKED EXAMPLE 19.16

1 Use Figure 19.26 to find the latitude and longitude for:

  a A

   Point A on the map is 30°N, 30°W

  b B

   Point B on the map is 18°S, 120°W

  c C

   Point C on the map is 70°N, 120°E

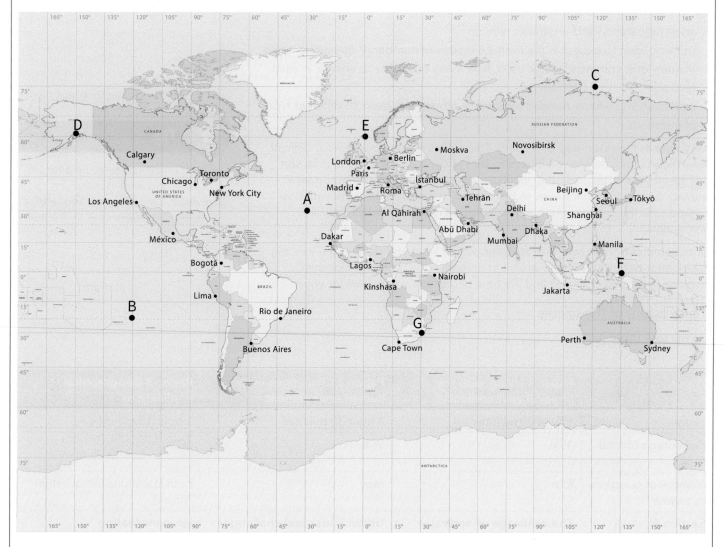

Figure 19.26 A world map with lines of latitude and longitude.

ACTIVITY 19.20

1 Use Figure 19.26 to find the latitude and longitude for:

  a D

  b E

  c F

  d G

20 Preparing for an assessment

20.1 Structure of the GCSE Geography exams

Luck with the questions set has only a small part to play in exam success. These are much more important:

- good preparation and organised revision
- knowing what to expect in the exams (in terms of number of questions, types of questions, command words used, how many questions to answer in each section)
- understanding what is needed to produce effective answers to different types of questions.

Everyone has their own way of revising. This makes giving good advice difficult. Nevertheless, well before you intend to start revising, it is a good idea to sort through all your notes and other work. Arrange them in order of importance. Estimate the amount that needs to be done to go through the work you have covered over the last two years, and how long it is likely to take to revise thoroughly.

Various studies have shown that the brain retains what you learn in the evening much better before sleep than what is crammed in the morning. However, everyone is different so try to find the time of day that works best for you.

Tip

Don't underestimate the amount of time and work needed for revision. Don't leave it too late!

The structure of the exams

You will take **three** papers.

Paper	Paper 1 Living with the physical environment	Paper 2 Challenges in the human environment	Paper 3 Geographical applications
Length	1 hour 30 minutes	1 hour 30 minutes	1 hour 15 minutes
Percentage of total marks	35%	35%	30%
Number of questions to answer	four	four	both questions (no question choice)
Section A	**The challenge of natural hazards** Question 1 Answer all the questions 33 MARKS	**Urban issues and challenges** Question 1 Answer all the questions 33 MARKS	**Issue evaluation** 37 MARKS All the questions will be based on the resources booklet, which your teacher will give to you 12 weeks before the date of Paper 3. You will be able to look at and study the resources, and discuss what they show with your teacher in advance

Paper	Paper 1 Living with the physical environment	Paper 2 Challenges in the human environment	Paper 3 Geographical applications
Section B	**The living world** Question 2 Answer all the questions 25 MARKS	**The changing economic world** Question 2 Answer all the questions 30 MARKS	**Fieldwork** 39 MARKS These questions are about the two fieldwork enquiries you will have carried out during the course. One piece of fieldwork will be on physical geography, for example a river study, or a coast study. One piece of fieldwork will be about human geography, for example studying land use in a town.
Section C	**Physical landscapes in the UK** Choose **two** questions to answer 15 + 15 MARKS Question 3 – Coastal landscapes in the UK Question 4 – River landscapes in the UK Question 5 – Glacial landscapes in the UK	**The challenge of resource management** Answer Question 3 – Resource management 14 MARKS Choose one more question to answer 11 MARKS Question 4 – Food Question 5 – Water Question 6 – Energy	

20.2 Types of questions

There are a number of types of questions that can be used in assessments:

- multiple-choice
- short answer
- extended prose

Multiple-choice

A list of possible answers is given in the question and you have to identify the correct answer. Sometimes you are asked to identify more than one correct answer. The question will always make clear how many answers you must choose.

Examples:

1 (List of four grid references referring to an OS map) – Which shows the steepest slope?
2 (List of four continents referring to a world map of birth rates) – Which continent has most of the high birth rates?

Occasionally a longer list of answers is given when there are two or more correct answers.

Example:

1 The world map shows rates of deforestation in different parts of the world. Identify the two correct statements about the pattern of deforestation shown on the map.

Short answer

These are questions worth one, two or three marks. Short answer questions can be answered in a few sentences, rather than several paragraphs (an extended response).

Extended prose

Some questions require more extensive responses. For example, questions allocated 4 marks and above generally require longer and more detailed answers.

As well as making relevant geographical points, you should also aim to provide:

- balanced coverage between different aspects of the question
- specific information about a relevant named example or case study
- good summary comments about the relative importance of different factors.

20.3 Answering questions

There are some general techniques that can be helpful to follow in assessments:

- read all of the questions carefully before beginning to write
- start with the question you feel you can answer best
- write all of your answers in the spaces provided in the examination paper
- you do not need to write out the question in your answer
- keep an eye on the time.

Strong and weak answers

Here are two examples of answers to a question. Which do you think is better?

Question Outline one strategy used to manage tropical rainforests sustainably. (2 marks)

Student sample answer 1

One strategy used to manage tropical rainforests sustainably is selective logging. Another strategy used to manage tropical rainforests is replanting new trees.

Student sample answer 2

Selective logging involves cutting down only the tall fully grown trees. This leaves smaller trees to keep on growing and allows the rest to fill the gaps and repair itself.

Both answers include a strategy – in fact, answer 1 provides two strategies! However, answer 2 has included more information to explain what the strategy is.

Command words – what they mean, why they are important

All exam questions have a command word (or words) and a theme (or geographical topic). Command words are important because they tell you what to do. It is helpful to pick them out. Some are simple and need no further comment, such as:

- Name
- Give
- State

- What is?
- How many?
- What is meant by?

However, other command words can be more complicated.

Describe

Describe is a very important command word in questions that include resources, such as graphs, photographs and tables of data. It is also very precise.

The question 'Describe what Figure 1 shows. (2 marks)' commands you to write about what is there, what is shown or what can be seen. It is not asking you to give an explanation.

Describing a line graph

To describe what a **line graph** shows, you could:
- Describe the general trend (increase, decrease, fluctuations).
- Look for the speed of change (variations in line steepness).
- State and use values (values at start and end, the difference between them, peaks and troughs, range (difference between highest and lowest).

Describing a distribution or pattern from a map

To describe a **distribution** or a **pattern** from a map, you could:
- State where there are many/most/highest values.
- State where there are few/values are lowest.
- End with a comment summarising the main features.

For describing from a world map, it is helpful to know:

- continents and oceans
- major world regions

- hemispheres
- world zones (tropical, temperate, polar)

Similarly, it is helpful to learn what to look for from other resources frequently used in geography, such as photographs, climate graphs, bar graphs, pie charts, scatter graphs and population pyramids.

Explain

Questions asking for **explanation** include:

- Explain why
- Suggest reasons for
- Outline why

- Give reasons for
- Why does?

It can be helpful to mention examples, such as case study information.

Compare

Compare requires that you make a direct comparison of the similarities and differences between two sets of information or places, such as between: two map distributions; two countries (e.g. rich and poor); two climate graphs; an old and a recent photograph.

Note that it does **not** mean two separate accounts. Using link words like 'whereas' or 'similarly' helps to make it more obvious that you are comparing.

20.4 Approaching extended response questions

Command words used in extended response questions require you to go beyond description and explanation. Analysis and evaluation become much more important. You often need to write about your opinion and use evidence to support it. Typical question commands include:

- To what extent ...?
- Assess how effective ...
- Assess the extent to which ...
- Evaluate the effectiveness of ....
- Which is more important ...?

For example, you might have a question like the following.

Assess your progress

command words

Assess the extent to which it is possible for people to

reduce the effects of tropical storms.   9 MARKS
                    question theme

Before answering the question, it can be useful to:

- Underline or highlight the key elements in the question – the command words and question theme.
- Make notes about what you might include in your answer. This can help you to work out a structure and an order for answering. Make sure that you cross out any work that you don't want marking before submitting your paper.

For example, you might start with the theme, in this case – ways to reduce the effects.

> o Monitoring storms using weather satellites – predicting speed and direction of movement
> o Weather forecasts as warnings For example, once the tropical storm has formed these are ways to know where it is going and to be prepared
> o Build shelters Stock them with emergency supplies Undertake practice drills
> o Educate people about tropical storms and their effects i.e. For example, what can be done in advance

You might also want to note down any details about any relevant examples or case studies.

Once you know what geographical information can be included, you could think about what you are going to do with it to answer the question. In other words, how to obey the command words.

Assess the extent could be anything from 0 per cent (totally impossible to reduce any effects) to 100 per cent (all effects stopped). In geography such extremes are highly unlikely.

Think about what stops the answer being 100 per cent and note them down – things like:

> *Storms are unpredictable and change direction*
> *countries too poor/lack organisation to build shelters*
> *poor tropical countries lack the technology/communications*
> *effects more likely to be reduced in rich countries like the USA/Japan*

Once you've made some notes, you will hopefully feel more ready to begin writing your answer. Use your notes to help you. For this question, it is important to include a summary that addresses the command words 'assess the extent'.

Extended response questions with the command word 'justify'

Some questions ask a direct question like these:

- Do you agree with this statement?
- Do you think the project should go ahead?
- In your opinion, should the new dam be built?

These questions are asking you to make a decision and then to justify it.

For example:

Next you must tick a box – yes or no. The nine mark question is '**Justify** your choice.'

One example of a statement is made such as:

'The poorer the country, the greater the effects from natural hazards (such as tropical storms, earthquakes and volcanoes).' Do you agree?

The following actions can be helpful when answering this type of question, or you could devise your own.

1. Underline/highlight the key elements in the question.
2. Before writing, make notes on rough paper. It can be helpful to include two headings, e.g. 'Yes/agree' and 'No/disagree', and then create two lists of points for and against.
3. Looking at your lists, make your decision about which view to take.
4. Explain your choice.

 Tip

You might feel that neither option is completely correct. You can explain your thoughts in your justification.

Assess to progress question bank

These questions provide additional practice.

Chapter 2 Tropical storms

 Assess your progress

1 Give **one** factor that is needed for a tropical storm to occur. `1 MARK`

2 Using the information in Table 2.1, give **three** kinds of damage caused by a category 4 tropical storm. `3 MARKS`

Category	Wind speed	Type of damage
1	119–153 km/h 74–95 mph	Very dangerous winds – some damage to well-constructed houses, roofs and gutters. Large branches will snap. Power outages with damage to power lines and poles.
2	154–177 km/h 96–110 mph	Extremely dangerous winds – extensive damage to roofs and houses. Shallow trees will be uprooted and block roads. Near-total power loss for a minimum of several days.
3 (major)	178–208 km/h 111–129 mph	Devastating damage to well-built houses. Many trees uprooted and blocking numerous roads. Electricity and water unavailable for days, if not weeks.
4 (major)	209–251 km/h 130–156 mph	Catastrophic damage – severe damage to houses, especially roofs and walls. Fallen trees and power lines will leave residential areas isolated with no power for weeks or months.
5 (major)	252 km/h + 157 mph +	Catastrophic damage – widespread destruction with a high proportion of homes and infrastructure destroyed. Most of the area uninhabitable for weeks or months.

3 Describe two features of a storm surge. `4 MARKS`

4 For either an HIC or an LIC, describe the primary and secondary effects of a tropical storm. Use a example in your answer. `6 MARKS`

Chapter 5 Tropical rainforests

 Assess your progress

1 Evaluate the success of the Plan Pacifico in the Chocó forests of Colombia in conserving the rainforest while allowing some economic development to support the local population. `9 MARKS + 3 SPaG`

 Tip

- Consider what 'evaluate' means.
- You need to give the advantages and disadvantages of the Plan Pacifico.
- You should also say how successful you think the projects have been.
- Add your views as to why, or why not, the Plan has been a success.

Chapter 10 Glacial landscapes in the UK

 Assess your progress

1 Define these geographical terms.
   a Glacier   d Snout
   b Interglacial period e Tundra
   c Permafrost   `5 MARKS`
2 Identify one landscape feature in an upland glaciated area in the UK such as the Lake District, Snowdonia or the Scottish Highlands
   a Describe two features of the landform.   `4 MARKS`
   b Evaluate the use of this landscape feature for tourist use. `6 MARKS`

Chapter 17 Energy resources

 Assess your progress

1 Study the data table

Country	Total energy consumption per person per year in kg of oil equivalent		
	2005	2008	2015
Australia	5956	5762	5505
Bangladesh	153	164	205
Brazil	1074	1114	1371
Canada	7985	8411	7303
China	896	1242	2029
Czech Republic	4049	4460	4138
Egypt	737	783	978
France	4487	4547	3868
India	515	531	614
Japan	4099	4173	3610
Kenya	500	506	480
Malaysia	2168	2279	2639
Thailand	1235	1524	1790
UK	3982	3906	2987
USA	7996	7921	7032
Vietnam	495	611	697

Changing global energy use 2005 – 2015. *Source: The Economist Pocket World in Figures* (2005, 2008, 2015).

Choose one country for each of these categories and describe:

a i a post-industrial economy   iii an LIC

   ii an NEE

b Describe the changes in energy consumption per person between 2005 and 2015.   `3 MARKS`

2 Explain why some countries' energy consumption per person is increasing at the moment, but in others it is decreasing.   `6 MARKS`

World map for Chapter 12 The development gap

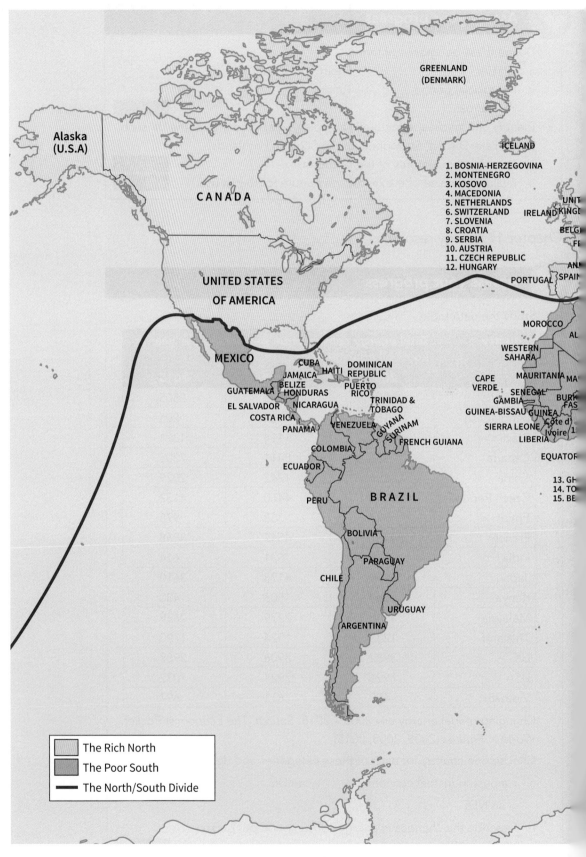

Figure 12.5 The North/South divide.

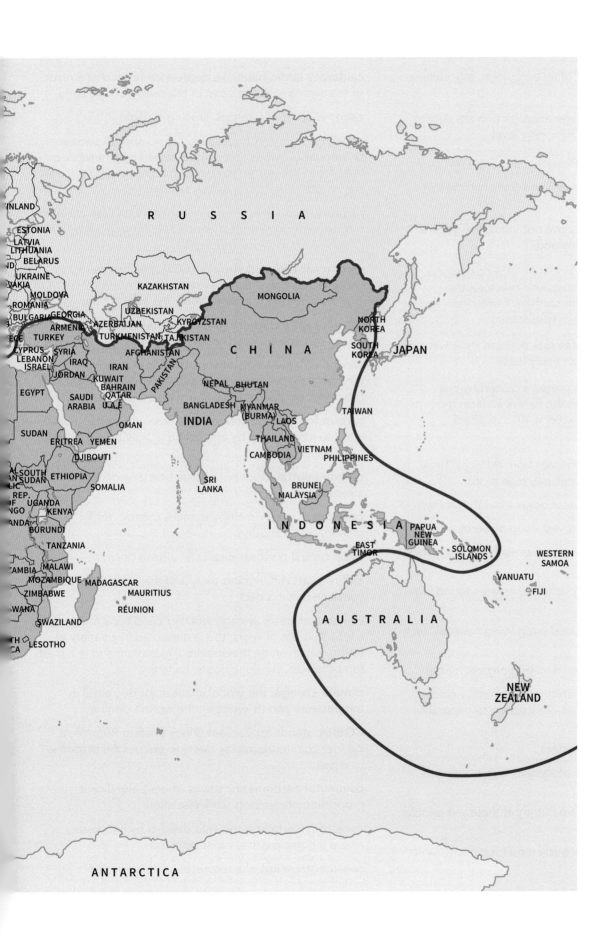

FINLAND
ESTONIA
LATVIA
LITHUANIA
BELARUS
UKRAINE
AKIA
MOLDOVA
ROMANIA
BULGARIA GEORGIA
ARMENIA
ECE TURKEY
CYPRUS SYRIA
LEBANON IRAQ
ISRAEL
JORDAN
EGYPT

RUSSIA

KAZAKHSTAN

UZBEKISTAN

AZERBAIJAN
TURKMENISTAN TAJIKISTAN

AFGHANISTAN

KYRGYZSTAN

MONGOLIA

CHINA

NORTH
KOREA
SOUTH
KOREA

JAPAN

KUWAIT
BAHRAIN
QATAR
U.A.E.
SAUDI
ARABIA
OMAN

IRAN

PAKISTAN

NEPAL BHUTAN

BANGLADESH MYANMAR
(BURMA) LAOS

INDIA

TAIWAN

SUDAN
ERITREA YEMEN
DJIBOUTI
AL SOUTH
AN SUDAN
LIC
OF ETHIOPIA
GO UGANDA
ANDA KENYA
BURUNDI
TANZANIA
AMBIA MALAWI
MOZAMBIQUE MADAGASCAR
ZIMBABWE
WANA
SWAZILAND
CA LESOTHO

THAILAND
CAMBODIA

VIETNAM
PHILIPPINES

SRI
LANKA

BRUNEI
MALAYSIA

INDONESIA

PAPUA
NEW
GUINEA

EAST
TIMOR

SOLOMON
ISLANDS

WESTERN
SAMOA

VANUATU
FIJI

MAURITIUS
RÉUNION

AUSTRALIA

NEW
ZEALAND

ANTARCTICA

Glossary

abiotic: the non-living part of an ecosystem, e.g. climate and soils

adult literacy: the percentage of adults who are able to read and write to a basic functioning level

afforestation: planting trees

agribusiness: highly-intensive, large-scale commercial farming

anomaly: a data point that does not follow the general pattern or that seems slightly odd

anticyclone: a large-scale circulation of winds around a central region of high atmospheric pressure; the circulation is clockwise in the northern hemisphere and anti-clockwise in the southern hemisphere

aquifer: a body of permeable rock that can store water and through which water can easily move

atmospheric pressure: the pressure caused by the weight of air at any point on the Earth's surface; the average air pressure at sea level is 1013 millibars

baby-boomer: someone who was born during a 'baby boom', which is a time of unusually high births

backwash: movement of water down a beach (gravity)

bedload: larger particles moved along a riverbed

bedrock: solid rock which lies underneath any loose rock deposits or soil

biodiversity: the range of plants and animals found in an area

biofuel: a fuel that is produced using living material, such as plants

biogas: energy derived from decaying organic matter

biomass: organic material which comes from living or recently living organisms that can be used to generate electricity

biome: a global-scale ecosystem

biotic: the living part of an ecosystem, e.g. plants and animals

birth rate: the number of live births per thousand people per year

brownfield sites: land previously used for industry, which has fallen into decay

bunds: low-lying rows of stones which reduce runoff and allow rainwater to infiltrate the soil

caldera: a large, basin-like depression formed as a result of the explosion or collapse of the centre of a volcano

canopy: cover of treetops, limiting sunlight

capacity: in terms of transport, the number of people or amount of goods that can be carried by a network or hub, for instance, the number of people passing through a particular airport in a certain time

capital intensive: an economic system with high inputs of money, expensive equipment and technology; people's physical labour is less important

carbon capture and storage (CCS): the process of capturing carbon dioxide emissions and storing them in a way that they are unable to affect the atmosphere.

carbon capture: where carbon dioxide produced by power stations is stored underground

carbon dioxide emissions: carbon dioxide being emitted into the atmosphere as a byproduct of industrial processes or other economic activities

carbon footprint: the amount of carbon dioxide released into the atmosphere by an economic activity (e.g. person, business or event)

carbon neutral: does not add carbon dioxide to the atmosphere

cash crops: growing crops to make money, not for personal consumption

CBD: central business district (in a town or city)

cliff retreat: cliff eroding away and the position of the coastline moving back

climate: an area's average weather conditions measured over a number of years; this is measured in a variety of terms, e.g. average precipitation, maximum and minimum temperatures, sunshine hours, humidity

climate change: the global increase (or decrease) in temperature and its effect on the world's climate

COBRA: stands for Cabinet Office Briefing Room A; a cabinet committee which meets to discuss the response to crises

commuter settlements: towns where a significant proportion of residents work elsewhere

cone of uncertainty: the area defined by forecasters where a hurricane may cause damage

consumption: use of a resource

contours: lines on a map that show places of equal height

conurbation: extensive urban areas resulting from the expansion of several towns or cities so that they merge together but maintain their separate identities. For example, the West Midlands conurbation includes the cities of Birmingham, Coventry and Wolverhampton, as well as many large towns, including Sutton Coldfield, Dudley, Walsall and West Bromwich.

convection currents: circular movements of heat in the mantle; generated by radioactive decay in the core

convectional storms: heavy rain falling as a result of high temperatures

core: (in physical geography) dense hot rock at the centre of the Earth

core: (in human geography) the strongest regions of a country economically; these are usually main cities and their surroundings, such as London and the South East

Coriolis effect: the effect, caused by the rotation of the Earth, which deflects winds to the right in the northern hemisphere and the left in the southern hemisphere

correlations: relationships between two sets of data

cross profile: a cross-section drawn across the river valley

crust: the outermost layer of the Earth

culturally diverse: having a variety of cultural/ethnic groups within a society

deciduous woodland: trees that lose their leaves in winter

decommissioned: closed down and safely dismantled

decomposers: mosses, lichens and bacteria which decompose and recycle dead vegetation

deficit: not having enough of something

de-industrialisation: the decline of traditional forms of industry, often accompanied by increased unemployment

Demographic Transition Model (DTM): the five stages through which a country passes in terms of birth, death rate and natural change; as a country passes through the DTM its economy becomes more sophisticated

depression: (in meteorology) where warm and cold air meet, usually at mid-latitudes over the UK – the warm air is less dense than the cold air and so rises above it, creating low pressure on the ground; weather associated with a depression includes rain and strong winds

dereliction: land and buildings that have fallen into disuse

desalination: the process of creating fresh water by removing salts and minerals from seawater

desalinisation: industrial process to remove the salt from seawater

desertification: a reduction in the biological productivity of the land which leads to desert-like conditions (as defined by the United Nations in 1977)

development gap: differences in level of people's total well-being and happiness, physical standards of living and national wealth (Physical Quality of Life Index, GDP per head) between countries

discharge: the volume of water at a given point in a river (measured in cumecs)

disposable income: the amount of money a person has left over to spend freely after paying for the essentials like food, housing and taxes

diurnal: showing the altering conditions between day and night

dormant: (in relation to volcanoes) has not erupted in living memory, but it could become active in the future

dormant: (in relation to plants) when a plant's metabolism slows to a point where its growth and development are temporarily stopped

drainage basin: the area of land drained by a river system

drilling rig: a floating or fixed structure for removing oil from deep within the seabed

drip tips: leaves that are shaped in a way that allows excess water to run off them

drip-feeding: pipes laid across a cropped area have regular small holes to distribute water. Slow flow limits wastage through evaporation

drought: long periods of time without rainfall

eastings: numbers found along the x-axis of a map

economic development: a change in the balance between primary, secondary, tertiary and quaternary economic production. Poorer countries rely on primary (raw materials) production. As countries develop economically there is a move towards secondary activities (manufacturing) and then to tertiary (services) and quaternary (research and development activities). As a country develops economically, people's standards of living increase

electricity grid: a network for transporting electricity to consumers

emergent trees: the very tallest trees that grow higher than the rainforest canopy

emigrate: leave a country to live in another, with the intention of remaining at least a year

endemic: when a species is only found in a particular place

energy mix: the combination of energy sources that make up the total supply

energy security: availability of affordable energy; not being reliant on countries who may cut supply for political reasons (e.g. Russian gas supply to Ukraine in the 2000s)

enhanced greenhouse effect: the increase in the effects of global warming due to human activities

equilibrium: the balance between all parts in a system

erosion: the breaking up of rocks that is the result of movement

estuary: wide part of a river where it nears the sea

European Economic Community (EEC): the name for an organisation that links European countries through trade and, more recently, political agreements. In 1993 it was renamed the European Union (EU)

eutrophication: a form of water pollution, caused when nutrients from sewage or agricultural fertiliser run off into waterways and cause algal blooms to grow. These use up all the available oxygen. Fish and other marine life then suffocate

evaporation: the effect of the Sun heating liquid water on the Earth's surface and converting it into a gas in the atmosphere

exploited: to use something

extinct: has not erupted in historic times, in the last 10 000 years

extracting: removing something, e.g. oil

extreme events: hurricanes, drought, earthquakes, volcanic eruptions – physical events causing serious impacts on life and economy

eye wall: the towering banks of cloud bearing heavy rainfall which surround the eye

eye: the centre of a tropical storm where sinking air causes relatively calm, clear conditions

flood plain: area of flat land which is prone to flooding

flash floods: rapidly rising river levels leading to a rapidly developing flood situation

food insecurity: having no reliable access to affordable and nutritious food

food miles: the distance food is transported between producer and consumer

food security: having a reliable, affordable food supply

fossil fuels: energy from plant and animal remains, such as coal, oil and natural gas, including shale gas

fracking: extracting gas from rocks by injecting water, sand and chemicals into them at high pressure

freeze-thaw: weathering of rocks by continued freezing and thawing of moisture in cracks

freight: goods being transported by road, rail, container ship or plane

genetically modified (GM): specific changes made to DNA to improve them, e.g. to create disease resistance

geology: study of the Earth, especially rocks

geothermal energy: where the heat from within the Earth is harnessed to generate electricity; geothermal power stations are usually found near areas of volcanic activity

glacial: a cold period where average global temperatures decrease to around 11 °C

glaciers: slow moving rivers of ice formed from compacted snow

global financial institutions: financial institutions which cover the world in terms of borrowing and lending – the two main ones are the World Bank and the IMF – wealthier countries contribute and any country can borrow or be funded by these institutions

global warming: the general increase in global temperature

globalisation: the integration of economic, financial, social and cultural ideas and contacts between countries; increased trade and labour migrations are a big part of this

gradient: the change in height of a landscape

green space: parks and vegetated areas and walkways

greenhouse effect: the gases in the atmosphere which trap outgoing radiation and keep the Earth at a temperature at which humans can live

grey water: wastewater from people's homes can be recycled and put to good use; treated grey water can also be used to irrigate both food and non-food producing plants

grid references: coordinates on a map that are used to locate places

gross national income (GNI) per head: in simple terms, the total value of a country's goods, services and overseas investments, divided by the number of people in that country

Gross National Income (GNI): the total value of goods and services produced within a country, usually measured per year; it is a measure of wealth and standard of living

ground cover: low-growing plants

ground source heat pump: a system for converting heat from the ground into heating systems within homes

groundwater management: regulation and control of water levels, pollution, ownership and use of groundwater

groundwater: water found underground in soil, sand or rock

Gulf States: the seven Arab states which border the Persian Gulf

habitats: the environments that plants and/or animals live in

hazard risk: the probability or chance that people will be seriously affected by a natural hazard

headland: a narrow piece of land extending out into the sea, sometimes called a promontory

HIC: a higher income country is defined by the World Bank as a country with a gross national income per capita above US$12 735 in 2014

hub: in terms of airports it means one of the most important globally; millions of passengers change planes at global hubs to reach their destinations

Human Development Index (HDI): development indicator combining life expectancy at birth, education and income

hunger: prolonged or frequent undernourishment, where a person has insufficient calories to carry out light activities

hydroelectric power (HEP): electricity generated by turbines that are driven by moving water

hydro-electricity: electricity that is generated from running water

hydroponics: a method of growing plants using mineral nutrient solutions, in water, without soil

hydropower: electricity generation from moving water: damming rivers, wave and tidal power

hypothesis: an idea or explanation that is tested through study and investigation

ice caps: smaller mass of glacial ice which are less than 50 000 km²

ice cores: samples of ice, often many metres thick, gained by drilling through the ice cover in cold environments

ice sheets: large masses of glacial ice which are over 50 000 km²

immigrate: enter a new country with the intention of living there at least a year

indigenous: native people who originate from a particular place

infant mortality: the average number of deaths of babies under 12 months old per 1000 live Births

infrastructure: the framework of transport and energy networks, including roads, railways, ports and airports, plus energy distribution

in-migrant: someone moving into another region of their country

integrated transport system: where all parts of the transport system link together to make journeys more efficient

intensive agriculture: farming that requires large inputs of labour, chemicals, capital and so on, to produce as many crops or animals as possible on the available land

interdependence: two or more things that depend on each other, in this case countries in terms of trade

interglacial: a warm period where average global temperatures increase to around 15 °C

Intergovernmental Panel on Climate Change (IPCC): an international body of scientists which assesses all of the evidence of human-induced climate change

International Monetary Fund (IMF): an organisation of 188 countries that aims to promote global economic stability

intertidal habitat: land exposed at low title and covered at high tide

interval data: data where the difference between each value is equally split

irrigation: taking water from a store such as an aquifer or river and distributing it across areas of landscape to make the land suitable for growing crops

jet stream: strong winds (around 200 mph) that circle the Earth between 5 and 10 miles up in the atmosphere

kerosene: a flammable oil that is made using petroleum

Kyoto Protocol: an international treaty to reduce greenhouse gas emissions; the original treaty ran from 1997 to 2012 but the Doha Amendment extends the protocol until 31 December 2020

labour intensive: an economic system dependent on people's physical labour – there is little technology, but that does not mean this economy cannot thrive

lagoons: stretches of salt water separated from the sea by a barrage or beach

land reclamation: gaining land from the sea

landlocked country: a country without a coast, so without access to the sea, affecting trade

landslide: the movement of earth or rock from a slope as a result of it becoming unstable (usually from heavy rainfall)

lateral erosion: erosion of the sides of a valley

latitude: imaginary lines that run from east to west around the globe

leaching: when nutrients wash out of soil

LIC: a lower income country is defined by the World Bank as a country with a gross national income per capita below US$1 045 in 2014

life expectancy: the average number of years a person is expected to live from birth in a particular society at a certain time

line of best fit: a straight line drawn through the points on a scatter graph

litter: fallen plants and leaves on the forest floor

long profile: a line showing the gradient of a river from source to mouth

longitude: imaginary lines that run from north to south around the globe

magma: heat from the Earth's core is hot enough to melt rock in the mantle; this molten (liquid) rock is called magma

mantle: a layer of rock between the core and crust made of molten rock

manufacturing: the process of making a product

marginal land: areas which can only be farmed when conditions (e.g. rainfall) are very good

mb (millibars): a unit of atmospheric pressure

meander: a large bend in a river

megacity: a city that has 10 million or more people

meltwater: water which is formed by the melting of snow and ice

Mercalli scale: a scale used to describe the damage caused by an earthquake

meridians: another name for lines of longitude

meteorologist: a scientist who studies the causes of particular weather conditions

micro-loans: very small loans with low interest payment

migration: the movement of people from one place to another

mixed land use: a mixture of uses such as business, leisure, residential

monoculture: the farming of a single crop

monsoon: seasonal wind system in the Indian Ocean which controls the rainfall pattern

mouth: the area where a river flows into the sea

mudflats: a stretch of muddy land which is uncovered at low tide

multiplier effect: the 'snowballing' of economic activity, for example, if new jobs are created this gives people more money to spend which means that more workers are needed to supply the goods and work in the shops

nature reserve: a conservation area which is set aside to preserve plants and animals

networks: interconnecting patterns of roads or railways, etc.

newly emerging economies (NEEs): countries in the process of changing from an LIC (lower income country) to a highly developed, more complex economy

NGOs: non-governmental organisations such as charities

non-renewable: a resource that is limited in supply and cannot be replaced

northings: numbers found along the y-axis of a map

nutrients: chemicals in the soil (often from decaying vegetation) that help plants to grow

ocean: large body of salt water

opencast: mining that takes place on the surface of the Earth by scraping away soil and rock

out-migrant: Someone leaving one part of a country to move to another

over-abstraction: when water is taken from rivers and other sources more quickly than it is being replaced.

overcultivation: the excessive use of farmland to the point where productivity falls due to land degradation

parallels: another name for lines of latitude

peninsula: a piece of land almost surrounded by water

pension: money paid to someone who has retired from work

periphery: the weakest regions of a country economically; they tend to be physically remote, making them more difficult to attract industry, for example, the Scottish Highlands

permafrost: ground that has been frozen for two or more years; in the northern hemisphere over 19 million km² is covered in permafrost, most of which is found in Siberia, Alaska, Greenland, Canada and northern Scandinavia

pesticide: chemicals used on crops to kill pests and diseases

photosynthesis: the chemical process where plants convert carbon dioxide to oxygen photovoltaic (PV) panel: a panel used to convert sunlight into electricity

plate margins: the place where tectonic plates meet and the Earth is particularly unstable

polar front: the boundary between a Polar cell and Ferrel cell

population census: an official count of the population. In the UK there is a census every 10 years, the last census was in 2011

population density: number of people per square kilometre (a measure of 'crowdedness')

post-industrial economy: a period of growth in an industrialised economy in which the relative importance of manufacturing decreases and the relative importance of services, information and research increases

poverty line: the minimum amount of money needed to be able to live

power stations: places that produce electricity

precipitation: any form of water, both liquid and solid, which falls from the sky; rain, snow, sleet and hail are all examples of precipitation

primary data: data that you have collected yourself

primary effects: the immediate damage caused by a tectonic hazard. It can include death and destruction of property

primary forest: native trees that are undisturbed

primary goods: raw materials – mining, oil/gas extraction, renewable energy, farming, fishing, forestry. In poorer countries these employ most people

primary hazards: those caused directly by the hazard, such as lava flows, ash falls as a result of a volcano erupting

primary products: raw materials from the Earth, including fossil fuels, uranium (fuel for nuclear power), metal ores, crops, fish and timber

producer: plants that provide food for herbivores at the beginning of the food chain

public transport: shared methods of travelling, such as buses, trams and trains

qualitative: information that is in the form of words

quantitative: information that is numeric

Quaternary period: a period in geologic time which stretches from about 2.5 million years ago to the present; it is divided into two main sections: the Pleistocene (2.5 million years ago to 11.7 thousand years ago) and the Holocene (11.7 thousand years ago to the present)

rapids: fast flowing river over an uneven riverbed

rate of urbanisation: the percentage increase in the urban population

raw materials (primary products): unprocessed material collected from the Earth: examples include fossil fuels, metal ores, agricultural produce

regeneration: improving the economic, social and environmental conditions of previously run-down areas

renewable: a resource that can be replaced or replenished over time

reservoir: a man-made lake, or natural one which has been adapted, used for collecting and storing water supply

Richter scale: a scale used to measure the magnitude of earthquakes

runoff: all precipitation that reaches a river

Saffir-Simpson scale: the five-point scale used to classify tropical storms according to their strength

salinisation: the accumulation of soluble salts in the soil, making the soil infertile

salinity ingress: when salt water from the sea invades water supplies on the land; this can happen due to sea level rise

salt flats: areas of flat land covered with a layer of salt

salt marsh: an area of coastal grassland regularly flooded by seawater

sand dam: a sand dam is a concrete wall built across a seasonal sandy riverbed

satellites: space stations that orbit the earth

satnav: satellite navigation systems; these link to satellites to provide the best routes for drivers

scree: angular rock fragments that are usually found at the base of mountains

sea: a region of water within an ocean or partly enclosed by land

secondary data: data that someone else has collected, but that you are going to use

secondary effects: the unforeseen consequences of tectonic hazards such as fires, spread of disease and food shortage

secondary forest: replanted trees

secondary hazards: hazards caused as an indirect result of the primary hazard; for volcanoes these include landslides and tsunamis

sediment: material moved and deposited in a different location

seismic activity: movement within the Earth's crust

services: another term for tertiary industries, the sector of the economy providing services for individuals, other industries and the community

shifting cultivation: when the land is farmed for a period of time and then farmers move to another location and the land is left to recover

shrub layer: bushes under woodland

silt: very fine material deposited by water

site: the precise area where the enquiry is taking place

situation: the general area or surroundings

slab pull theory: a theory that outlines how large and dense tectonic plates sinking into the mantle at ocean trenches drives tectonic plate movement.

slash and burn: where forest is cut down and cleared by burning; the ash from the fire provides nutrients to the soil

smog: a mixture of smoke and fog that can be dangerous to human health

soft engineering: working with the environment in order to reduce the risks of flooding and erosion

source: the starting point of a river

spatial: linked to a location or place

squatter settlements: illegal settlements where people have no legal rights over the land on which they live

storm surge: when low-pressure storm conditions cause the sea to rise

sub-aerial processes: processes that affect the face and top of cliffs

subduction: the process by which one tectonic plate moves under another tectonic plate and sinks into the mantle

subsidise: when the government pays towards the cost of producing something

suburbs: largely residential (housing) areas in the outer part of a city

succulents: plants with a thickened and fleshy structure which allows them to retain water

sulphur scrubbing: removing sulphur from flue gases before they escape from the power station chimney

sunspot: dark spots on the surface of the Sun that sometimes discharge big bursts of heat given out by the Sun

supply and demand: supply means how much of something is available; demand is how much of it people would like

surplus: having more of something than is needed

sustainability: meeting the needs of today without harming future needs

Sustainable Development Goals: United Nations goals that came into force at the end of 2015; they are international development targets to be achieved by 2030

sustainable management: management that meets the needs of the present without compromising the ability of future generations to meet their own needs; it takes into account the environment, the needs of present and future generations and the economy

sustainable water supply: meeting the present-day need for safe, reliable and affordable water, which minimises adverse effects on the environment, while enabling future generations to meet their requirements

sustainable: a method of using a resource so that it is not permanently damaged

swash: movement of waves up a beach

tap roots: long roots which extend far down in the soil to find water

tariffs: extra taxes put on goods imported, making them more expensive to buy

tectonic hazards: natural hazards caused by movement of tectonic plates (including volcanoes and earthquakes)

tectonic plate: a rigid segment of the Earth's crust which floats on the heavier, semi-molten rock below

theory: one or more statements that explain a situation or a course of action. It has to be proved by evidence from an investigation

tonnes of oil equivalent: the same amount of energy from any source compared with the equivalent amount of energy from oil

trading policies: rules decided by a country or group of countries to control imports and exports

transect: a straight line on the ground along which measurements are taken

transnational corporations (TNCs): TNCs link several countries together in the production and marketing of goods

transpiration: evaporation of water from plant leaves

transport hub: connecting point for transport links

tundra: vegetation found in cold environments which is mainly composed of shrubs, grasses, mosses and lichens – the low temperatures mean that it is difficult for trees to grow

UK (United Kingdom): the UK comprises the countries of England, Wales, Scotland and Northern Ireland

unsorted: deposits where all different sizes of rock – from tiny particles to massive boulders – are mixed together

urban slums: poor areas lacking in services; they are often called 'shanty towns' but also have localised names such as 'favelas' in Brazil, 'barriadas' in Peru and 'bustees' in India

urban sprawl: the expansion of an urban area into surrounding, less populated areas

urbanisation: an increase in the proportion of people living in towns and cities

velocity: (relating to a river) speed of flow, usually measured in metres per second

vertical erosion: downward erosion of a riverbed

virtual water: the 'hidden' volume of water that is used in the agriculture and industry.

water conflict: disputes between different regions or countries about the distribution and use of fresh water; water conflicts arise from the gap between growing demands and diminishing supplies

water conservation: strategies that use water more efficiently

water deficit: this exists where water demand is greater than supply

water quality: a measure of the chemical, physical, and biological content of water; high levels of bacteria or suspended material can result in poor quality water and pose a health risk for people

water security: having a safe and regular supply of a key resource; in this case, water

water stress: water stress occurs when the demand for water exceeds the available amount during a certain period or when poor quality restricts its use

water surplus: water surplus exists where water supply is greater than demand

water table: the level at which water is found under the Earth's surface

water transfer schemes: systems of pipelines and aqueducts to move water long distances from areas with plentiful supply to those of high demand

waterfall: a steep fall of water along the course of a river

weather: the day-to-day condition of the atmosphere, for example, the weather may be sunny, windy, rainy and so on

weathering: the breaking up of rocks that occurs *in situ* (the same place) with no major movement taking place

wilderness: a large area of land that has been relatively unaffected by human activity; these areas tend to have very low population densities and inhospitable environments

wildlife corridor: stretches of land that connect areas of native vegetation

Index

aeroponics 274
afforestation 92, 93
agribusiness 251, 252
agriculture: dry farming 92;
    glacial landscapes 157; and
    greenhouses gases 57, 58;
    large-scale 276–7; and
    volcanic activity 18
    see also irrigation
air pollution 185, 190, 223, 233, 257,
    263, 310
Alaska, development of 102–5
Antarctic Treaty 106
Antarctica 106–7, 146
anticyclones 35, 47
appropriate technology: and
    increased food supply 275; and
    water resources 297
aquifers 86, 88, 253, 282, 283, 289,
    292, 297
arêtes 153, 159
ash flows, volcanic 16t
assessment of course 356–63
atmospheric circulation, three cell
    model 34–5
atmospheric pressure 34

bars, coastal 122, 123
beaches, types of 116, 117
Beijing 167, 190t, 294, 295, 310fig
bergschrund 152
biodiversity: agribusiness 252;
    hot deserts 84;
    hydropower 212;
    tropical rainforests 67, 72, 74, 76;
    tundra 97, 98
biofuels 76, 270, 310
biomes, global distribution of 63fig
biotechnology 220, 275
Birmingham 180–7, 192–5
Boscastle 142–5
Bournemouth 240–3
Brazil: deforestation 71, 75fig;
    development 212, 213, 228fig;
    economic growth 225;
    energy 300fig, 305t;
    traffic congestion 190t, 191;
    water resources 282t, 284
bunds 92

caldera, volcanic 18
carbon capture and storage (CCS) 233
carbon footprint 233, 251, 316
Care International 214, 215
chalk streams 112, 113fig
Chile 21, 87t, 89fig, 305
China: hydropower 212;
    prediction of seismic activity 30–3;
    recorded earthquakes 33fig;
    South-to-North Water Transfer
    Project 294–5;
    Tianjin Eco-city 188–9;

trade with Africa 209;
    urban migration 167;
    water consumption 285
Chocó rainforests 76–7
climate change 54–9;
    adaptation 59;
    carbon capture 58;
    'carbon sinks' 58;
    causes of 56–7; and
    development gap 207;
    effects of 58;
    evidence for 54–5;
    and food supply 270;
    impact on coastlines 126;
    impact on tropical storms 38–9;
    mitigation of 58, 59; and
    UK extreme weather 51; and
    volcanic activity 16t; and
    water management 59; and
    water supply 289
coal 260–3, 310
coastal landscapes, UK 114–31;
    coastal deposition 122–5;
    coastal management 128;
    coastal system 118;
    managed retreat 128–9;
    marine erosion 119–21;
    mass movement 121;
    protection of 126–31;
    weathering 118, 119
cold environments 96–107;
    building and infrastructure 100, 101;
    characteristics of 96–7;
    distribution of 96fig; food
        chains 101;
    indigenous people 100, 101;
    precipitation 98;
    protection of 106–7;
    transport 100
collision zones 12
colonialism 206, 224
The Commonwealth 238, 239
conservative plate margins 13
constructive plate margins 12, 13fig
constructive waves 115–17
convection currents 11
convectional storms 66
Coriolis effect 35
corries 152, 153, 159
cycle of poverty 208fig
cyclones see tropical storms

dams 71, 141, 212, 213, 292, 312
deforestation: causes of 70, 71;
    Chocó rainforests 76–7; and
    ecosystems 61, 62; and
    greenhouse gases 57, 76;
    impact of 72; and
    income 72; and
    Madagascar 78–81;
    Malaysia 223;
    tropical rainforests 70–3, 76–7

Democratic Republic of Congo:
    calorie intake 267;
    hydropower 212;
    sustainable food production 280–1;
    rain forests 74
Demographic Transition Model see DTM
deposition: coastal 122–5;
    glacial 150–1, 155–6, 160;
    river 131fig, 135–6, 138–9
depressions 35, 46, 47, 52
desalination 59, 94, 95, 292–3
desertification 34, 90–3
deserts, hot 82–95;
    agriculture 86;
    animals 84, 85;
    characteristics of 82;
    causes of 90–2;
    distribution maps 83fig;
    ecosystem 84;
    environmental protection 89;
    inaccessibility 89;
    minerals 86, 87, 89;
    plants 84;
    population 86;
    precipitation 82, 83, 90;
    solar energy 86;
    temperature 82, 89;
    tourism 87;
    water supply 88–9
destructive plate margins 12, 14
destructive waves 115, 116fig
development gap 196–223; and
    aid 213, 214; and
    climate 206, 207; and
    colonialism 206;
    cycle of poverty 208fig;
    demographic transition model 204–5;
    and education 209; and
    extreme weather events 207; and
    Fairtrade 213;
    hydropower 212, 213; and
    inequality 210–11;
    intermediate technology 214;
    investment 214, 215; and
    location 206;
    microfinance 215; and
    migration 210, 211;
    political factors 209;
    reduction of 212–3; and
    trade 209;
    village associations 215; and
    war 206;
    water quality 208
development indicators 198–200, 202–3
disease: and air pollution 263;
    earthquakes 23, 28; and
    life expectancy 210; and
    malnutrition 272;
    regional differences UK 236fig; and
    sanitation 297;
    tropical areas 207; and
    tropical storms 40, 41;

urbanisation 175, 184; and
    volcanic eruption 17t,
    waterborne 289, 290
drinking water 88, 112, 248, 253, 284, 285,
    288–91
drumlins 155
DTM (Demographic Transition
    Model) 204–5

Earth: core 10;
    crust 10, 11t;
    magma 10, 12;
    mantle 10;
    rotation of 35, 36;
    structure of 10fig
earthquakes 20–33; and
    buildings 23, 31, 32; and
    conservative plate margins 13;
    global distribution 20fig;
    impact and population density 23;
    measurement of 21;
    planning for 24; and
    plate margins 20–1;
    prediction of 24;
    preparation for 24, 25;
    primary effects 22;
    protection 24, 25;
    recent 21fig;
    reduction of risk 24–5;
    secondary effects 23; and
    volcanic eruption 18
economic change, UK 224–43; change in
    industry 226, 228, 229;
    de-industrialisation 226;
    foreign suppliers 226, 227;
    globalisation 224–5;
    infrastructure 230–2;
    post-industrial economy 229
ecosystems 61–5
ecotourism 74
Egypt 86, 87, 88, 190t, 291, 311
energy insecurity 310–11
energy resources 300–19; and
    climate 308;
    coal 301;
    conservation of 316;
    cost of exploitation 257, 259, 308;
    deficit 306, 307;
    gas 302; global consumption 249,
        300fig, 304–5;
    impact of corruption 309;
    nuclear power 303;
    renewable 303, 306, 309, 312, 318–19;
    surplus 306;
    sustainability 316–17; and
    technology 308, 309; and
    war 308;
    ways to increase 312–13;
    wind farms 305, 312
erosion, river valleys 134, 135fig, 136–9
erratics 156
estuaries 139
European Union (EU) 238, 239
extreme weather, UK 8, 46–53; and
    climate change 51;
    COBRA committee 52;
    frequency of 50, 51;

Great Storm of 1703 50;
    heatwaves 48, 50;
    jet stream 51;
    reduction of 49;
    snowstorms 48;
    storms 48, 52–3

Fairtrade 213
famine 206, 222, 272
Ferrel cells 34, 35fig
fieldwork 322–9
financial services, UK 229
fires 23, 26fig
fisheries 278, 279
flooding 139, 140–5, 163
food miles 59, 251, 279
food resources: deficit 270, 271;
    food security 23, 265, 267–8, 271–3;
    global calorie consumption 266–7;
    increase 274–5;
    origins 264fig;
    production 246;
    seasonal 279;
    specialist 250;
    sustainable 278–81;
    uneven distribution 268;
    wastage and loss 279; and
    water insecurity 291
    see also famine
fossil fuels 56, 257, 306, 307figs, 308,
    313–15
fracking (shale gas extraction) 259, 308

gas clouds, volcanic 16t
genetic modification 275
glacial landscapes, UK 146–63;
    agriculture 157;
    deposition 150, 151, 155–6;
    erosion 149, 152, 153, 154, 155fig;
    formation 148;
    management of 157;
    movement 150;
    till 151;
    tourism 157;
    troughs 154
glacial retreat 55
glaciation 16t, 112, 113fig
global positioning systems (GPS) 128
global warming 54, 58
globalisation 197, 224–5, 250, 251
gorges 137
Green Revolutions 274
greenfield sites, development of 184,
    240, 243
greenhouse gases 56–7, 58, 72, 76,
    246, 309
groundwater management 88–9, 248, 253,
    270, 282, 283, 297

halocarbons 57
hard engineering 126, 127, 130, 131, 141
hazard risk equation 19
headland erosion 120
heatwaves 47, 48, 50
Human Development Index 199, 201t,
    203, 206
hurricanes see tropical storms

hydroponics 274, 296
hydropower 80–1, 212, 213

ice age 146, 147
Iceland 12, 18, 146, 306, 311
igneous rocks 110, 137
India: desert 86;
    economic growth 225;
    energy 301;
    food resources 264, 268;
    Himalayas 12;
    hydropower 212;
    urbanisation 167, 172–7, 190t;
    water resources 284, 285, 289
industrialisation: and energy supply 310,
    311; and
    water consumption 286
industry, sectors of 228fig
Intergovernmental Panel on Climate
    Change (IPCC) 58
interlocking spurs 136
international aid 29, 222–3
irrigation 86, 88, 92, 270, 274, 284

Japan 18, 19, 26–7, 222
Japan Trench 26
jet stream 51
Jordan 87, 290t

Kenya 298–9
Kyoto Protocol 58

lahars, volcanic 16t, 17fig
Lake District 154, 155fig, 158–63
landlocked countries and development
    gap 206
landslides 16t, 23, 28, 40, 118, 121,
    130–1
lava flows, volcanic 15t, 16t
levées 139
Los Angeles 13, 190
luxury goods 225

Madagascar 78–81
Malaysia 204, 205fig, 218–21, 223
malnutrition 272
managed retreat (coastal realignment) 126,
    128, 129
meanders 138
meat production 279
megacities 170–7
Mercalli scale 21, 22fig
metamorphic rocks 110
Milankovitch cycles 56
moraines, types 155
Mount Everest 28
Mount Fuji 18, 19fig
multiplier effect 102, 216, 221,
    232, 314
Mumbai 172–7

Nepal 23, 24
Nicaragua 18
Nigeria 167, 210
Nile river 86, 88
nuclear power 27, 87, 257, 258, 260,
    303, 313

ocean currents, and climate change 56
oil reserves: Alaska 102fig, 103, 104, 105;
    hot deserts 86;
    North Sea 314–15;
    Qatar 94
Okhotsk plate 26
organic farming 278
ox bow lakes 138

permaculture 278
permafrost 98, 100
pesticides 57, 58, 76, 89, 253, 270, 290
Philippines 38, 44–5
photosynthesis 57
physical landscapes, UK 109–13
plate margins 11, 24
plate tectonic theory 11
Polar cells 34, 35fig
polar environments 96, 97fig, 98
polar front 34
population increase, global: and
    desertification 90; and
    energy consumption 304; and
    food insecurity 247, 268, 280; and
    water consumption 285, 286
precipitation 48, 63, 64, 82, 98, 141,
    250, 283
pyramidal peak 153
pyroclastic flows, volcanic 16t

Qatar 94–5
Quaternary period 54

Red Sea 13
refugees 91, 211, 271
reservoirs 112, 212, 253–5, 283, 292
resource management 245–9
ribbon lakes 154, 155fig
Richter scale 21
rift valleys 12, 13
river landscapes, UK 132–45;
    flooding 140–5;
    landforms 136–9;
    river valleys 132–5
rural-urban migration 164–5, 167–9, 173,
    220, 280
Russia 97fig, 103, 225, 257, 258, 282t,
    302, 311

Saffir-Simpson scale 40, 42
Sahara desert 34, 63, 82, 83fig, 86, 89, 93;
    Sahel 90, 91
sand dams 298
sand dunes 122
sanitation 23, 168fig, 175, 176, 177, 214,
    290, 297
SDGs
    see Sustainable Development Goals
sea level rise 38, 58
sea-ice, Arctic 55
seasonal changes 55
sedimentary rocks 110
seismic gap model, earthquake
    prediction 24
seismometers 21
service industries 226, 228, 229
shale gas extraction see fracking

skills practice 330–55;
    atlas 354–5;
    cartographic 346–53;
    graphical 330–6;
    numerical 336–9;
    statistical 327–7, 340–5
slab pull theory 11
slash and burn 71, 79
snowstorms 48
Spain 86, 91
soft engineering 126, 128, 141
soil erosion 92, 93, 272
spits 122
storm hydrograph 140
subduction zones 12
Sudan 86, 88, 206, 207, 214, 271,
    291, 306
sunspot activity 56
surface wind patterns 35
Sustainable Development Goals (SDGs)
    (UN) 74, 309

Tanzania 216–7, 276–7, 318–19
tarns 152, 153
technology: and energy conservation 317;
    and food supply 270
tectonic plate 11
tourism: Alaska 102;
    Antarctica 106;
    Boscastle 143;
    Bournemouth 240, 241;
    glacial landscapes 157; and
    heatwaves 48;
    hot deserts 87;
    Lake District National Park 160–3;
    Malaysia 221;
    Qatar 94; Tanzania 216–17
trade 89, 173, 206, 209, 216, 232, 238,
    239
    see also globalisation
traffic congestion 160, 185, 188, 190–1,
    223, 230
transnational corporations (TNCs) 220–21
transport, sustainable 316, 317
transportation by rivers 135
tropical rainforests 66–81;
    biodiversity 67;
    conservation 74; and
    debt reduction 75;
    deforestation 70–3;
    importance of 74;
    location of 66fig;
    management strategies 74–5;
    rainforest layers 68–9;
    secondary forests 74;
    sustainability of 74–5
tropical storms 34–45;
    cone of uncertainty 42;
    eye of 37; eye wall 37;
    frequency and energy 38;
    impact of 40;
    impact of climate change 38–9;
    measurement of 40;
    monitoring 41;
    planning 42, 43;
    prediction 42;
    protection 42, 43;

responses to 41; and
    severe air instability 36;
    structure of 37fig;
    worldwide distribution of 36fig
truncated spurs 153
tsunamis 16t, 23, 26–7
tundra environments 96, 97, 98, 99
Typhoon Haiyan 38, 44–5
typhoons see tropical storms

UK: airports 232;
    climate 46fig;
    culture 238;
    electronic communications 238;
    energy mix 256–9;
    food production and demand 250–2;
    geological map 111fig;
    global trade 238;
    global transport 238;
    industry and environmental
    sustainability 233;
    lowland areas 112;
    most populous areas 178fig, 179;
    North/South divide 236–7;
    place in world 238–9; ports 232; railway
    links 180, 231;
    research and development 229;
    rivers 112, 113;
    road network 230;
    rural areas and change 234;
    upland areas 112
unemployment 175, 184, 210, 219, 314
undernutrition see famine
urban farming 278, 281
urban greening 188
urban issues 164–95;
    megacities 170–7;
    sanitation 23, 168fig, 175, 176, 177,
    214, 290, 297;
    sustainability 188–9; traffic
    congestion 190–1;
    urbanisation 166–9
urban landscape, UK 178–87;
    commuters 185;
    housing 184;
    population characteristics 179fig;
    urban sprawl 185
urban sprawl 169, 185
urbanisation 164–5, 166–9; and
    demand for water 247; and food
    supply 268;
    Kinshasa 280;
    Malaysia 220;
    Mumbai 173;
    rates of 167
USA: calorie intake 267;
    cold environments 104–5;
    desertification 90;
    energy 301–6, 308, 309, 310;
    earthquakes 13;
    fracking 259;
    hot deserts 87t;
    income distribution 210;
    irrigation 246;
    pesticides 270;
    population 204;
    solar energy 86;

traffic congestion 190;
trade 221, 225;
tropical storms 38, 39;
volcanoes 18, 62;
water resources 248, 292

virtual water 286
volcanic bombs 16t
volcanic islands 12
volcanoes 14–19; and
    climate change 56;
    composite 14, 15t;
    dormant 14;
    extinct 14;
    global distribution of active 14fig;
    primary and secondary effects 17t;
    shield (basic) 14, 15t;
    volcanic hazards 16–17

war, and food supply 271
wastewater use 297
water insecurity 290–1
water pollution 253, 289, 290, 291
water resources 282–9; agricultural use 284; domestic use 284; efficient consumption of 287fig; global freshwater supplies 282; global surplus and deficit 283; and hot deserts 88–9, 92, 94; impact of climate 288, 289; impact of geology 289; increase

of 292–3; industrial use 284; and infrastructure 289; over-extraction 289; quality 208, 253; supply and demand 247, 248, 253, 254–5, 288; sustainability of 296–7; in UK 253, 254–5; water conflict 291; water conservation 296, 298–9 see also drinking water
water stress 270, 288
water table 83
water transfer schemes 254, 255fig, 293
waterfalls, and erosion 137
waves 114–17
weathering: coastal 118, 119, 120, 121, 130, 131; freeze-thaw 149, 153; river valleys 134; of volcanic rock 18
women; development gap 198, 215; economic migration 211; urban farming 281; water access 297, 298
war, and food supply 271
wastewater use 297
water insecurity 290–1
water pollution 253, 289, 290, 291
water resources 282–9; agricultural use 284; domestic use 284; efficient consumption of 287fig; global freshwater supplies 282; global surplus and deficit 283; and

hot deserts 88–9, 92, 94;
impact of climate 288, 289;
impact of geology 289;
increase of 292–3;
industrial use 284; and infrastructure 289;
over-extraction 289;
quality 208, 253;
supply and demand 247, 248, 253, 254–5, 288;
sustainability of 296–7; in UK 253, 254–5;
water conflict 291;
water conservation 296, 298–9 see also drinking water
water stress 270, 288
water table 83
water transfer schemes 254, 255fig, 293
waterfalls, and erosion 137
waves 114–17
weathering: coastal 118, 119, 120, 121, 130, 131;
    freeze-thaw 149, 153;
    river valleys 134;
    of volcanic rock 18
women: development gap 198, 215;
    economic migration 211;
    urban farming 281;
    water access 297, 298

Acknowledgements

The author and publishers acknowledge the following sources of copyright material and are grateful for the permissions granted. While every effort has been made, it has not always been possible to identify the sources of all the material used, or to trace all copyright holders. If any omissions are brought to our notice, we will be happy to include the appropriate acknowledgements on reprinting.

Text

Extract on page 171. from 'China's Pearl River Delta overtakes Tokyo as world's largest megacity' by Nick Mead, *The Guardian*, 28/01/2015; Data in Table 12.2 on page 205. from *The Economist Pocket World in Figures*, Economist Books, 2005, 2008, 2015 editions; Extract on page 227. from 'M&S switch to foreign suppliers threatens 16,000 UK textile jobs' by Fran Abrams, *The Independent*, 02/11/1999; Extract on page 267. from 'World entering era of global food insecurity with malnutrition and obesity side by side within countries, says leading food expert' by Steve Connor, *The Independent*, 11/07/2015; Quotation on page 317. Reproduced by kind permission of Sustrans (http://www.sustrans.org.uk/).

Images

Front cover © Andrea Pucc/Getty images; Fig. S1.1 plainpicture/Bildhuset/Per Klaesson; Fig. Chapter 1 header SDubi/Shutterstock.com; Fig. 1.5 Joao Virissimo/Shutterstock.com; Fig. 1.6 Fredy Thuerig/Shutterstock.com; Fig. 1.10 solomonjee/Shutterstock.com; Fig. 1.11 Vadim Petrakov/Shutterstock.com; Fig. 1.12a Photovolcanica.com/Shutterstock.com; Fig. 1.12b Brisbane/Shutterstock.com; Fig. 1.13 © Tom Uhlman/Alamy Stock Photo; Fig. 1.14 © Kees Metselaar/Alamy Stock Photo; Fig. 1.15 PATRICK LANDMANN/SCIENCE PHOTO LIBRARY; Fig. 1.16 PavelSvoboda/Shutterstock.com; Fig. 1.20 © The Photolibrary Wales/Alamy Stock Photo; Fig. 1.21 © Better Late Images/Alamy Stock Photo; Fig. 1.23 Morenovel/Shutterstock.com; Fig. 1.26 © WENN UK/Alamy Stock Photo; Fig. 1.27 © Horizon Images/Motion/Alamy Stock Photo; Fig. 1.29 Niranjan Shrestha / AP/Press Association Images; Fig. 1.30 ROBERTO SCHMIDT/Staff/Getty images; Fig. 1.31 NICOLAS ASFOURI/Staff/ Getty images; Fig. 1.33 © John Henshall/Alamy Stock Photo; Fig. 1.34 © imageBROKER/Alamy Stock Photo; Fig. 1.35 KPG Payless2/Shutterstock.com; Fig. Chapter 2 header Graphithèque/Fotolia; Fig. 2.3 Chet Clark/Shutterstock.com; Fig. 2.6 Harvepino/Shutterstock.com; Fig. 2.7 Richard Whitcombe/Shutterstock.com; Fig. 2.10 meunierd/Shutterstock.com; Fig. 2.12 GOLFX/Shutterstock.com; Fig. 2.14 Niar/Shutterstock.com; Fig. 2.15 Chris Warham/Shutterstock.com; Fig. Chapter 3 header Roger Coulam/Getty images; Fig. 3.3 bdomanska/Shutterstock.com; Fig. 3.4 © Derek Croucher/Alamy Stock Photo; Fig. 3.5 allou/Getty images; Fig. 3.6 Jens Meyer/AP/Press Association Images; Fig. 3.7 England's Great Storm (engraving) (b/w photo), English School/Private Collection/Bridgeman Images; Fig. 3.8 © Dorset Media Service/Alamy Stock Photo; Fig. 3.11 stockphoto mania/Shutterstock.com; Fig. 3.12 Jason Salmon/Shutterstock.com; Fig. 3.13 Philip Bird LRPS CPAGB/Shutterstock.com; Fig. 3.14 Rui Saraiva/Shutterstock.com; Fig. Chapter 4 header Sondem/Fotolia; Fig. 4.1 Mohamed Shareef/Shutterstock.com; Fig. 4.2 Thomas Ramsauer/Shutterstock.com; Fig. 4.3 Public domain; Fig. 4.4 Public domain; Fig. 4.6 stocker1970/Shutterstock.com; Fig. 4.7 © BrazilPhotos.com/Alamy Stock Photo; Fig. 4.8 © jon gibbs/Alamy Stock Photo; Fig. S2.1 NASA; Fig. S2.5 Alison Rae; Fig. Chapter 5; header apiguide/Shutterstock.com; Fig. 5.3 Trevor Worden/Getty images; Fig. 5.6 © jeremy sutton-hibbert/Alamy Stock Photo; Fig. 5.7 guentermanaus/Shutterstock.com; Fig. 5.8 A.S. Zain/Shutterstock.com; Fig. 5.9 ©Hector Fernandez/Fotolia; Fig. 5.10 Dr. Morley Read/Shutterstock.com; Fig. 5.11 Fedorov Oleksiy/Shutterstock.com; Fig. 5.12 Rich Carey/Shutterstock.com; Fig. 5.13 © Nelson_A_Ishikawa/Thinkstock; Fig. 5.14 JulieHewitt/Getty images; Fig. 5.16 SUWIT NGAOKAEW/Shutterstock.com; Fig. 5.17 Kjersti Joergensen/Shutterstock.com; Fig. 5.18 Aleksey Stemmer/Shutterstock.com; Fig. 5.20 Pierre-Yves Babelon/Shutterstock.com; Fig. 5.21 © Suwan Waenlor/Alamy Stock Photo; Fig. 5.25 Rizzo Associates; Fig. Chapter 6; header Aiisha/Fotolia; Fig. 6.3 OFFFSTOCK/Shutterstock.com; Fig. 6.4 poppit01/Shutterstock.com; Fig. 6.5 EcoPrint/Shutterstock.com; Fig. 6.6 Arno Dietz/Shutterstock.com; Fig. 6.7 Yuval Helfman/Shutterstock.com; Fig. 6.8 Ilia Torlin/Shutterstock.com; Fig. 6.9 SSSCCC/Shutterstock.com; Fig. 6.10 muznabutt/Shutterstock.com; Fig. 6.14 abogdanska/Shutterstock.com; Fig. 6.16 DiversityStudio/Shutterstock.com; Fig. 6.18 Wolfgang Zwanzger/Shutterstock.com; Fig. 6.20 Unknown source; Fig. 6.22 Fitria Ramli/Shutterstock.com; Fig. Chapter 7; header David Tipling/Getty images; Fig. 7.3 © Steve Morgan/Alamy Stock Photo; Fig. 7.6 Vladimir Kovalchuk/Shutterstock.com; Fig. 7.7 © Global Warming Images/Alamy Stock Photo; Fig. 7.8 Marina Riley/Shutterstock.com; Fig. 7.9 Wikipedia; Fig. 7.12 US Coast Guard Photo/Alamy Stock Photo; Fig. 7.14 Public domain; Fig. 7.15 Sam Chadwick/Shutterstock.com; Fig. 7.16 Wolfgang Kaehler/Contributor/Getty images; Fig. S3.1 plainpicture/NaturePL/Ross Hoddinott; Fig. S3.2 © Scott Hortop

Travel/Alamy Stock Photo; Fig. S3.3 © Tony Watson/ Alamy Stock Photo; Fig. S3.4 © David Noton Photography/ Alamy Stock Photo; Fig. S3.5 © PearlBucknall/Alamy Stock Photo; Fig. S3.6 © Philip Enticknap/Alamy Stock Photo; Fig. Chapter 8 header Matt Cardy/Stringer/Getty images; Fig. 8.1 © Design Pics Inc/Alamy Stock Photo; Fig. 8.4 © Picture Hooked/Lisa Jacobs/Alamy Stock Photo; Fig. 8.6 © Kevin Britland/Alamy Stock Photo; Fig. 8.8 © Kate Eastman/Alamy Stock Photo; Fig. 8.9 © David Lichtneker/ Alamy Stock Photo; Fig. 8.11 © Jack Lane/Alamy Stock Photo; Fig. 8.12 © David Noton Photography/Alamy Stock Photo; Fig. 8.14 © Steve Morgan/Alamy Stock Photo; Fig. 8.17 © darryl gill/Alamy Stock Photo; Fig. 8.18 © robertharding/Alamy Stock Photo; Fig. 8.19 © A.P.S. (UK)/ Alamy Stock Photo; Fig. 8.20 © Doug Houghton/LGPL/ Alamy Stock Photo; Fig.8.20 © Crown copyright and database rights 2015 OS (100049945); Fig. 8.22 Matt Cardy/Stringer/Getty images; Fig. 8.24 © robert harrison/ Alamy Stock Photo; Fig. 8.25 OLI SCARFF/Stringer/Getty images; Fig. 8.26 © Mick House/Alamy Stock Photo; Fig. 8.27 © Mike McEnnerney/Alamy Stock Photo; Fig. 8.28 © Terry Whittaker/Alamy Stock Photo; Fig. 8.31 © Marc Hill/Alamy Stock Photo; Fig. 8.32a © David Bagnall/ Alamy Stock Photo; Fig. 8.32b © Pix/Alamy Stock Photo; Fig. Chapter 9 header Westend61/Getty images; Fig. 9.1 Crown copyright?; Fig. 9.2 AC Rider/Shutterstock.com; Fig. 9.3 Dave Willis/Getty images; Fig. 9.4 © Paul White North East England/Alamy Stock Photo; Fig. 9.8 © Chris Howes/Wild Places Photography/Alamy Stock Photo; Fig. 9.9 © Joseph Clemson/Alamy Stock Photo; Fig. 9.11 © Motoring Picture Library/Alamy Stock Photo; Fig. 9.13 © Peter J. Hatcher/Alamy Stock Photo; Fig. 9.16 Dave Head/ Shutterstock.com; Fig. 9.17 Banks Group; Fig. 9.20 2016 John Strain/fotoLibra; Fig. 9.21 © Paul White - UK Industries/Alamy Stock Photo; Fig. 9.23 © Paul White Aerial views/Alamy Stock Photo; Fig. 9.24 © allOver images/ Alamy Stock Photo; Fig. 9.26 © Mark Pearson/Alamy Stock Photo; Fig. 9.27 © Roy Riley/Alamy Stock Photo; Fig. 9.29 © Roger Bamber/Alamy Stock Photo; Fig. Chapter 10 header Martin Fowler/Shutterstock.com; Fig. 10.2 © Washington Stock Photo/Alamy Stock Photo; Fig. 10.8 richsouthwales/Shutterstock.com; Fig. 10.12 © Ashley Cooper/Alamy Stock Photo; Fig. 10.13 Brendan Howard/ Shutterstock.com; Fig. 10.15 © Jon Sparks/Alamy Stock Photo; Fig. 10.16 © Stewart Smith/Alamy Stock Photo; Fig. 10.18 © Anna Stowe Landscapes UK/Alamy Stock Photo; Fig. 10.20 © Ashley Cooper/Alamy Stock Photo; Fig. 10.21 © David Robertson/Alamy Stock Photo; Fig. 10.22 stocksolutions/Shutterstock.com; Fig. 10.23 crazychris84/Shutterstock.com; Fig. 10.24 Crown copyright and database rights 2015 OS (100049945); Fig. 10.25 Stewart Smith Photography/Shutterstock.com; Fig. 10.26 © Gordon Shoosmith/Alamy Stock Photo; Fig. S4.1 The India Today Group/Contributor/Getty images; Fig. Chapter 11 header Greg Fonne/Contributor/Getty images; Fig. 11.5 Andrea Izzotti/Shutterstock.com; Fig. 11.8 © Image Source Plus/Alamy Stock Photo; Fig. 11.9 Bethany Clarke/

Contributor/Getty images; Fig. 11.10 © Julio Etchart/ Alamy Stock Photo; Fig. 11.12 Aleksandar Todorovic/ Shutterstock.com; Fig. 11.13 © Gordon Dixon/Alamy Stock Photo; Fig. 11.14 Dezeen magazine; Fig. 11.18 Chris Hepburn/Getty images; Fig. 11.19 Public domain; Fig. 11.20 Public domain; Fig. 11.21 © Robin Weaver/ Alamy Stock Photo; Fig. 11.22 © John James/Alamy Stock Photo; Fig. 11.23 Public domain; Fig. 11.24 © A.P.S. (UK)/ Alamy Stock Photo; Fig. 11.25 Crown copyright?; Fig. 11.26 © Marmaduke St. John/Alamy Stock Photo; Fig. 11.27 © syrnx/Alamy Stock Photo; Fig. 11.28 © LOOK Die Bildagentur der Fotografen GmbH/Alamy Stock Photo; Fig. 11.29 © Geoffrey Morgan/Alamy Stock Photo; Fig. 11.30 © Daryl Mulvihill/Alamy Stock Photo; Fig. 11.32 © Images of Birmingham Premium/Alamy Stock Photo; Fig. S5.1 plainpicture/Cultura/Monty Rakusen; Fig. Chapter 12 header Grigvovan/Shutterstock.com; Fig. 12.1 © Andrew Twort/Alamy Stock Photo; Fig. 12.2 bikeriderlondon/Shutterstock.com; Fig. 12.3 © Stephen Barnes Photography/Alamy Stock Photo; Fig. 12.4 © Jake Lyell/Alamy Stock Photo; Fig. 12.7 TonyV3112/Shutterstock. com; Fig. 12.10 Peteri/Shutterstock.com; Fig. 12.12 © Jake Lyell/Alamy Stock Photo; Fig. 12.13 © Boaz Rottem/ Alamy Stock Photo; Fig. 12.14a John Wollwerth/ Shutterstock.com; Fig. 12.14b Morphart Creation/ Shutterstock.com; Fig. 12.15 nicolasdecorte/Shutterstock. com; Fig. 12.16 © David Burton/Alamy Stock Photo; Fig. 12.19 Credit: Baciu/Shutterstock.com; Fig. 12.20 Reproduced under Creative Commons 4.0 licence; Fig. 12.22 © Aurora Photos/Alamy Stock Photo; Fig. 12.25 szefei/Shutterstock.com; Fig. Chapter 13 header Sharon Vos-Arnold/Getty images; Fig. 13.2 © David Bagnall/ Alamy Stock Photo; Fig. 13.3 Galyamin Sergej/Shutterstock. com; Fig. 13.6 © J Marshall - Tribaleye Images/Alamy Stock Photo; Fig. 13.10 © David Gee 2/Alamy Stock Photo; Fig. 13.12 Mike Hewitt/Staff/Getty images; Fig. 13.13 © Ashley Cooper/Alamy Stock Photo; Fig. 13.14 Alison Rae; Fig. 13.15 stocker1970/Shutterstock.com; Fig. 13.18 DavidGraham86/Shutterstock.com; Fig. 13.20 Google; Fig. S6.1 © dpa picture alliance archive/Alamy Stock Photo; Fig. S6.5 © Universal Images Group North America LLC/ Alamy Stock Photo; Fig. Chapter 14 header Majeczka/ Shutterstock.com; Fig. 14.3 Dick Kenny/Shutterstock.com; Fig. 14.4 Valdis Skudre/Shutterstock.com; Fig. 14.6 Platslee/Shutterstock.com; Fig. 14.11 Randi Sokoloff/ Shutterstock.com; Fig. 14.14 © Graeme Peacock/ Alamy Stock Photo; Fig. 14.18 © The Photolibrary Wales/ Alamy Stock Photo; Fig. Chapter 15 header Zeljko Radojko/ Shutterstock.com; Fig. 15.2 Renate Wefers/Fotolia; Fig. 15.3 Natalia Mylova/Shutterstock.com; Fig. 15.4 HeinzTeh/Shutterstock.com; Fig. 15.9 dailin/Shutterstock. com; Fig. 15.10 Ministr-84/Shutterstock.com; Fig. 15.11 Richard Jary/Shutterstock.com; Fig. 15.12 Jim Parkin/ Shutterstock.com; Fig. 15.13 John Wollwerth/Shutterstock. com; Fig. 15.14 Byelikova Oksana/Shutterstock.com; Fig. 15.17 Sura Nualpradid/Shutterstock.com; Fig. 15.18 JAKKRIT SAELAO/Shutterstock.com; Fig. 15.19